FATIGUE OF FIBROUS COMPOSITE MATERIALS

A symposium
sponsored by ASTM
Committee D-30 on High
Modulus Fibers and Their
Composites and Committee
E-9 on Fatigue
AMERICAN SOCIETY FOR
TESTING AND MATERIALS
San Francisco, Calif., 22–23 May 1979

ASTM SPECIAL TECHNICAL PUBLICATION 723
K. N. Lauraitis
Lockheed-California Company
symposium chairman

ASTM Publication Code Number (PCN)
04-723000-33

AMERICAN SOCIETY FOR TESTING AND MATERIALS
1916 Race Street, Philadelphia, Pa. 19103

Library of Congress Catalog Card Number: 80-67399

NOTE

The Society is not responsible, as a body,
for the statements and opinions
advanced in this publication.

Printed in Baltimore, Md.
January 1981

Foreword

This publication on Fatigue of Fibrous Composite Materials contains papers presented at a symposium held 22–23 May 1979 at San Francisco, California. The symposium was sponsored by the American Society for Testing and Materials through its Committees D-30 on High Modulus Fibers and Their Composites and E-9 on Fatigue. K. N. Lauraitis, Lockheed-California Company, served as symposium chairman.

Related ASTM Publications

Commercial Opportunities for Advanced Composites, STP 704 (1980), $13.50, 04-704000-33

Composite Materials: Testing and Design (Fifth Conference), STP 674 (1979), $52.50, 04-674000-33

Advanced Composite Materials—Environmental Effects, STP 658 (1978), $26.00, 04-658000-33

Fatigue of Filamantary Composite Materials, STP 636 (1977), $26.50, 04-636000-33

Composite Materials; Testing and Design (Fourth Conference), STP 617 (1977), $51.75, 04-617000-33

Fatigue of Composite Materials, STP 569 (1975), $31.00, 04-569000-33

A Note of Appreciation to Reviewers

This publication is made possible by the authors and, also, the unheralded efforts of the reviewers. This body of technical experts whose dedication, sacrifice of time and effort, and collective wisdom in reviewing the papers must be acknowledged. The quality level of ASTM publications is a direct function of their respected opinions. On behalf of ASTM we acknowledge with appreciation their contribution.

ASTM Committee on Publications

Editorial Staff

Contents

Introduction

This symposium, the second co-sponsored by ASTM Committees D-30 and E-9 focusing on the fatigue of fiber-reinforced composite materials, was held on the 22 and 23 May 1979, in San Francisco, California. It was a product of the same momentum that set the first such conference in motion two and a half years earlier. Composites had come of age. They had moved from the laboratory into the shop and were ready for their next step into service in critical structure—perhaps. With this last step imminent, durability and damage tolerance inevitably forced themselves into view. Therefore, our energies and efforts over the last seven years have been funneled into studying fatigue and environmental effects. The works published herein exemplify our considerable progress in the field and are a statement of our position today. A position which to me produces a feeling of *déjà vu.* We have explored the use of the dominant flaw approach in composites; tried to guarantee minimum life through proof testing, attempted statistical descriptions of the fatigue process and evaluated various cumulative damage models. While reminding ourselves to *think composites,* we have followed the well-trodden path of those who have *thought metals* before us. Through attempts to emphasize the differences, we have discovered the similarities; and, so find ourselves now, as do our metals colleagues, at a point where "despite all this progress in detail we are still faced with considerable uncertainties when attempting to design a component or structure to avoid the occurrence of fatigue failures."[1] Yet, major advances in our understanding are apparent in reviewing the papers presented at this conference, especially compared to ten years ago when the word fatigue was hardly linked with the word composites. Our data base has been expanded considerably. We have taken our studies to the microlevel and explored the sequences of events and have had some success in mathematically modeling cracking/delamination states.

However, "we [are not] yet able to separate and then integrate the individual aspects of the process."[1] In this quote from Professor Dolan, Professor Le May possibly brings forth the key to converting our knowledge to practical wisdom. The noteworthy words here are separate, integrate, and process. The last of these is probably most important since the first two follow from the recognition of and focus on fatigue as a *process*. It is dynamic—a horse race. And, to date, as Professor Morrow[2] has noted we have been taking snapshots of the horses. This exercise has been necessary, good, fulfilling, and progressive, but we need only one trip along that circle

[1]LeMay, I., "Symposium Summary and an Assessment of Research Progress in Fatigue Mechanisms," *Fatigue Mechanisms, ASTM STP 675,* American Society for Testing and Materials, 1979, pp. 873–888.

[2]Morrow, J., "General Discussion and Concluding Remarks," *Fatigue Mechanisms, ASTM STP 675,* American Society for Testing and Materials, 1979, pp. 891–892.

and must take a step forward and up before we circle again, thereby always spiraling ahead. As we remove our composite blinders, we must not trade them for those labeled *metals, plastics, fibers,* or even *materials.* We must behold the field as a whole. Recognize that we are dealing with a system created by the physical (material) mechanical, chemical, thermal, and electrical interactions. With this point of view, we will necessarily cease to break down the fatigue process to its separate parts and will, through the concentration of our energies and attention, proceed to integrate and synthesize our knowledge so it may be utilized in design.

We have not been investigating fatigue as a process but have been involved in the description of its effects. Fatigue has become the cause of failure rather than a word used to describe the systematic interactions occurring as a result of repetitive load applications. We as researchers desire and do intend to have the designer in mind. Let us indeed approach the problem from a consideration of the designer's needs, something we all try to do, but let it be the needs as he sees them. Often what the designer requires is for the purpose of meeting certain requirements, which, though important, play no active part in the design stage. For the fact remains that aircraft and other dynamically loaded large-scale structures have been designed and built and have functioned successfully for lifetimes in excess of 20 years despite our inability to predict fatigue life. Perhaps we have been unable to find the answers because we have been asking the wrong questions and the wrong people.

The *fatigue problem* is not necessarily one of determining some underlying principle, useful for life prediction, but instead one of determining how to use our descriptive knowledge in the design process. We may be able to assure safe structures without actually predicting fatigue life. Design of structures has been primarily based on stiffness and static strength. Thus, if the design is correctly done, can we determine if fatigue will be a problem? Such questions constitute a future research direction.

Many have contributed to the success of this symposium and, I am certain, success of this publication. I am most grateful for the assistance of the Session Chairmen, K. T. Kedward, K. L. Reifsnider, G. L. Roderick, and J. T. Ryder, through whose efforts the sessions progressed without fault. I also extend my gratitude to the keynote speaker, D. W. Hoeppner, whose words gave us pause to think, and most sincerely to the authors without whose contributions there could not have been an ASTM Special Technical Publication. Nor would this volume or symposium have existed without the considerable efforts of the ASTM Staff whom I thank wholeheartedly.

K. N. Lauraitis

Rye Canyon Research Laboratory, Lockheed-California Company, Burbank, Calif. 91520; symposium chairman.

J. E. Rhodes[1]

Effect of Post Buckling on the Fatigue of Composite Structures

REFERENCE: Rhodes, J. E., **"Effect of Post Buckling on the Fatigue of Composite Structures,"** *Fatigue of Fibrous Composite Materials, ASTM STP 723,* American Society for Testing and Materials, 1981, pp. 3–20.

ABSTRACT: This paper discusses the physics involved in shear and compression post buckling and compares their forced displacement forms. First level, simplified mathematical treatise are presented. The more complex rigorous mathematics are avoided, with emphasis being placed on the qualitative aspects. The objective is to contribute to a fundamental understanding of post-buckling behavior that will help establish practical design limits.

It is shown that the surface strains and substructure separation forces can be assessed, with reasonable accuracy, once the displacement shapes are established. Test panel deflections and strain measurements are compared with predicted values. Static and fatigue test results on panels subject to loadings in the post-buckled range are presented.

KEY WORDS: fibrous composites, shell structures, post buckling, compression, shear, forced displacement, fatigue test panels, moiré patterns, displacement strain, fatigue (materials), composite materials

The criterion of "no buckling up to ultimate load" was generally applied during the design of most fibrous composites hardware. Primary airframe structural application was generally limited to wing and empennage torsion boxes. Ultimate strength tests on these structures often demonstrated a static strength capability in shear and compression well above the initial buckling loads. A potential buckling capability, although not used, was demonstrated. Since the maximum applied fatigue loading for all applications was well below the initial buckling level, the fatigue tests on these structures contributed little to an understanding of post-buckled repeated loading.

The potential use of composites has more recently been explored in applications other than thin wing and empennage torsion boxes. In these structures, particularly fuselage shells, a lower load intensity range is

[1]Senior research and development engineer, Lockheed-California Company, Burbank, Calif. 91520.

encountered. In the past, these shell structures have generally been fabricated from aluminum alloy sheet supported by formed frames and stiffeners with the skin thicknesses ranging from 0.64 to 2 mm (0.025 to 0.080 in.). Examples of these types of structures are shown in Fig. 1. The unit weight of these lightly loaded structures is relatively low, however, the total surface area of this type of shell structure on some aircraft is very high so the total weight is appreciable.

In applying fibrous composites to these structures, a nonbuckling criterion would result in configuration optimizations different from those that would be selected if post buckling were allowed. Devices for increasing effective skin thicknesses, such as various forms of sandwich construction, would be favored over monolithic construction. It is, therefore, necessary to establish allowables for post buckling of monolithic composite panels.

Related Aluminum Experience

There is a tendency to emphasize the differences between metallic and composite structures rather than their similarities. There is a distinct similarity in the behavior of shell structures. It is, therefore, worthwhile to review the ground rules applied to aluminum shell structures.

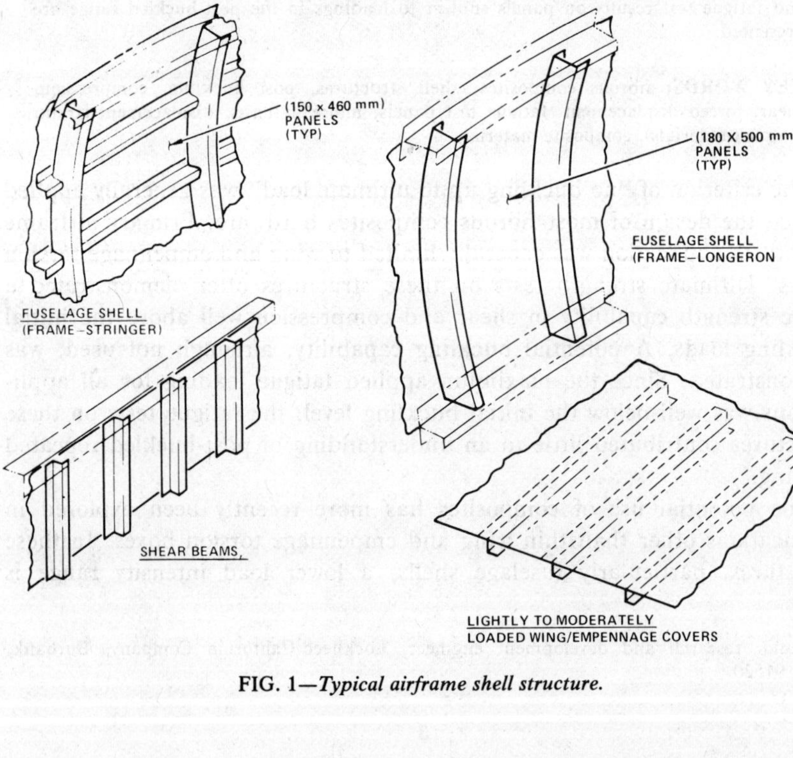

FIG. 1—*Typical airframe shell structure.*

Post buckling in aluminum shell structures was generally limited by one of the following:

1. No shear buckling below a given fraction of ultimate load.

$$\frac{qult}{qcr} < 5.0 \text{ (to minimize fatigue)}$$

2. Aerodynamic surface smoothness.
3. Aesthetic—no pillowing or buckling in the static ground loading.
4. Aeroelastic stiffness requirements.
5. Acoustic fatigue and noise transmissibility.
6. Service handling and damage.

The limits established were somewhat arbitrary. In some aircraft, buckling was allowed below the 1 g level flight loads. Some aircraft clearly showed buckles just sitting on the ground.

Shear and compression panel tests of representative structures were conducted to finalize the design. Proof tests on the complete structures (fuselages, wings, tails, etc.) were conducted prior to first flight, and the appropriate modifications made. There was no concern about skin delamination because the shear and transverse properties relative to the longitudinal strength were high. The differences in this respect between the interlaminar properties of graphite/epoxy composites and solid metal sheet are shown in Table 1.

TABLE 1—*Shear and transverse properties comparison.*

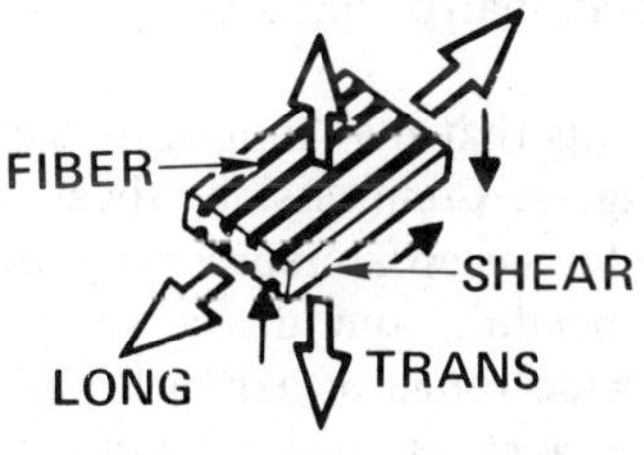

	Strength, MPa (ksi)			Strength Ratio	
Material	Longitudinal Tension	Shear	Transverse Tension	Shear ÷ Longitudinal	Transverse Tension ÷ Longitudinal Tension
Aluminum sheet	448(65)	276(40)	448(65)	0.62	1.0
Graphite/Epoxy Quasi-isotropic	621(90)	31(4.5)	41(6.0)	0.05	0.07

The allowable post-buckling level was usually established by a local mechanical attachment failure or stiffener column-crippling. Failures usually occurred at intersection areas of panel support structure.

The basic data available for initial sizing came from many sources: Stress Memo Manuals, data sheets, and NACA reports typified by NACA TN 2661.[2] Curves for initial buckling in terms of panel length to width (a/b) and panel width to thickness (b/t) had common usage. Post-buckling or crippling allowable prediction methodology or both varied from company to company, and for the most part was semi-empirical in nature with constants introduced to fit the test data bank. Static strength was the prime issue, fatigue rarely entered the picture except as a general judgment factor.

Composite Fatigue Considerations

Durability requirements for non-buckled structures have been covered in a general way by limiting the gross area strain at ultimate load. This is similar to the approach used in aluminum structures. In these structures, an ultimate load, gross area stress cut-off is established, which is consistent with the fatigue quality that can be achieved in the numerous structural details. Ultimate load strain cut-off is a candidate for establishing post buckling and membrane limits for the design of composite shell structures.

A generally held view is that we do not have the same tension-tension fatigue problem in graphite/epoxy composites as we have in aluminum structures. This view is based primarily on thick sheet, non-buckled test specimen experience and is not necessarily applicable to thin sheet, post-buckled shells. It is expected that matrix initiated failures will establish fatigue life for thin shells.

The forced displacements that may induce matrix cracking and stiffener peeling are shown in Fig. 2. Compression, shear, and pressure pillowing are illustrated. An angular sweep as represented by θ in Fig. 2 shows the existence of combined bending and direct strains for compression and tension existing over a wide range of azimuth positions. It seems logical, therefore, to assume for a worst case assessment that the most critical surface fiber orientation exists irrespective of the actual stacking sequence.

A review of the many existing reports on fatigue tests does not yield definitive data for thin laminates. The myriad combinations of fibers, matrices, stacking sequences, fiber volume, and quality levels limit meaningful comparisons between data, particularly where strain levels, or properties to convert to strain levels, are not quantified. Most of the composite fatigue data is for thick sheets often with holes. Application to thin

[2]Kuhn, P. L. and Peterson, J. P., "Summary of Diagonal Tension," AFML Advanced Composites Design Guide-TN 2660, Air Force Materials Laboratory, May 1952.

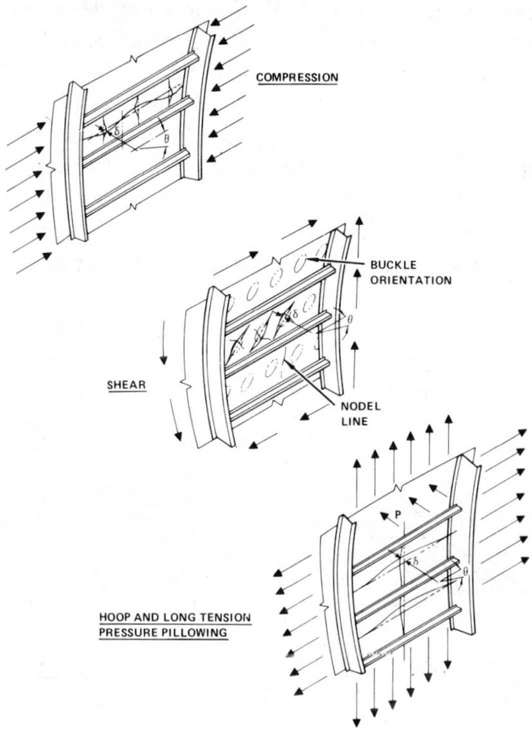

FIG. 2—*Fuselage—forced displacement.*

sheets of the level of 1 mm (0.040 in.) thick is at best intuitive. Several failure function hypotheses are given by Hahn.[3] Strain is shown to be a more sensitive parameter than stress in establishing allowables. A tensile-tensile strain versus applied cycles relationship for matrix cracking is suggested from the results of repeated torsion load applied to a ±45-deg fiber orientation tube.[4] While the results of these tests are given in shear strain, an equivalent tensile strain across the fibers can be determined. Fatigue tests on thin sheet panels should help establish the failure modes and point the way to a design allowable format.

Analysis

A simple strip theory approach is suggested for a post-buckling evaluation. The limits of this analysis, particularly for anisotropic plates, is

[3]Hahn, H. T., Symposium on Fatigue Behavior and Life Prediction of Composite Laminates, American Society for Testing and Materials, 20 March 1978.

[4]Fujezak, R. R., "Torsional Fatigue Behavior of Graphite Epoxy Cylinders," American Institute of Aeronautics and Astronautics presentation.

recognized. It is an analysis tool that requires engineering judgment. Post buckling, however, is a result of a forced displacement, and a smooth deflected form has to be established almost independent of the sheet elastic properties. An example of this procedure as applied to panel compression is shown in Fig. 3. The edge member represented by the stringer or longeron is assumed to remain stable up to a given ultimate strain value. After initial panel buckling, the center panel no longer deforms under direct compression and continued forced strain increases the depth of the buckles in the panel. The forced displacement shape at the center

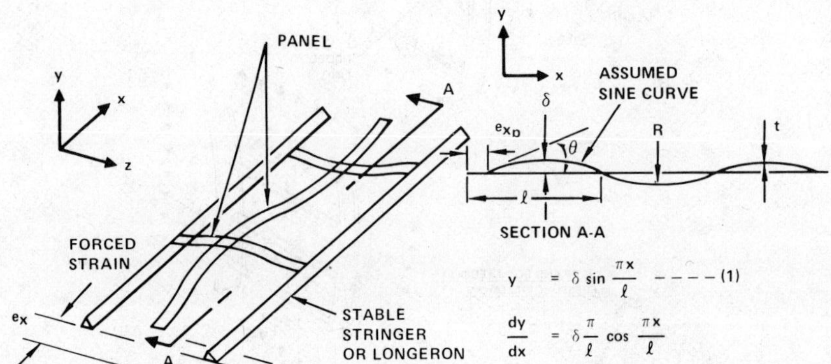

$$y = \delta \sin \frac{\pi x}{\ell} \quad ----(1)$$

$$\frac{dy}{dx} = \delta \frac{\pi}{\ell} \cos \frac{\pi x}{\ell}$$

$$\frac{-d^2y}{dx^2} = \frac{\delta \pi^2}{\ell^2} \sin \frac{\pi x}{\ell} = \frac{1}{R} \quad \text{FOR SMALL DEFLECTIONS}$$

WHERE:
t = PANEL THICKNESS
e_x = TOTAL STRAIN e_{xp}, POST BUCKLED STRAIN R = RADIUS OF CURVATURE

$$\text{LENGTH OF 1/2 WAVE} = S = 2\int_0^{\ell/2} \sqrt{dx^2 + dy^2}$$

$$\text{LENGTH OF 1/2 WAVE} = 2\int_0^{\ell/2} \left[1 + 1/2\left(\frac{dy}{dx}\right)\right]^2 dx \quad \left[\begin{array}{l}\text{BINOMIAL}\\ \text{THEOREM}\\ \text{SIMPLIFICATION}\end{array}\right]$$

$$e_{xp} = \frac{2\int_0^{\ell/2} \left[1 + 1/2\left(\frac{dy}{dx}\right)\right]^2 dx}{\ell} - 1 = \left(\frac{\delta\pi}{2\ell}\right)^2 \quad ----(2)$$

$$\delta = \frac{2\ell}{\pi}(e_{xp})^{1/2} \quad ----(3) \qquad \theta_{x=0} = \frac{\delta\pi}{\ell} \quad ----(4)$$

MAX BENDING STRAIN @ ℓ/2

$$e_{sb} = \frac{t}{2R} = \frac{\delta\pi^2}{\ell^2}\frac{t}{2} \quad ----(5)$$

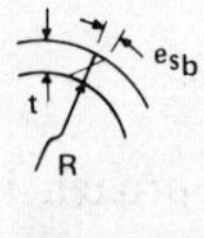

$$e_{sb} = \frac{\pi t}{\ell}(e_{xp})^{1/2} \quad ----(6)$$

$$\text{TOTAL SURFACE STRAIN COMPRESSIVE} = (e_x - e_{xp}) + \frac{\pi t}{\ell}(e_{xp})^{1/2} \quad ----(7)$$

FIG. 3—*Strip representation compressive forced displacement.*

of the panel is assumed to be represented by a sine curve, and the displacement magnitude is expressed in terms of the post-buckled strain. It can be observed that the assumed displacement shape results in:

1. The maximum deflection of the half wave being a direct function of the wave length and the square root of the edge member strain after panel initial buckling.
2. The deflection to wave length ratio remaining constant at a given post-buckled strain.
3. Induced bending surface strain in the compression strip being proportional to the thickness of the panel.

Analyses of the induced strains normal to the loading are somewhat more complex. The induced strains are dependent on the surrounding restraints. A strip representation that includes stiffener attachment peeling forces is shown in Fig. 4. The analyses assume a complete lateral restraint of the panel in the local buckled area. Tensile strains calculated using this simplified analysis would represent the upper limit of the strains that could be realized. The lateral restraint on a compression test panel depends on the fixture, edge number, and cross member stiffeners. The induced lateral membrane force causes a sharpening of the radius of curvature at the stiffener that is dependent on the level of stiffener restraint.

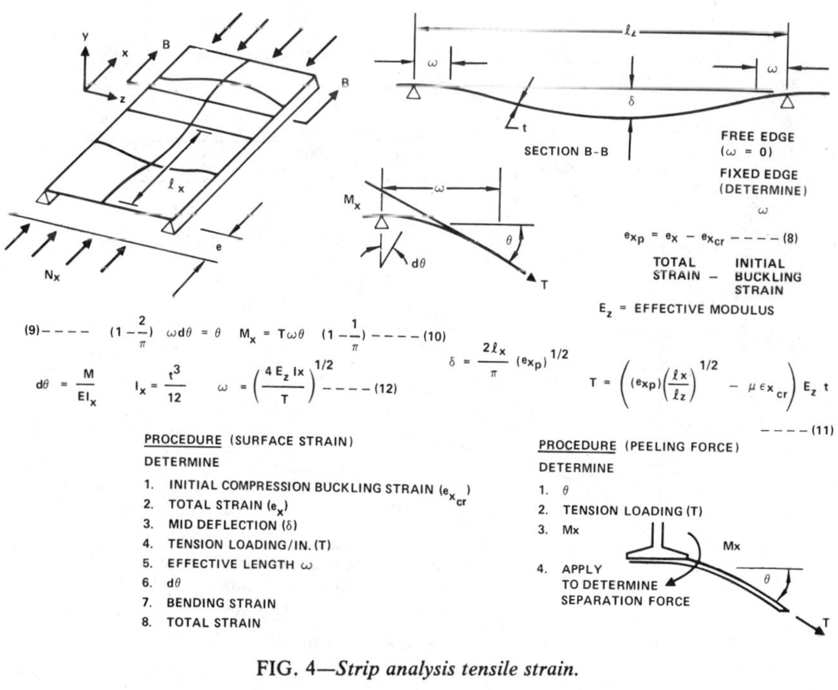

FIG. 4—*Strip analysis tensile strain.*

Application of the strip analysis given in Fig. 3 to an idealized 152.4 mm (6-in.) wide compression panel yields the compression surface strains versus applied strains shown in Fig. 5. To highlight the effect of width to thickness ratio (b/t), four thicknesses are shown.

Compression Panel Test Results

In 1976 at Lockheed, a compression panel of the dimensions shown in Fig. 6 was tested to destruction. The object was to verify by test the stability of a typical T300/5209 graphite/epoxy hat stiffened panel containing three stiffeners and a single rib attachment. Shadow moiré techniques were employed to define the buckle wave forms, and strain gages were used to measure the panel response to compressive loading. A schematic diagram of the shadow moiré test arrangement is also given in Fig. 6. A photo of the pattern near failure load is given in Fig. 7. Out-of-plane panel skin surface displacements derived from the pattern are given in Fig. 8. Displacements calculated using the strip analysis and the measured values are compared. Failure occurred at the 0.004 strain level. A strip analysis assessment of surface strains and peeling indicates failure could be expected at this level.

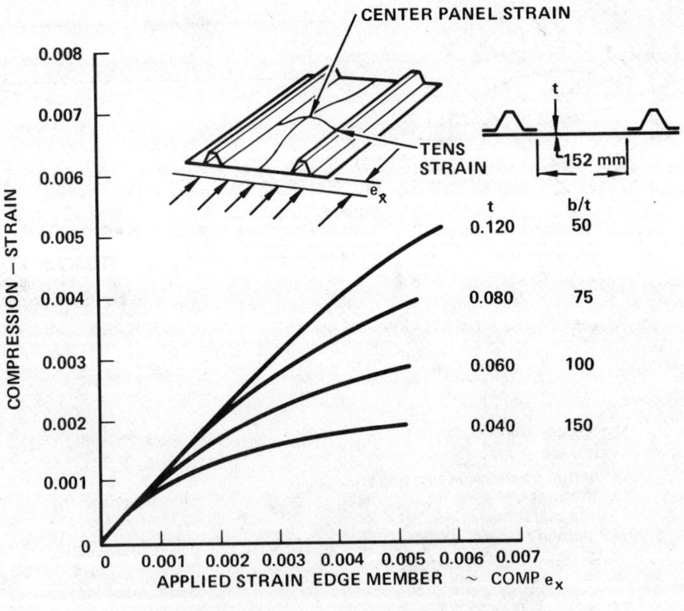

FIG. 5—*Longitudinal strain—post buckled panel.*

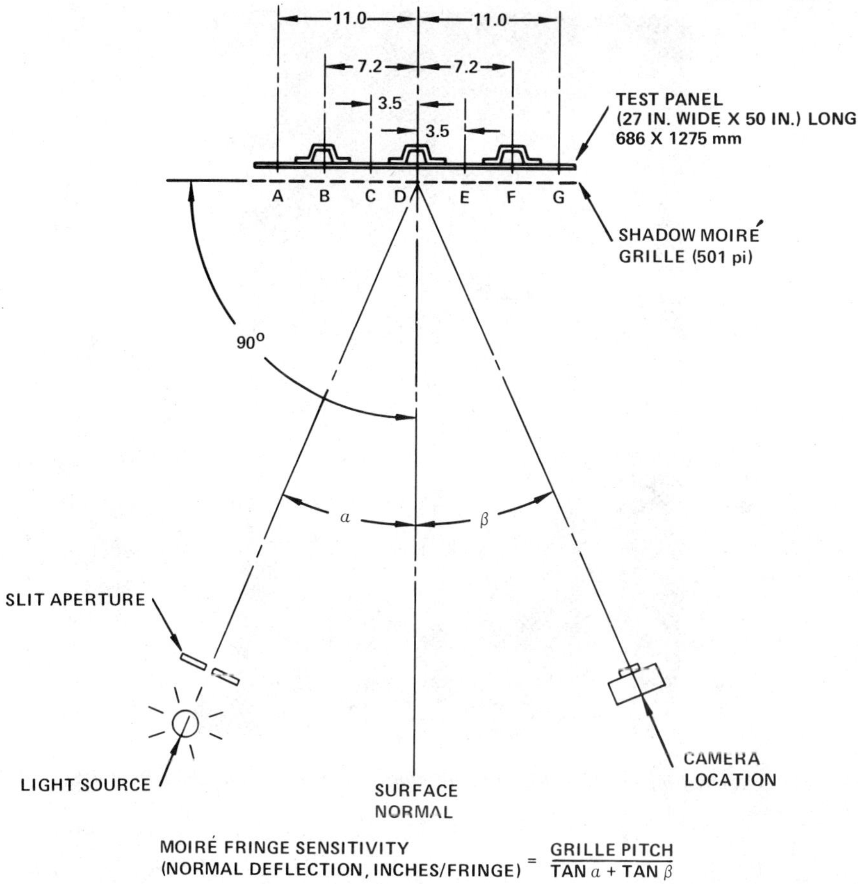

FIG. 6—*Shadow moiré test arrangement.*

Shear Post Buckling

Under independent research, a program was initiated at Lockheed to determine the post-buckling behavior of thin graphite/epoxy sheet composite panels in shear. The program had the following objectives.

1. Determine the post-buckling behavior of thin sheet composite panels in shear up to ultimate strength.
2. Explore the fatigue capability of panels subject to repetitive buckling.

One 8-ply, 1-mm (0.040-in.) T300/5208 graphite/epoxy shear panel and one 12-ply panel were statically tested to ultimate. One 8-ply panel was fatigue cycled in shear under a random fatigue spectrum representing

FIG. 7—*Moiré fringe patterns depicting buckling near failure. Moiré pattern at 415 000 N (93.3 kips).*

three lifetimes (based on a fuselage side panel loading). After this test, no damage was detected.

Shadow moiré techniques were used to define the skin buckling wave form to static ultimate load.

Formed blocks with specific contour depths were used to calibrate fringe patterns. The initiation of buckling was determined from the computer plot of applied load versus back-to-back strain gage readings. Electrical resistance foil gages matched for the coefficient of expansion of graphite were used. The fatigue cycled test panel installed in a closed loop, load controlled MTS hydraulic test machine is shown in Fig. 9. Views of the test equipment and data retrieval units are shown in Fig. 10. The extent of panel skin buckling at a shear flow of 153 N/mm (874 lb/in.) is shown in Fig. 11. The moiré pattern after panel failure 157 N/mm (899 lb/in.) shear flow is shown in Fig. 12. A view of the failed static shear panel with the shadow moiré grille screen removed is shown in Fig. 13. The maximum out-of-plane skin surface displacements for the center bay were determined from the moiré fringe and are shown in Fig. 14. For

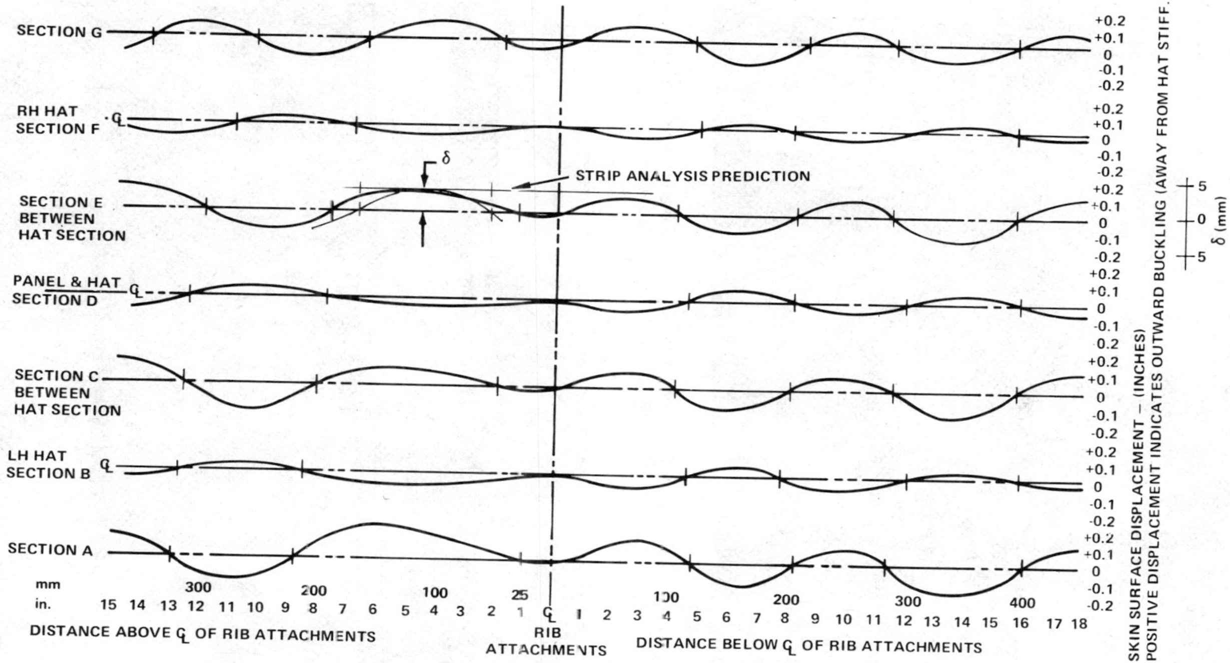

Analysis: Eff length = 4.7 e_{cr} = .0025 e_x = .004
e_x-e_{cr}=.0015 δ = .12

FIG. 8—*Out-of-plane surface displacements at an axial compression load of 415 000 N (93.34 kips); T300/5209 graphite/epoxy hat stiffened panel.*

FIG. 9—*Shear panel test set up in MTS machine for fatigue cycling in shear random fatigue spectra.*

FIG. 10—*Test equipment visicorder for monitoring load frequency modulus (FM) tape load signal generator and ancillary equipment.*

FIG. 11—*Moiré fringes depicting skin buckling at a shear flow of 153 N/mm (874 lb/in.) in static test panel.*

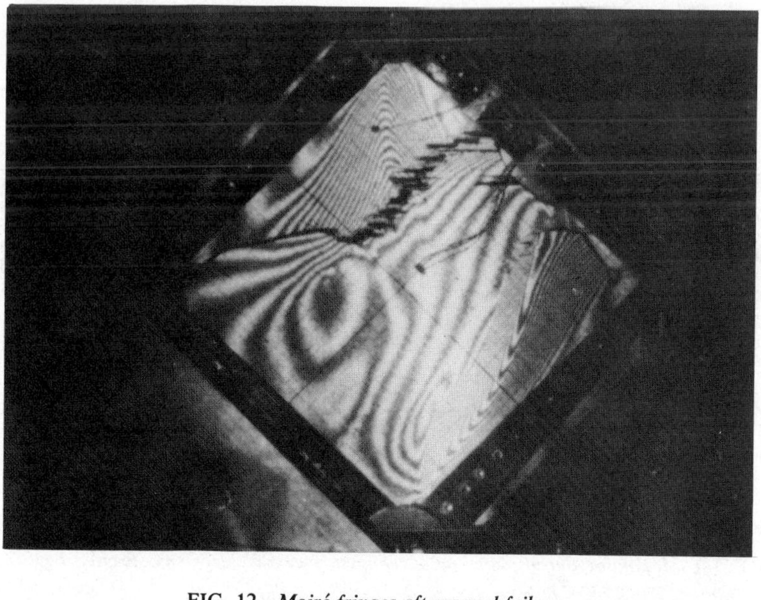

FIG. 12—*Moiré fringes after panel failure.*

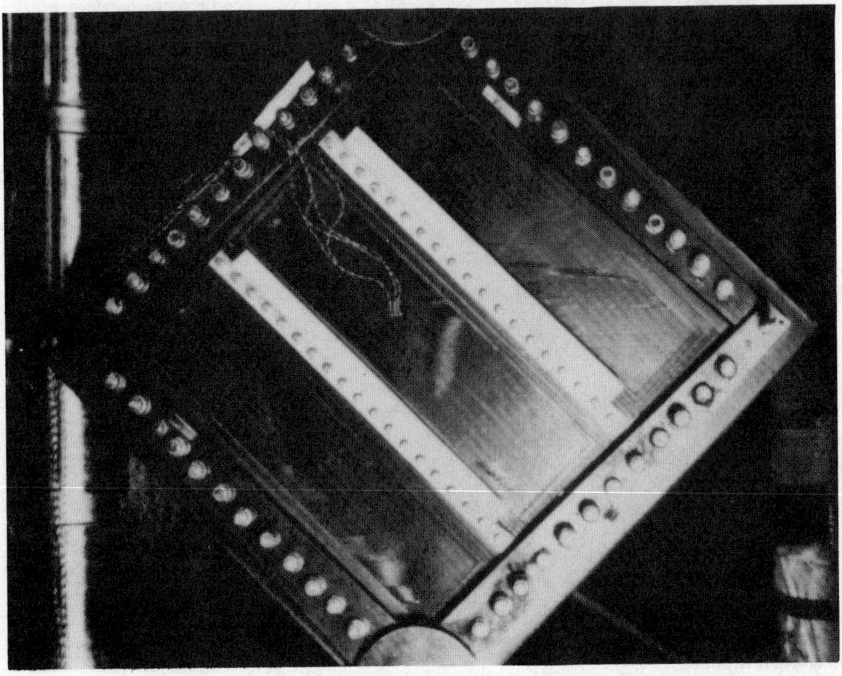

FIG. 13—*Stiffener side of panel after failure.*

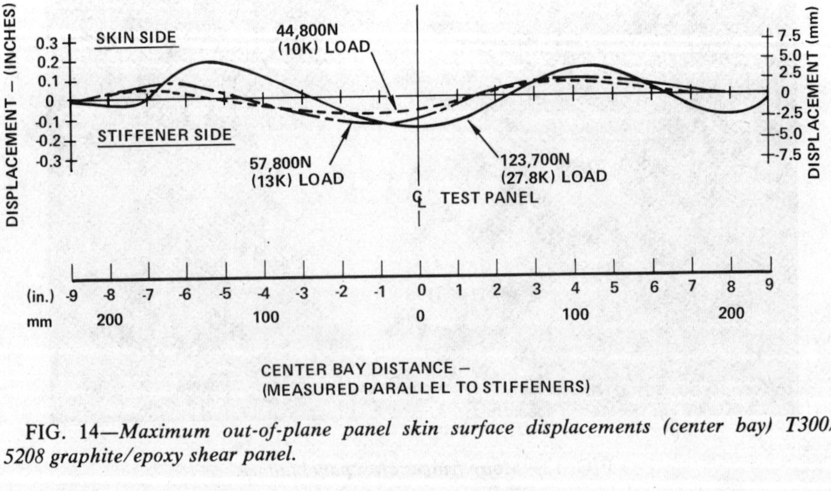

FIG. 14—*Maximum out-of-plane panel skin surface displacements (center bay) T300/5208 graphite/epoxy shear panel.*

analysis comparison, the wave should be plotted against the buckle minimum and maximum axes.

Application of the strip analysis technique to shear post buckling is shown in Fig. 15. A comparison between the deflection and strain determined from the analysis and the 8-ply shear panel test results are shown

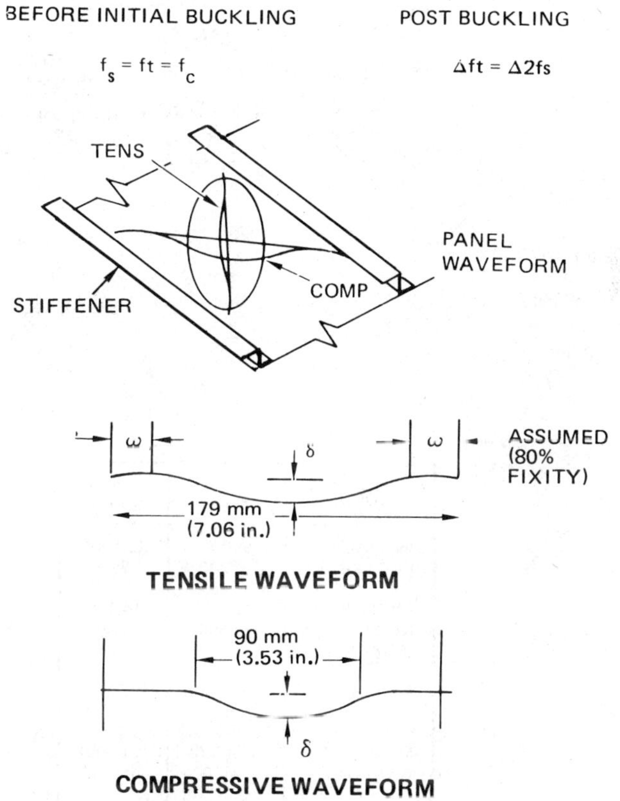

ANALYSIS PROCEDURE

- DETERMINE INITIAL BUCKLING STRAIN

$\frac{ft}{E} = 0.785\ Pcr/E,$ Pcr (lbs), E (psi) $\frac{ft}{E} = 1.217\ Pcr/E$ SI (units) Pcr (N), E (MPa)

DETERMINE POST BUCKLED STRAIN
(P-Pcr) (2) (.785/E)

- CALC δ
- CALC ω
- DETERMINE BENDING PLUS DIRECT STRAINS

FIG. 15—*Shear analysis.*

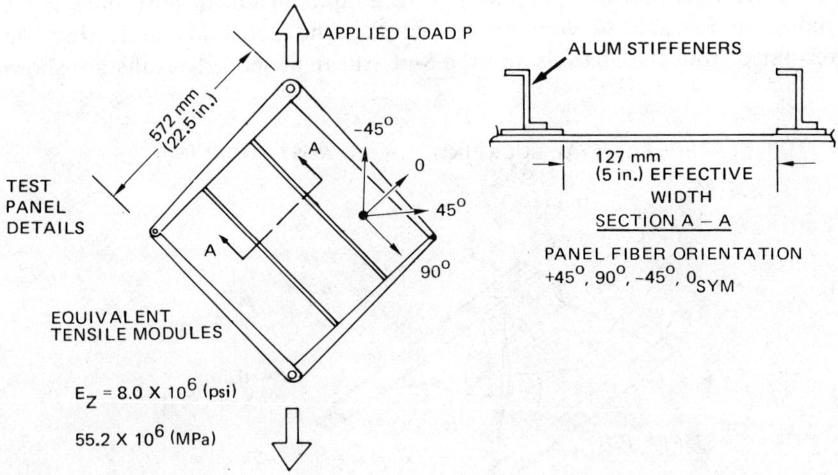

STATIC TEST RESULTS

COND.	P N (lb)	fs MPa (psi)
INITIAL BUCKLING	(4,000) 17800	(3,100) 21.7
EXTENSIVE NOISE (CRACKING)	(18,000) 80000	(14,130) 97.4
FAILURE	(28,400) 126000	(22,300) 153
MOIRE GRID DATA POINT	(27,800) 124000	(21,800) 150.8

MOIRÉ GRID DATA POINT ANALYSIS ~ TEST COMPARISON	STRIP ANALYSIS	MEASURED TEST
δ mm (in.)	3.9 (0.150)	3.95 0.155
MID STRAIN TENSILE	0.0055	0.006
MID STRAIN COMPRESSIVE	–0.003	–0.003

FIG. 16—*Analysis-test results comparison.*

in Fig. 16. Typical strain gage results for the 12-ply panel along with analysis values are shown in Fig. 17.

Fatigue Results

The fatigue spectrum applied to the 8-ply panel is shown in Fig. 18. Panel tension strains versus applied cycles are presented. Also shown is a tangent theory summation.

Conclusions

From the limited test results and the first level simplified analysis assessment the following conclusions are drawn.

1. A post-buckling static capability in compression and shear for flat panels has been demonstrated.
2. A post-buckling repeated loading capability in shear for flat panels has been demonstrated.

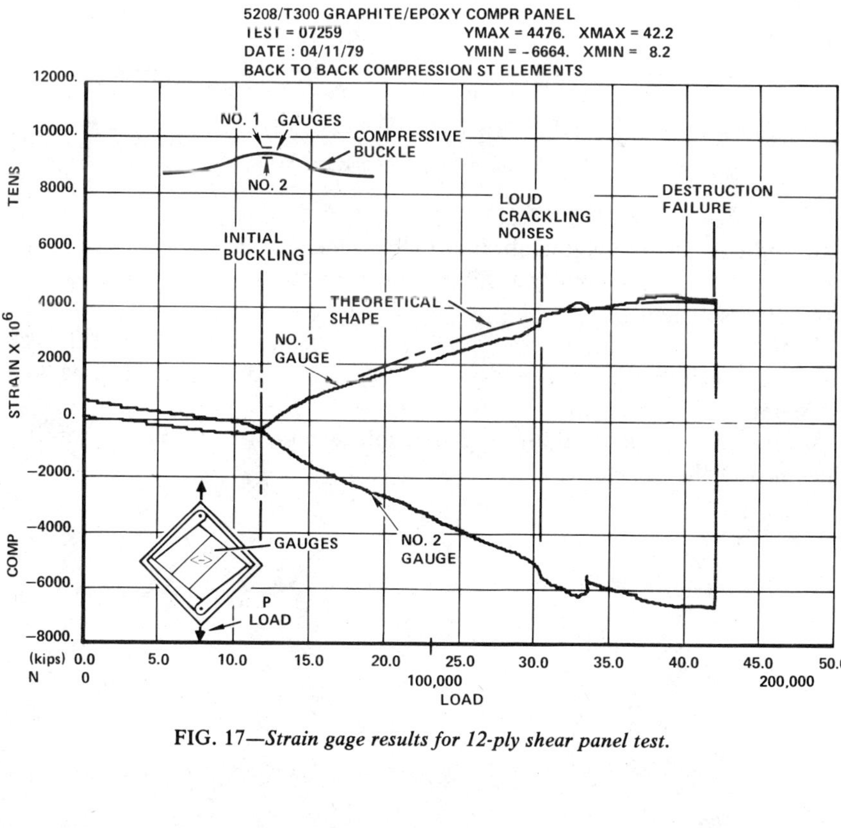

FIG. 17—*Strain gage results for 12-ply shear panel test.*

8 PLY PANEL ~ 1 mm (0.040) SHEAR PANEL FATIGUE TEST

TEST RESULTS
THREE LIFETIME EQUIVALENT LOADINGS (78000 HRS) NO APPARENT DAMAGE
ONE LIFETIME WITH CENTER HOLE ADDED (NO DAMAGE)

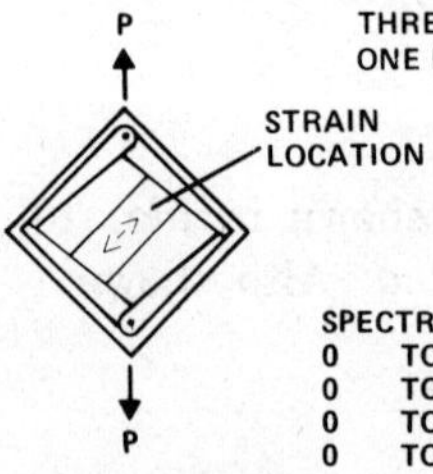

SPECTRUM	$\frac{q}{q_{cr}}$	P N (LB)	APPLIED CYCLES	TANGENT Σ THEORY	e MAX APPLIED TENSILE STRAIN
0 TO 0.52 LIMIT	1.56	32,500 (7,300)	395,200	436,000	0.002
0 TO 0.75 LIMIT	2.25	45,800 (10,300)	39,500	41,268	0.0035
0 TO LIMIT	3.0	62,300 (14,000)	1,560	1,768	0.004
0 TO 1.25 LIMIT	3.75	77,800 (17,500)	208	208	0.006
			Σ 436,000		

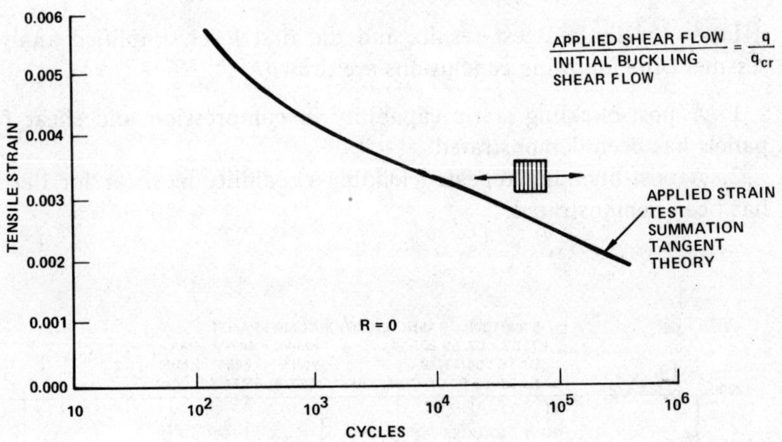

FIG. 18—*Shear panel fatigue test data.*

3. Induced bending membrane displacement can be assessed with simple strip analysis, and the induced surface strains provide a basis for strength assessments.

4. High induced strains and peeling forces can exist at the stiffeners and are amenable to analysis assessment provided edge restraints are evaluated.

5. The strains induced by forced displacement are a function of the thickness of the panels.

C. Y. Kam[1]

Bolt Hole Growth in Graphite-Epoxy Laminates for Clearance and Interference Fits When Subjected to Fatigue Loads

REFERENCE: Kam, C. Y., **"Bolt Hole Growth in Graphite-Epoxy Laminates for Clearance and Interference Fits When Subjected to Fatigue Loads,"** *Fatigue of Fibrous Composite Materials, ASTM STP 723,* American Society for Testing and Materials, 1981, pp. 21-30.

ABSTRACT: This paper presents the results of an experimental program that was conducted to evaluate the damage to the bolt hole as related to the fit of the bolt to the hole in the graphite-epoxy laminate when the joint is subjected to a fatigue load spectrum. The experimental program was conducted using a double-lap bolted-joint test specimen. Three types of bolt fit were used in the experiment: interference fit, clearance fit, and clearance fit with wet sealant. The bolts were torqued to values representing standard installation torque. The graphite-epoxy laminate used for the test specimens was pseudo-isotropic laminate of $(0/90, \pm45)_{4S}$. The prepreg was T300/5208 and the laminate was cured in the standard Narmco 5208 cure cycle.

The test program consisted of applying cyclic loadings to produce bearing stress levels of 397×10^6 N/m^2 (+57 600 psi) for 50 000, 100 000, 200 000, and 500 000 cycles. The bolted specimens were disassembled and the bolt hold diameters measured for hole growth. Preliminary results indicated that hole growth will occur when the bearing stresses are above 50 000 psi. Bolts installed in clearance holes with the PR1422 sealant had the same hole growth as bolts installed without the sealant.

KEY WORDS: fatigue tests, graphite-epoxy, bolted joints, fatigue (materials), composite materials

The use of graphite-epoxy material in structural components of transport aircraft appears to be rapidly increasing. New and derivative airplanes are being developed that will have control surfaces, vertical and horizontal tails, floor beams, gear doors, trailing edges of the wing, and vertical and horizontal stabilizers manufactured from composite materials. Large development efforts support this proliferation of composite structures. The most visible support is the National Aeronautics and Space Administration's (NASA)

[1] Unit chief-Design, Structural Composites Technology, Douglas Aircraft Co., Long Beach, Calif. 90846.

Aircraft Energy Efficient (ACEE) Composite Structures Program. The graphite-epoxy structural components under development and in flight service use a variety of mechanically fastened joint configurations in the assembly of the component and in the installation of the structure. The installation method used for installing the mechanical fasteners was adopted from methods already established for metal-alloy structures; for example, the use of titanium bolts in a clearance hole. Recent tests indicate that the use of bolts in an interference fit hole may be beneficial in advanced composite structures.

The design of structural components for revenue-producing aircraft is usually dominated by the requirement for long service life, which means that the structure must be tolerant to fatigue load cycles. It has been well documented that commercial airplanes in service can easily accumulate 60 000 or more flight hours. The long service life requirement also results in a large number of landings that generally produces the high-fatigue stresses. It is of interest to compare the service life of a passenger-carrying aircraft to the various types of aircraft used by the military. This comparison, shown in Table 1, notes that some military transport/cargo airplanes can log 50 000 flight hours; however, the number of landings is only about one-half as many as for civil transports. Thus, it can easily be deduced that the design of bolted joints for civil transports must be as fatigue-resistant as possible.

Failure Modes of Composite Joints

Examinations of many bolted joints often indicate that the predominant load transfer in the joint is by shear in the attachments rather than by tension. Thus, depending on the joint geometry and the relationship of bolt diameter, bolt spacing, edge distance, and laminate thickness, the failure of the joint may occur in the specific modes shown in Fig. 1. However, tests have shown that the particular failure mode is directly influenced by the joint geometry. Net-tension failures will occur when the bolt hole is a large fraction of the bolt spacing. Shear-out failures will occur when the bolt is located too close to the edge of the laminate in the direction of the load. Shear-out failures can also occur in laminates that are highly orthotropic even when the

TABLE 1—*Service life of aircraft.*

Types	Flight Hours	Number of Landings	Usage, hour/year
Commercial transports	60 000 (minimum)	20 000 to 60 000	3000
Military cargo/transports	15 000 to 50 000	20 000 to 25 000	500
Trainers	15 000 to 25 000	10 000 to 40 000	1500
Bombers	15 000	5000	300
Fighters	6000 to 8000	8000 to 10 000	300

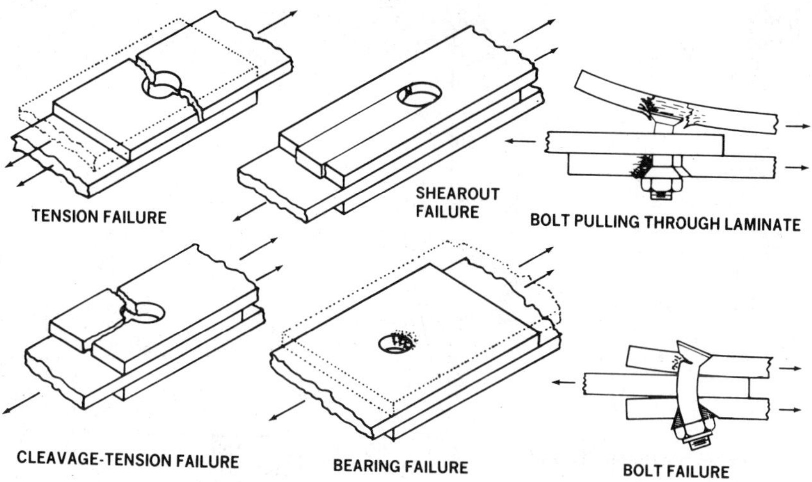

FIG. 1—*Possible failure modes for bolted joints in advanced composites.*

edge distance is very large. Cleavage failure will usually occur when the edge distance of the bolt hole to the laminate is small and the laminate has low bending strength for resisting the load applied by the bolt to the thin strip of laminate. Bearing stress failures will occur when the bolt diameter is a fraction of the bolt spacing.

Two other failure modes should be mentioned: the pulling of the bolt head through the laminate, and the failure of the bolt itself by the bending loads imposed on the bolts. In general, most of these failure modes can be prevented by proper selection of the joint geometry and laminate-layup configuration. However, bearing failures can be induced in the composite laminates when repeat loads are applied to the bolted joint even though the bearing stress was acceptable for the static-load condition.

Since fatigue-load cycles are a dominant design parameter, and the results of recent tests show that interference-fit fasteners can result in higher composite laminate-failure loads, an experimental program was initiated to explore the fatigue damage of the bolt hole when the bolt fit is either a clearance fit or an interference fit.

Test-Specimen Preparation

A test-specimen configuration was selected that was bearing-stress critical and that permitted the application of a fatigue-load ratio of $R = -1.0$. The composite laminate was made from T300 biwoven cloth, and impregnated with Narmco's 5208 epoxy resin system. The layup configuration (Fig. 2) selected was pseudo-isotropic, $(0/90, \pm 45)_{4S}$, which resulted in an average

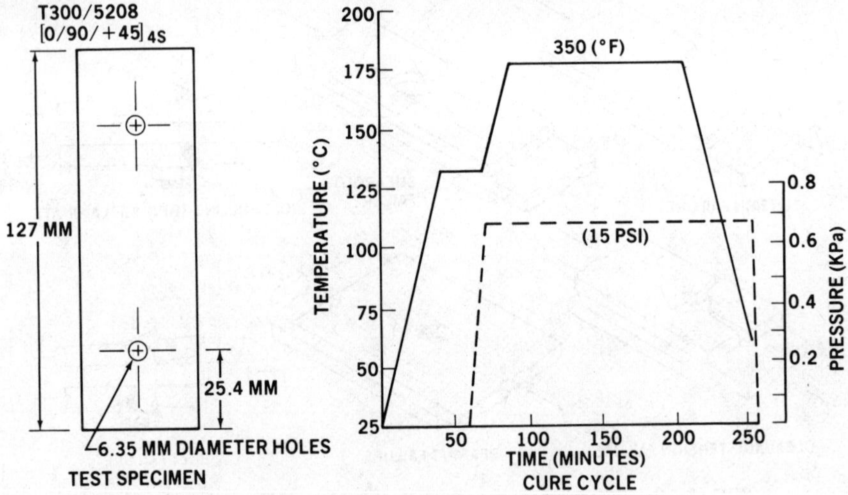

FIG. 2—*Composite specimen configuration and cure cycle.*

laminate thickness of 4.32 mm (0.17 in.). The test coupon was 127 by 44.5 mm. The cure cycle is also given in Fig. 2.

After the laminates were cured, the panel was subjected to C-scan for check of any delaminations or high-porosity areas. Resin and void content measurements from the cured panel showed a resin content of 30.2 percent by weight and a void content of 0.33 percent. The individual test specimens were then machined from the panel and the 6.35 mm (0.25 in.) diameter holes were drilled into the composite specimens. The completed test specimen was assembled as shown in Fig. 3. The 6.35 mm-diameter Hi-lok fasteners were installed and torqued to 7.9 N·m (70 in-lb).

Fatigue Tests

The fatigue-load cycles were applied with an MTS machine at the rate of 3 Hz and at the maximum load level of 13 789 N (3100 lb). The load was cycled at full reversal to apply a bearing stress of 397 $\times$ 10^6 N/m^2 (57 600 psi) to each side of the bolt hole.

The test matrix was selected to evaluate the effect of the fit of the bolt to the hole diameter, such as (1) the bolt in a clearance hole, (2) the bolt in a clearance hole and installed with a wet sealant, and (3) the bolt in an interference fit hole. The specimens were subjected to 50 000, 100 000, 200 000, and 500 000 loading cycles. After the load cycles, the specimens were dismantled and the hole diameter in the graphite-epoxy laminate was measured. The test data are noted in Tables 2 through 4.

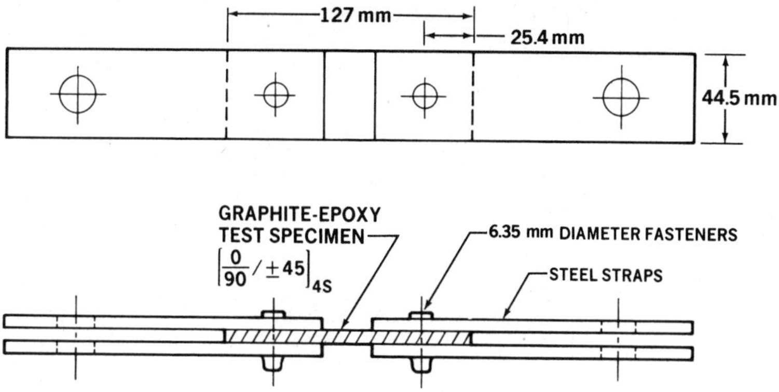

FIG. 3—*Test specimen configuration.*

Results and Discussion

The tests show that when clearance-fit bolted-joint specimens are subjected to a full reversal cyclic load that results in a bearing stress level of 397×10^6 N/m^2 in a graphite-epoxy laminate, the hole diameter appears to have a constant growth with relation to the number of cyclic loads. The hole diameter growth versus number of cyclic loads is shown in Fig. 4. In addition to the clearance hole fit, tests were also run on specimens that were assembled using a standard wet sealant. The application of wet sealant appears to have no effect on the hole diameter growth. These data points are also shown in Fig. 4.

One interesting observation was noted from Test Specimen No. 2-20A that showed an abnormal amount of hole growth. The measured growth was 0.246 mm (0.0097 in.) after 500 000 cycles. This large amount of hole growth was attributed to the bolt shank slipping back and forth in the hole and thus pounding on the graphite laminate at each load cycle. The other half of the test specimen, No. 2-20B, showed that the hole diameter growth was about as expected. Visual examination of Test Specimen 2-20A hole showed a layer of powdered graphite-epoxy sticking to the sides of the hole. This large amount of hole growth and powdering is similar to some earlier tests where the bolt was observed to be sliding back and forth in the hole.

The effect of hole diameter growth when interference bolts were used was also investigated. The results of these tests are shown in Fig. 5. In addition to the apparent hole diameter growth, Fig. 5 also shows the amount of the interference fit as indicated by the solid symbols. Although the data show some amount of hole diameter growth after the fatigue test cycles, in no case did the amount of growth exceed the amount of the interference fit. It was clearly noted that the interference fit bolts did not have any hole growth greater than the initial hole expansion due to the interference fit because each bolt had to

TABLE 2—*Hi-lok bolt in clearance hole.*

Specimen No.	Shank Diameter		Initial Hole Diameter		Number of Cycles	Final Hole Diameter	
	A	B	A	B		A	B
2-17	6.330[a] (0.2492)[b]	6.325 (0.2490)	6.350 (0.2500)	6.345 (0.2498)	50 000	6.353 (0.2501)	6.350 (0.2500)
2-18	6.327 (0.2491)	6.325 (0.2490)	6.355 (0.2502)	6.355 (0.2502)	100 000	6.363 (0.2505)	NO DATA
2-19	6.327 (0.2491)	6.327 (0.2491)	6.345 (0.2498)	6.350 (0.2500)	200 000	6.350 (0.2500)	6.358 (0.2503)
2-20	6.327 (0.2491)	6.327 (0.2491)	6.375 (0.2510)	6.350 (0.2500)	500 000	6.622 (0.2607)	6.350 (0.2505)

[a] Millimetres.
[b] Inches.

TABLE 3—*Hi-lok bolt in clearance hole installed with wet sealant.*

Specimen No.	Shank Diameter		Initial Hole Diameter		Number of Cycles	Final Hole Diameter	
	A	B	A	B		A	B
2-21	6.327[a] (0.2491)[b]	6.327 (0.2491)	6.345 (0.2498)	6.347 (0.2499)	50 000	6.350 (0.2500)	6.350 (0.2500)
2-22	6.327 (0.2491)	6.327 (0.2491)	6.350 (0.2500)	6.350 (0.2500)	100 000	6.350 (0.2500)	6.360 (0.2504)
2-23	6.327 (0.2491)	6.327 (0.2491)	6.363 (0.2505)	6.373 (0.2509)	200 000	6.375 (0.2510)	BOLT FAILURE
2-24	6.327 (0.2491)	6.327 (0.2491)	6.363 (0.2505)	6.350 (0.2500)	500 000	BOLT WOBBLED (0.3705)	6.350 (0.2500)

[a] Millimetres.
[b] Inches.

TABLE 4—*Hi-lok bolt in an interference hole.*

Specimen No.	Shank Diameter		Initial Hole Diameter		Number of Cycles	Final Hole Diameter	
	A	B	A	B		A	B
2-25	6.330[a] (0.2492)[b]	6.330 (0.2492)	6.259 (0.2464)	6.261 (0.2465)	50 000	6.279 (0.2472)	6.327 (0.2491)
2-26	6.327 (0.2491)	6.327 (0.2491)	6.269 (0.2468)	6.261 (0.2465)	100 000	6.276 (0.2471)	6.276 (0.2471)
2-27	6.327 (0.2491)	6.327 (0.2491)	6.270 (0.2470)	6.274 (0.2470)	200 000	6.325 (0.2490)	6.302 (0.2481)
2-28	6.327 (0.2491)	6.327 (0.2491)	6.274 (0.2470)	6.274 (0.2470)	500 000	6.325 (0.2490)	6.289 (0.2475)

[a] Millimetres.
[b] Inches.

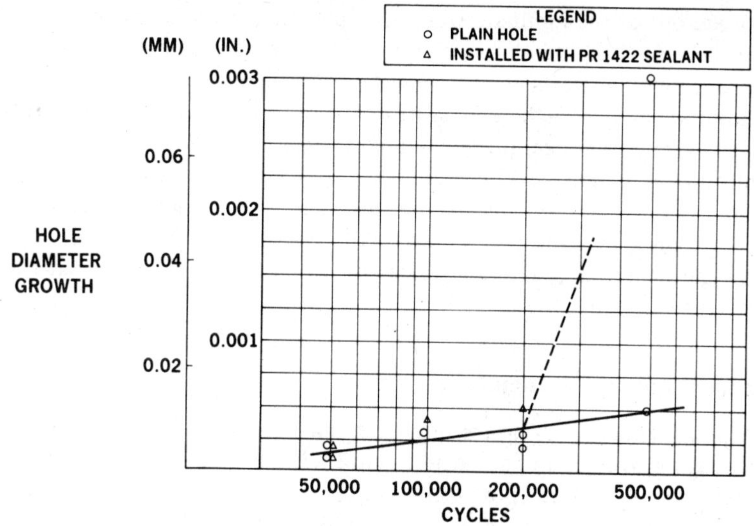

FIG. 4—*Effect of cyclic loading on clearance fit hole diameter.*

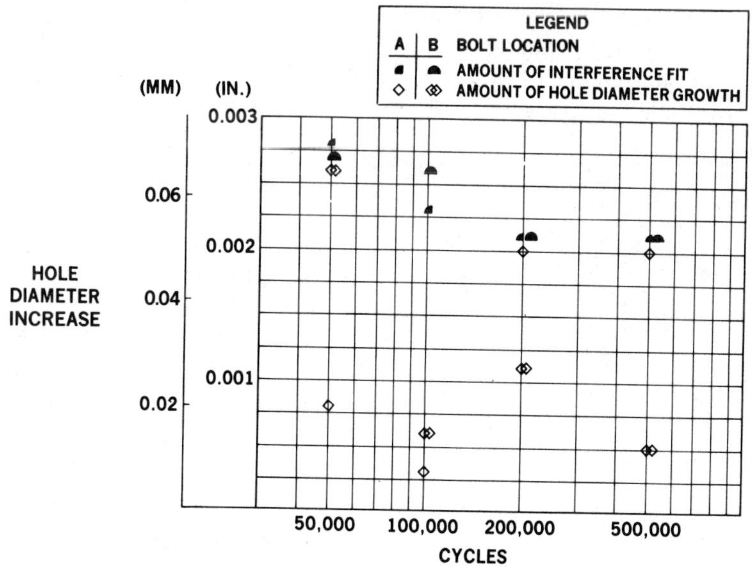

FIG. 5—*Effect of cyclic loading on interference fit hole diameter.*

be forced out of the laminate when the specimens were being disassembled. Since the bolts had to be pushed out of the holes, it was not certain whether the resulting hole size as measured was due to the impact of the bolt shank on the laminate or was the hole expansion due to forcing the bolt into and out of the drilled hole. On examination, the interference fit holes did not show the powdering noticed in Test Specimen No. 2-20A. It may also be surmised that the hole diameter expansion can be the result of the hole size being permanently expanded by the bolt shank or the hole elastically shrinking back to nearly the drilled-hole diameter after the bolts were removed from the test specimens.

Summary and Observations

This experimental program indicates that clearance-fit bolted joints with high bearing stresses and subjected to full reversal fatigue loadings can cause damage to the graphite-epoxy laminate. The damage is most likely caused by the back-and-forth impact of the bolt shank on the composites laminate. The use of interference fit bolts appears to eliminate the back-and-forth slipping of the bolt shank in the bolt hole. Although some hole diameter growth was detected after the fatigue tests on the interference fit bolts, the exact cause of the hole diameter growth could not be easily determined because of the interference fit condition existing after the test loadings. However, the following observations can be made:

1. Clearance fit bolts can result in hole diameter growth when the bolted joint is subjected to fatigue loadings.
2. Bolt slipping in the graphite-epoxy laminate can be prevented by the use of an interference fit of bolt to hole.
3. Multiple bolt patterns need to be investigated.

D. Schütz,[1] *J. J. Gerharz,*[1] *and E. Alschweig*[1]

Fatigue Properties of Unnotched, Notched, and Jointed Specimens of a Graphite/Epoxy Composite

REFERENCE: Schütz, D., Gerharz, J. J., and Alschweig, E., **"Fatigue Properties of Unnotched, Notched, and Jointed Specimens of a Graphite/Epoxy Composite,"** *Fatigue of Fibrous Composite Materials, ASTM STP 723,* American Society for Testing and Materials, 1981, pp. 31–47.

ABSTRACT: Within a continuing program on high tensile graphite/epoxy composite, stress-strain, axial fatigue, and compliance behavior of unnotched, notched (3-mm diameter hole), and jointed specimens made of $[0_2/\pm 45/0_2/\pm 45/\overline{90}]_s$ T300/914C laminates (177°C curing temperature) have been studied. In addition, the behavior of unnotched specimens cut from (1) the same laminate but with the longitudinal specimen axis now perpendicular to the zero-degree fiber direction, and (2) the high modulus fiber laminate with the same build-up was investigated.

Stress-strain curves, *S-N* curves, and increase-in-compliance versus percentage-of-total life curves were determined for all specimen types for stress ratios, *R*, ranging from $R = +5.0$ (compression-compression cycling, C-C) to $R = +0.1$ (tension-tension cycling, T-T).

An overall comparison of results from specimens with different stress raisers shows that the stress raisers diminish fatigue strength in the low-cycle range, but in the high-cycle range their influence has vanished. Effective stress concentrations were found to be different for compression and tension. During T-C cycling, increase of compliance was lowest for the fastener-filled no-load transfer joint and largest for the single-shear load transfer joint. The large compliance changes of the load-transfer specimens were attributed to increased bearing damage.

In general, the scatter in static and fatigue strength was found to be comparable with that for similar features in metals. When the plain material was loaded transversely instead of longitudinally, static and fatigue strength were lower by a factor of about 3.

KEY WORDS: composite materials, fatigue (materials), notches, notch sensitivity, joints, compliance, failure modes, scatter, *R*-value (influence of mean stress), high tensile strength fibers, high modulus fibers, loading direction

Within the course of development of advanced composites, only recently fatigue investigations were extended to tension-compression and compression-compression loading, thus giving a more complete picture of the fatigue

[1] Chief, Research Group for Fatigue Analysis and Fracture Mechanics, and fatigue research engineers, respectively, Fraunhofer-Institut für Betriebsfestigkeit (LBF), Darmstadt, Germany.

behavior of composites. Besides plain specimens [*1,2*], also notched [*2–5*] and jointed specimens [*5*] were included in these investigations. It was found that generally tension-compression loading is more critical than tension-tension or compression-compression loading [*1,2,6*].

To find characteristic correlations in critical tension-compression fatigue behavior between plain, notched, and jointed specimens and to establish design data, a research program was carried out at the Fraunhofer-Institut für Betriebsfestigkeit (LBF) using the same prepreg system (graphite/epoxy), the same laminate structure, and the same manufacturing procedure for all specimens. Special attention was given to jointed specimen design covering a wide range of practical joint configurations.

A detailed report has previously been given to the Society of Environmental Engineers Conference in London on 29 June 1977, concerning the fatigue behavior of the plain specimens investigated [*6*]. Results of the fatigue behavior investigations on notched and jointed specimens and how they compare to plain specimen fatigue behavior will now be discussed and, in addition, results of investigations into the effect of fiber material (high strength and high modulus fiber) and load direction will be presented.

Specimens and Testing Procedure

All specimens were made from the same prepreg system, consisting of 60 percent by volume of high strength fiber T300 from Toray and epoxy resin 914C from Ciba. From this prepreg material, the firms of Dornier, Messerschmitt-Bölkow-Blohm GmbH, and VFW-Fokker manufactured, in an autoclave, uniform plates of multidirectional composite from which the specimens were cut.

The plates consisted of 17 layers having a central 90-deg layer and equal contents of 0- and 45-deg layers.

The ratio of diameter of the hole to the width of the specimen was $d/w \approx 0.2$ for all notched and jointed specimens. The non-waisted notched specimens had a 3-mm hole. In the case of jointed specimens, 4-mm Hi-Lok bolts were used; these were installed as a press fit (interference, $\Delta d = 0.007$ to 0.037 mm) and with a clamping torque of 2 Nm. The three types of jointed specimen shown in Fig. 1 represent constructional details close to those encountered in practice, and differ in the load transfer (F_{LT}/F) and secondary bending (ϵ_b/ϵ_{ax}) parameters. While load transfer values were calculated, the secondary bending values given for jointed specimens of Type 3 were based on double strain gage measurements. For Type 2, an edge distance corresponding to $e/d = 3.75$ was chosen. With all specimens, extensometers carrying strain gages were used in the test; these recorded the longitudinal extension of the specimens in the vicinity of the notches during test loading (the gage lengths are given, for the individual specimens, in Fig. 1). In the case of the plain specimen, the notched specimen, and the

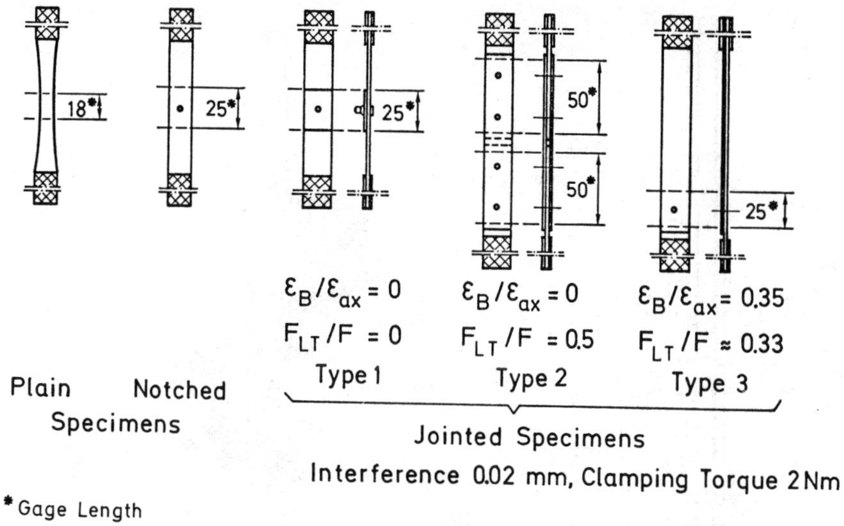

FIG. 1—*Forms of specimens, laminate structure:* $[0_2/\pm 45/0_2/\pm 45/\overline{90}]_s$.

Type 1 jointed specimen, changes in length were measured thereby, while in the case of jointed specimens of Types 2 and 3, the measurement taken was the displacement of jointed parts of the specimen in relation to each other, in accordance with MIL-STD 1312. For this purpose, the tips of the extensometer were placed on the sides of the jointed parts of the specimen. For clarification, Fig. 2 shows the positioning of extensometers on the jointed specimen of Type 2.

All specimens were secured in the hydraulic clamps of the servohydraulic test machine between smooth jaws using a pressure of about 50 N/mm[2].

Results and Discussion

Stress-Strain Behavior

Before fatigue strength tests were carried out, static tensile and compressive tests involving stress-strain measurements were performed. Figure 3 shows the stress-strain curves for the three different jointed specimens subjected to tension and compression; these are compared with the curve for the plain specimen. In addition, the figure gives, in tabular form, the mean values of tensile and compressive strengths and of deformations at failure. A striking feature is the considerable reduction in the strengths of notched and jointed specimens compared to values for plain specimens. Also note

[2] The italic numbers in brackets refer to the list of references appended to this paper.

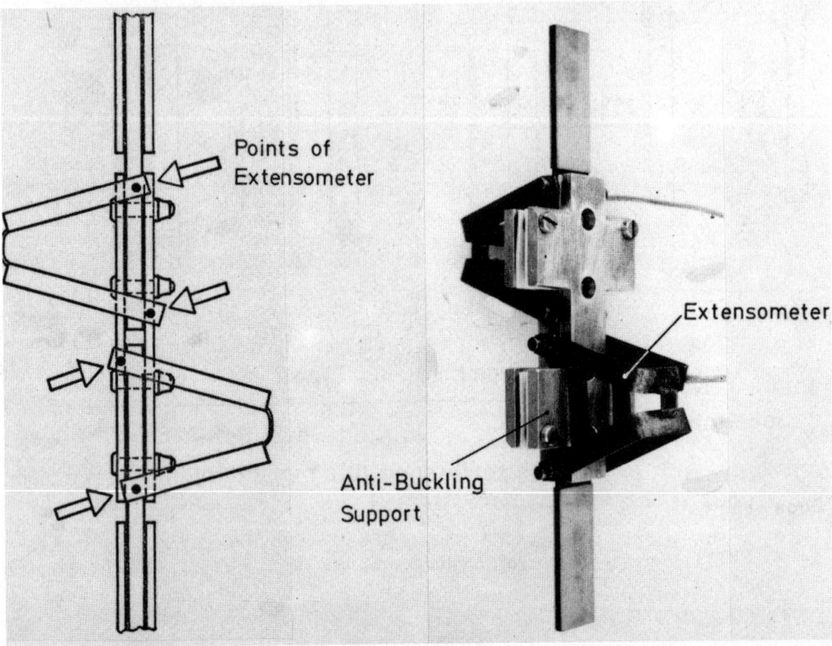

FIG. 2—*Jointed specimen with extensometers and antibuckling support.*

that the plain specimens have absolutely equal tensile and compressive strengths, whereas this is not so in the case of notched and jointed specimens. The greatest difference occurs with the Type 1 jointed specimen; here the compressive strength is almost double the tensile strength and even exceeds the compressive strength of the notched specimen (see table in Fig. 3). The installation of an interference fit bolt in the hole obviously reduces shear stresses when compression is applied. The stress-strain curves for jointed specimens of Types 2 and 3 have a well-defined, nonlinear pattern. In contrast, the stress-strain curve of the jointed specimen of Type 1 is almost linear and has the same slope initially as the stress-strain curve for the plain specimen. The sharply increasing deformation experienced by Type 2 jointed specimens shortly before the compressive strength is reached is caused by lengthening of the hole, as examination of the fractured specimen reveals.

Fatigue Strength Behavior

All fatigue strength results are presented in the form of *S-N* curves. In each case, these curves have been calculated according to a nonlinear regression method. The following equation was chosen to describe the test data mathematically

$$\bar{S} = S_f + \Delta S/\exp\,(\log N_f/A)^B \qquad (1)$$

where

$\bar{S}$ = average fatigue strength,
S_f = fatigue limit,
ΔS = range of fatigue strength,
N_f = number of cycles to failure,
A = scale parameter, and
B = shape parameter.

For $S_f > 0$, $\Delta S > 0$, $A > 0$, and $B > 0$, this equation has the potential of fitting the test results best [7].

Estimates of the parameters were determined by the maximum-likelihood method assuming log-normal distribution for the fatigue strength and constant scatter over the total life range. By this estimation procedure, run-out data as well as failure data were utilized in the statistical analysis, see Appendix. Thus, uncensored and censored data samples were considered [8]. The illustrations, for reasons of clarity, do not show individual results. Depending on the slope of the *S-N* curve, between 8 and 30 tests were carried out to establish each *S-N* curve. As an example, the fatigue test results of the Type 2 jointed specimens for $R = -1.66$ are presented in Table 1, including some results of the statistical analysis conducted.[3]

Type :	1	2	3	Plain	Notched
$\bar{S}_u^*$	369	303	348	860	450
$\bar{S}_{cu}^*$	-660	-434	-336	-850	-509
$\bar{v}_u$	5,2***	0,27**	0,28**	11,1***	6,9***
$\bar{v}_{uc}$	-11,2***	-0,49**	-0,28**	-15,3***	-8,4***

*N/mm²

**mm

***Strain at Fracture, б, ‰

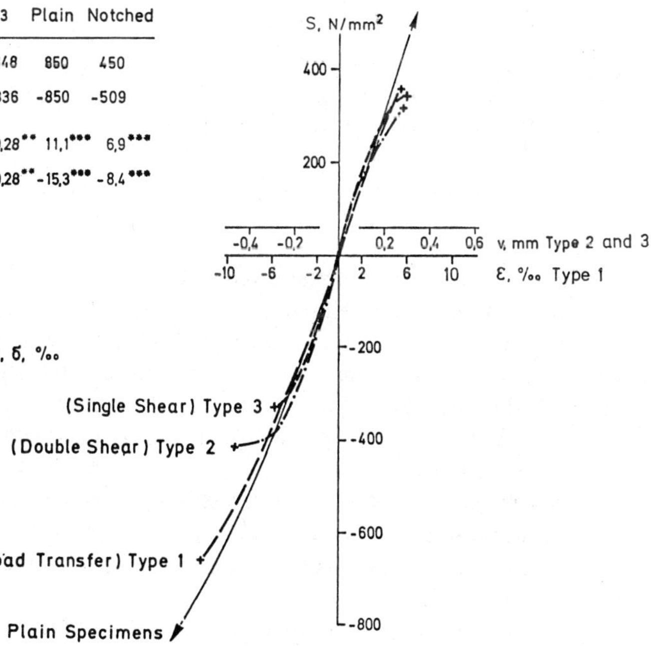

FIG. 3—*Stress-strain behavior of graphite/epoxy composite joints with Hi-Loks.*

[3] Additional tables with individual test results may be obtained directly from the authors.

TABLE 1—*Example of fatigue test data and results of analysis: Type 2 (double shear) jointed specimen,* R = *−1.66.*

Fatigue Test Data				
Gross Area, A, mm²	Load, F_{min}, N	Stress, S_{min}, N/mm²	Load Cycles, N_f	Analysis Results Equation of S-N Curve
44.02	19 105	−434[a]	1	
44.04	18 717	−425[a]	1	
43.99	18 124	−412[a]	1	
43.80	17 827	−407[a]	1	
43.85	20 346	−464[a]	1	
43.82	19 281	−440[a]	1	
44.02	20 205	−459[a]	1	
44.00	19 008	−432[a]	1	
44.09	16 357	−371	160	
44.11	16 232	−368	80	
44.06	15 862	−360	2 250	Average fatigue strength:
44.06	14 540	−330	720	$\lvert S_{min} \rvert = 84.4 + 349.3/\exp(\log N_f/9.42)^{1.0}$
44.00	14 520	−330	19 500	Standard deviation of fatigue strength:
44.06	14 540	−330	1 590	$s = 0.0224$ (log normal)
43.97	14 510	−330	46 300	
44.04	13 212	−300	18 730	
44.04	13 312	−300	3 430	
43.78	13 134	−300	16 200	
44.08	13 224	−300	71 140	
44.00	12 320	−280	133 500	
44.06	12 337	−280	449 220	
44.91	12 575	−280	14 750	
44.00	12 320	−280	420 380	
44.00	11 000	−250	1 684 000	

[a] Compressive strength.

In Fig. 4, *S-N* curves for plain and notched specimens are compared for different *R*-values. The reduction in fatigue strength resulting from the notch is greatest in the short life region. In contrast with metals, this reduction becomes smaller and smaller and can completely disappear as the endurance limit region is approached. The difference between the static and endurance limit strengths, both for plain and notched specimens, is greatest at the negative *R*-values of −1.0, −1.66, and −5.0. These *R*-values were therefore chosen for fatigue strength tests on jointed specimens.

Results of jointed specimen tests given in Fig. 5 show that, of these three *R*-values, the sharpest decrease in fatigue strength compared with the static strength occurs at $R = -1.66$.

Figure 6 shows the *S-N* curves at the critical *R*-value ($R = -1.66$) for the three types of jointed specimen. The position of the *S-N* curve is determined by the static compressive strength of the joints. In the load-cycle number

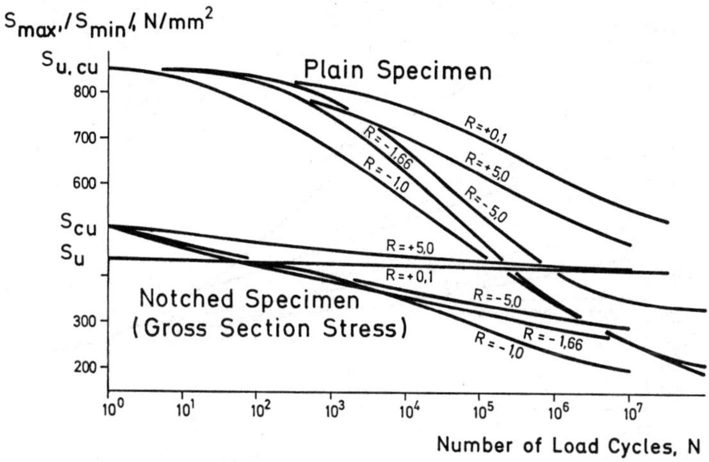

FIG. 4—S-N *curves of notched and plain specimens from a graphite/epoxy composite.*

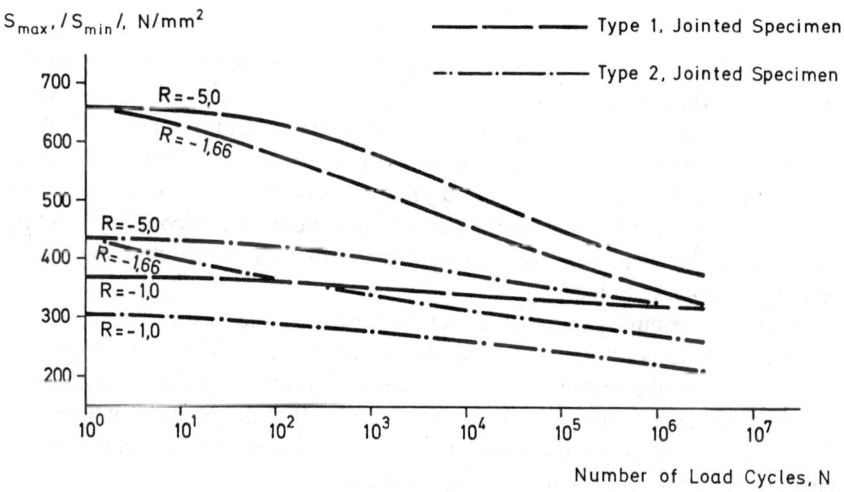

FIG. 5—S-N *curves for graphite/epoxy joints and various stress ratios,* R.

region $10^3 < N < 10^7$ that is of practical interest, Types 2 and 3 have almost the same fatigue strength, while the *S-N* curve for Type 1 in this region lies considerably higher. Also, for purposes of comparison, the *S-N* curves for plain and notched specimens of the $R = -1.66$ are shown in Fig. 6. The *S-N* curve for Type 1 jointed specimens lies above that for notched specimens, due to the higher compressive strength of the former.

It is clear here that the positions of *S-N* curves for the different specimens are determined by the static strengths of the specimens.

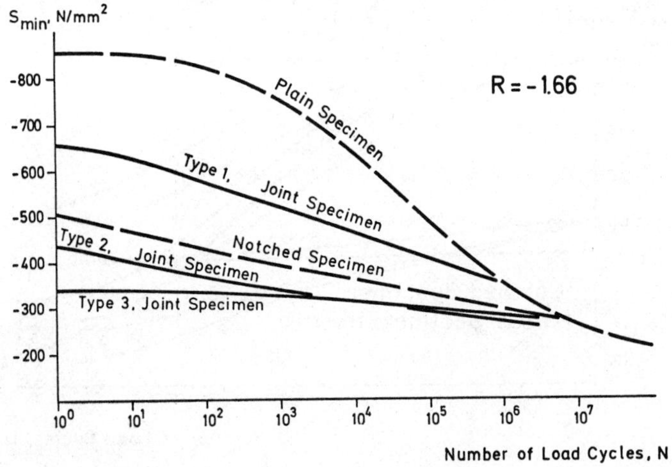

FIG. 6—S-N *curves of jointed specimens in comparison with* S-N *curves of notched and plain specimens of graphite/epoxy composite.*

In the literature, the ratio of the strength of the plain specimen to the strength of the notched specimen is frequently given for fiber composites. This is a good reference value for the stress concentration factor of fiber-controlled composites, since they exhibit brittle behavior. These ratios have been calculated for the notched and jointed specimens and are summarized in Table 2. With the exception of jointed specimens with secondary bending (Type 3), these effective stress concentration factors are greater for tensile than for compressive loading.

With the various specimens, different positions and types of fracture are encountered, depending on the number of load cycles. The most important differences are illustrated in Fig. 7. A typical feature of specimens with holes is the occurrence of longitudinal cracks originating at the hole, whereupon the fracture can run partly along these longitudinal cracks. These cracks are visible only for load cycle numbers greater than 10^3.

In the case of jointed specimens, there are essentially three types of failure: a fracture through the hole (as in the case of notched specimens, but without longitudinal cracks), a fracture ahead of the hole (as in the case of the static compression test), and failure of the inside wall of the hole (considerable lengthening of the hole). In the case of jointed specimens of Type 1, only a "fracture through the hole" occurred at $R = -1.0$. At $R = -1.66$ and -5.0, this type of fracture was encountered mainly in the range $N < 10^4$, while at larger numbers of cycles, fracture ahead of the hole predominated. In the case of Type 2, "fracture ahead of the hole" and "failure of hole walls" were chiefly involved. In fact, "fracture ahead of the hole" predominated in the range $N < 10^5$ and "failure of hole walls" in the

TABLE 2—*Influence of stress concentration on the static strength of carbon/epoxy composite,* $S_{u,plain}/S_{u,notched}$.

	Specimen with Hole	Jointed Specimens		
		Type 1	Type 2	Type 3
Tension	1.8	2.3	2.7	2.4[a]
Compression	1.7	1.3	2.0	2.6[a]

[a] First failure.

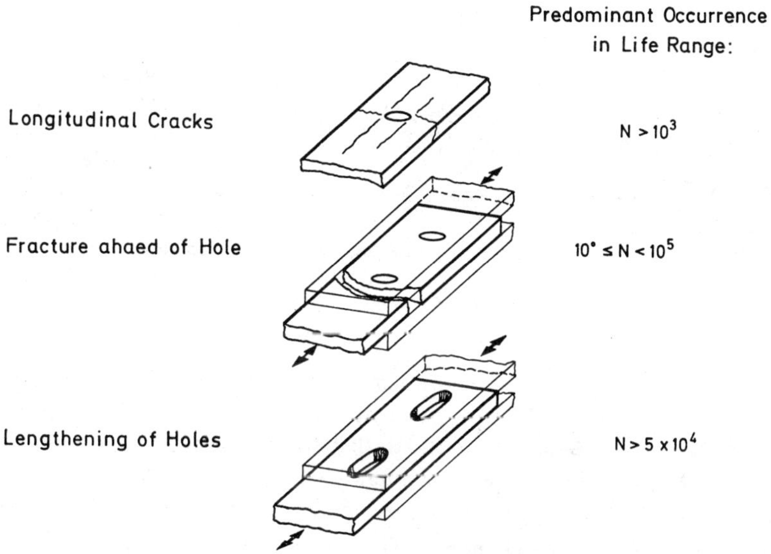

FIG. 7—*Typical locations and modes of failure for tension-compression loading.*

range $N > 5 \times 10^4$. With Type 3, all three types of fracture were encountered over the full range of load-cycle numbers, with a tendency towards the occurrence of a higher proportion of hole failure fractures at high load-cycle numbers. Obviously, multiple damage develops in different ways in the various specimens, depending on the number of load cycles.

Deformation Behavior Under Fatigue Loading

Apart from failure due to fracture, the development of excessive deformation may result in a component no longer fulfilling its function. Therefore, deformation produced during fatigue loading of specimens was registered continuously by the extensometers already mentioned and the increase in deformation was recorded. In Fig. 8, the increase in deformation is plotted

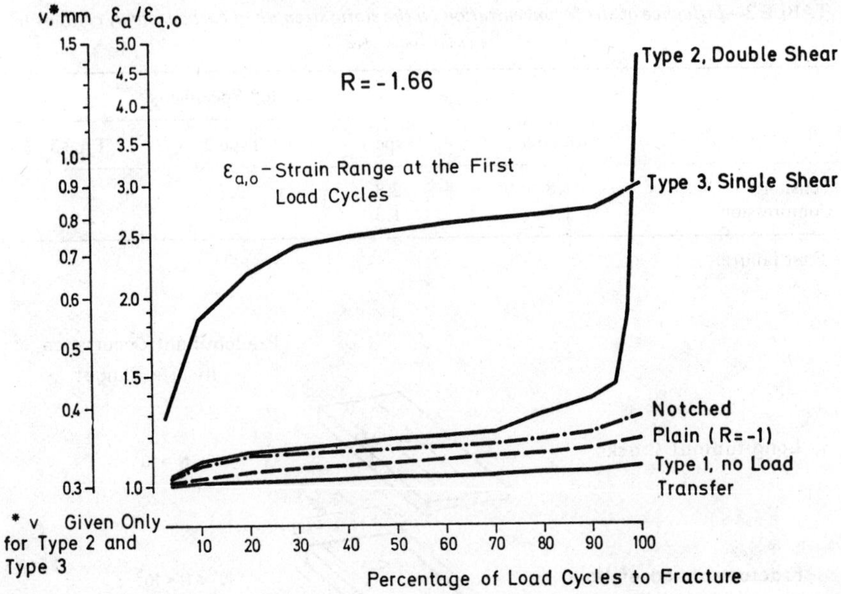

FIG. 8—*Increase in deformation during fatigue loading.*

against the percentage of the number of cycles relative to the number at failure. Curves represent averages of at least ten specimens and illustrate significant differences arising particularly between the types of jointed specimen. In the case of the jointed specimen Type 1, the increase in deformation is at its lowest; it even lies below the increase in deformation encountered with the plain and notched specimens. The deformation of jointed double-shear specimens having large areas of hole wall (Type 2) is not much greater than that of the notched and plain specimens for load cycle numbers up to 70 percent of the value at failure. The deformation increases sharply only towards the end of the endurance of the jointed specimens and just before fracture, it is five times as great as at the beginning of fatigue loading. Because of the test method used, the deformation measured in the case of jointed specimens Types 2 and 3 is the relative displacement of the parts connected by Hi-Lok bolts, lengthening of the hole has a great effect on the measured value. The deformation of jointed single-shear specimens of Type 3 with secondary bending increases sharply and uniformly at the beginning of fatigue loading and, at 40 percent of the number of load cycles at fracture, has risen by a factor of 2.5. Thereafter, the deformation still increases, but more moderately, and just before failure occurs, it is three times as large as at the beginning of fatigue loading. In this case, the bolt is loaded asymmetrically, in contrast to the case of the double-shear jointed specimen. The inclination of the bolt changes visibly during fatigue loading and gives rise to lengthening of the hole as the result of locally high stressing of the hole wall. With lengthening of the hole, load transfer and hence the stress on the hole wall decreases more and more with this type of joint, in contrast to Type 2. After this initial phase, the further, now considerably smaller, increase in deformation is caused predominantly by the axial and bending stress in the test region. A critical increase in deformation of about 0.4 mm is achieved considerably sooner in the case of a Type 3 joint than in the case of Type 2, where the corresponding deformation occurs only just before failure.

If the fatigue strength and deformation of all three types of joints are compared, then: (1) joints of Types 1 and 2 are critical as regards fatigue strength for useful life goals of $N > 10^5$, and (2) joints of Type 3 are critical as regards deformation over the region of the *S-N* curve of interest.

Scatter of the Strength Values

An important strength characteristic that must also be considered is the scatter observed in the critical values. Scatter values have been determined from the multitude of test results obtained by statistical analysis. The literature contains many examples of statistical analyses of static strength values, and the Weibull distribution is preferentially used since it can better describe the distribution of strengths. From the results of the static strength investigations, the scatter parameters of the Weibull distribution were calculated. For fatigue strength results, the standard deviations of the stress values were determined. The scatter parameters (weighted mean [*8*]) for the static strength and fatigue strength are summarized in Table 3, together with scatter values from the literature. For the same family of laminates, the scatter values found in the literature for static strengths of plain and notched carbon fiber composite specimens are greater than LBF test results. In comparison with metals, the scatter of static strengths and fatigue strengths of carbon fiber composites are only slightly larger.

Effect of the Laminate Structure

For all specimens on which test results have been reported, the longitudinal axis of the specimen and the axial loading lie parallel to the eight 0-deg layers of the composite (see laminate structure in Fig. 1). From this laminate, plain specimens were also cut, the longitudinal axis of which lay at right angles to the 0-deg fibers and hence parallel to the fibers of the one 90-deg layer. Results are given for some investigations on these 90-deg specimens under static and fatigue loading and are compared with results of investigations on 0-deg specimens (see Fig. 9). The mean values of the static strength, the *E*-modulus, and the elongation at rupture are also given in

TABLE 3—*Scatter of static and fatigue strength (weighted mean [8]).*

	Plain and Notched,		Jointed Specimens					
			Type 1		Type 2		Type 3	
	$1/\alpha$[a]	s[b]	$1/\alpha$	s	$1/\alpha$	s	$1/\alpha$	s
<u>Static Strength</u>								
Composite: LBF tests	0.035	0.02	0.027	0.016	0.046	0.027	0.07[c]	0.04[c]
Composite: literature	0.09 ÷ 0.04	...	...	...	...	...	...	...
Metals, literature	0.03	...	...	...	...	...	...	...
<u>Fatigue Strength</u>								
Composite LBF tests		0.03		0.016 ÷ 0.025		0.02 ÷ 0.036		0.03
Aluminum alloy literature		0.023 ($K = -6$)[d]						

[a] Parameter of Weibull distribution.
[b] Standard deviation of strength.
[c] First failure.
[d] Slope of *S-N* curve.

Fig. 9. The static strength, the E-modulus, and the fatigue strength over the whole range of load cycles (10 to 10^7) are lower than in the case of 0-deg specimens by a factor of about 3.

The degree of deformation under fatigue loading increases much more sharply for the 90-deg specimen than it does for the 0-deg specimen. Just before fracture occurs, it has increased by 125 percent of the initial value as compared with an increase of 20 percent for the 0-deg specimen.

According to investigations described in the literature, the differences recorded are characteristic when composites having a fiber-controlled behavior (0-deg specimens) are compared with composites having matrix-controlled behavior (90-deg specimens). A typical feature of fiber-controlled behavior is the increase in the slope of the S-ϵ curve under tensile loading, such as occurs for the 0-deg specimen, see Fig. 9. In contrast, the slope of the S-ϵ curve for 90-deg specimens decreases with increasing loading. A typical feature of matrix-controlled behavior is the appearance of discontinuities in the S-ϵ curve, such as seen for the 90-deg specimens. This is probably associated with failure of individual laminate layers running parallel to the longitudinal axis of the specimen.

Influence of Fiber Material

Besides the tests with the high-strength carbon fiber composite, some additional tests were run with an unnotched high-modulus fiber composite of the same laminate build-up. The high-modulus fiber is typified by Toray

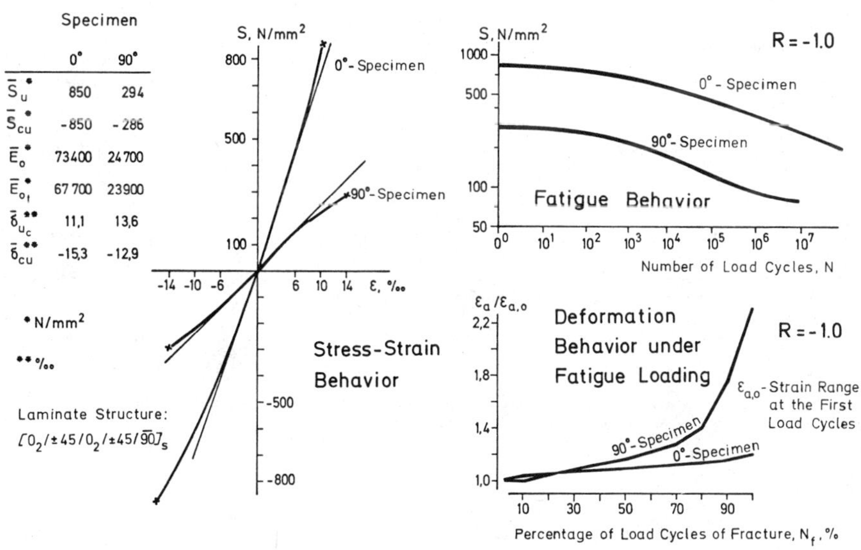

FIG. 9—*Influence of load direction.*

M40B and the epoxy resin is again 914C from Ciba. The stress-strain behavior as well as the fatigue behavior and the deformation behavior under $R = -1.0$ fatigue loading of high-strength and high-modulus fiber composites are compared in Fig. 10. The high-modulus fiber composite is stiffer, but the strain at fracture with tensile and compressive loading is less than half of that of the high-strength fiber composite. The mean values of Youngs modulus, the ultimate stresses, and strains resulting from three tests are given in the table on the left side of Fig. 10. The fatigue strength is remarkably lower for the high-modulus fiber composite in the low and medium cycle range of the *S-N* curve. In the region of the endurance limit, the strength values are similar. The increase of the deformation during fatigue loading that was measured on all fatigue specimens is somewhat less for the high-modulus fiber composite over the whole life to failure. In most cases, the stiffness loss was less than 10 percent shortly before failure.

Conclusions

Finally, the important results and trends will be summarized. It should be observed that information concerning the behavior of the joints is particular to the combination of the parameters involved in this study, that is, load transfer, secondary bending, the fit of the bolt, clamping forces, geometry, and so forth. Thus, for example, it is to be expected that if clearance bolts instead of interference-fit bolts were employed, not only the fatigue behavior but also the deformation behavior of jointed specimens would be greatly

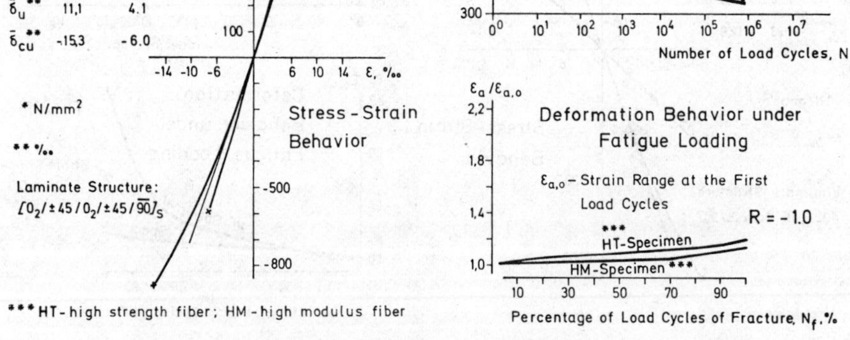

FIG. 10—*Influence of fiber material.*

changed under static and fatigue loading. The test results available show multiple relationships between the stress-strain and the fatigue strength behavior of specimens of graphite/epoxy composites. The results and trends found, bearing these relationships in mind, are:

1. Stress concentrations in specimens containing open holes and in joints reduce the static tensile and compressive strengths by comparison with plain specimens.
2. The absolute values of the tensile and compressive strengths can differ in magnitude for both notched and jointed specimens of graphite/epoxy composites.
3. Stress concentrations reduce the fatigue strength in the short-life region; in the long-life region, their effect disappears.
4. The position or slope of *S-N* curves for notched and jointed specimens can be evaluated from their static strength and from the "endurance limit" of plain specimens of the same laminate structure.
5. The stress concentration of an open hole is reduced by installing an interference-fit bolt.
6. The additional local stresses arising from load transfer and secondary bending result in nonlinear behavior and lead to the occurrence of different types of failure.
7. In the case of joints involving load transfer, progressive lengthening of the hole under fatigue loading is the greatest contributor to failure.
8. Under fatigue loading, lengthening of the hole sets in at an early stage in the case of single-shear joints, but not until just before failure in the case of double-shear joints.
9. Single-shear joints tend to fail earlier than double-shear joints, as the result of an excessive degree of deformation (lengthening of the hole).
10. The scatter in strength of specimens of graphite/epoxy composites has reduced and currently lies within the range of scatter encountered in metals.
11. The fatigue strengths of fiber- and matrix-controlled composites differ by about the same factor over the whole range of numbers of load-cycles.
12. The high modulus fiber composite material had lower static strength than the high-strength fiber composite material; but the fatigue strength of both does appear to be the same in the high-cycle fatigue range.

Acknowledgments

The research program on the "Fatigue Strength of Graphite/Epoxy Composite Material" was sponsored by the German Ministry of Defense Research Branch Rü Fo IV. The "Fatigue Strength of Graphite/Epoxy Composite Material" working party was entrusted with planning and monitoring the program; the working party was brought into being by the German Defense Ministry and consists of members of Dornier, Messerschmitt-

Bölkow-Blohm GmbH, VFW-Fokker, Deutsche Forschungs-und Versuchsanstalt für Luft-und Raumfahrt e.V.-Stuttgart, Bundesamt für Wehrtechnik und Beschaffung-Musterprüfstelle für Luftfahrtgerät, and Fraunhofer-Institut für Betriebsfestigkeit.

APPENDIX

With n_f failure observations (S_i, N_i) and n_g non-failure observations $(\hat{S}_j, \hat{N}_j)$ in the test sample, the support function of the maximum likelihood method is, corresponding to Ref *9* and with Eq 1

$$\text{Sup} = \sum_{i=1}^{n_f} [-(\log S_i - \log \bar{S}_i)^2/2s^2 - \ln s] + \sum_{j=1}^{n_g} \ln \int_{t_j}^{\infty} \exp(-u^2/2)du \qquad (2)$$

with

$$t_j = (\log \hat{S}_j - \log \bar{S}_j)/s$$

where

$S_i, \hat{S}_j$ = fatigue strength (stress level of the test),
$\bar{S}_i, \bar{S}_j$ = average fatigue strength from Eq 1 for N_i or $\hat{N}_j$, respectively,
s = standard deviation, and
u = integration variable.

The support function, Eq 2, has its largest value if

$$\frac{\partial\,\text{Sup}}{\partial S_f} = \frac{\partial\,\text{Sup}}{\partial \Delta S} = \frac{\partial\,\text{Sup}}{\partial A} = \frac{\partial\,\text{Sup}}{\partial B} = \frac{\partial\,\text{Sup}}{\partial s} = 0 \qquad (3)$$

Equation 3 therefore sets the maximum likelihood estimates of the parameters of Eq 1.

References

[*1*] Ryder, J. T. and Walker, E. K. in *Fatigue of Filamentary Composite Materials, ASTM STP 636,* K. L. Reifsnider and K. N. Lauraitis, Eds., American Society for Testing and Materials, 1977, pp. 3-26.

[*2*] Ramani, S. V. and Williams, D. P. in *Fatigue of Filamentary Composite Materials, ASTM STP 636,* K. L. Reifsnider and K. N. Lauraitis, Eds., American Society for Testing and Materials, 1977, pp. 27-46.

[*3*] Roderick, G. L. and Whitcomb, J. D. in *Fatigue of Filamentary Composite Materials, ASTM STP 636,* K. L. Reifsnider and K. N. Lauraitis, Eds., American Society for Testing and Materials, 1977, pp. 73-88.

[*4*] Rosenfeld, M. S. and Huang, S. L. "Fatigue Characteristics of Graphite/Epoxy Laminates Under Compression Loading," AIAA Paper No. 77-473, AIAA Conference on Aircraft Composites: The Emerging Methodology for Structural Assurance, San Diego, Calif., 24-25 March 1977.

[*5*] McCarty, J. E. and Johnson, R. W. "An Approach to the Development of Durable and Damage Tolerant Composite Commercial Aircraft Structures," AIAA Paper No. 77-471, AIAA Conference on Aircraft Composites: The Emerging Methodology for Structural Assurance, San Diego, Calif., 24-25 March 1977.

[6] Schütz, D. and Gerharz, J. J., *Composites,* Vol. 8, No. 4, Oct. 1977, pp. 245–250.
[7] Gecks, M. and Och, F. "Ermittlung dynamischer Festigkeitskennlinien durch nichtlineare Regressionsanalyse," MBB-Bericht UD-208-77, Messerschmitt-Bölkow-Blohm GmbH, Unternehmensbereich Drehflügler, Ottobrunn (Germany), 1977.
[8] Butler, J. P. in *International Conference on Structural Safety and Reliability,* A. M. Freudenthal, Ed., Pergamon Press, 1972, pp. 181–211.
[9] Spindel, J. E. and Haibach, E. "The Method of Maximum Likelihood Applied to the Statistical Analysis of Fatigue Data Including Run-Outs," S.E.E. International Conference, 3-6 April 1978, University of Warwick, Coventry, England, *Applications of Computers in Fatigue,* pp. 7.1–7.23.

J. D. Whitcomb[1]

Experimental and Analytical Study of Fatigue Damage in Notched Graphite/Epoxy Laminates

REFERENCE: Whitcomb, J. D., "**Experimental and Analytical Study of Fatigue Damage in Notched Graphite/Epoxy Laminates,**" *Fatigue of Fibrous Composite Materials, ASTM STP 723,* American Society for Testing and Materials, 1981, pp. 48–63.

ABSTRACT: Fatigue damage development in notched $(0/\pm45/0)_s$, $(45/0/-45/0)_s$, $(90/\pm45/0)_s$, and $(45/90/-45/0)_s$ graphite/epoxy laminates was investigated. Both tension and compression fatigue behaviors were studied. Most of the tests were conducted at load levels equal to two thirds of the ultimate tensile strength of the notched specimens. After fatigue loading, specimens were examined for damage type and location using light microscopy, scanning electron microscopy, ultrasonic C-scans, and X-radiography. Delamination and ply cracking were found to be the dominant types of fatigue damage. In general, ply cracks did not propagate into adjacent plies of differing fiber orientation. To help understand the varied fatigue observations, the interlaminar stress distribution was calculated with finite element analysis for the regions around the hole and along the straight free edge. Comparison of observed delamination locations with the calculated stresses indicated that both interlaminar shear and peel stresses must be considered when predicting delamination. The effects of the fatigue cycling on residual strength and stiffness were measured for some specimens of each laminate type. Fatigue loading generally caused only small stiffness losses. In all cases, residual strengths were greater than or equal to the virgin strengths.

KEY WORDS: composite materials, fatigue damage, stress analysis, graphite/epoxy composites, fatigue (materials)

Nomenclature

E_x, E_y, E_z	Extensional moduli
e_0	Specified axial strain
G_{xy}, G_{xz}, G_{yz}	Shear moduli
R	Ratio of minimum to maximum stress in fatigue cycle
S_{max}	Peak fatigue stress

[1]Research engineer, NASA Langley Research Center, Hampton, Va. 23665.

S_{ult}	Ultimate tensile strength
S_x	Gross axial stress
U, V, W	Functions used to describe u, v, and w
u, v, w	Displacements in x, y, and z directions, respectively
x, y, z	Cartesian coordinates
θ	Polar coordinate
ν_{xy}, ν_{xz}, ν_{yz}	Poisson ratios
σ_z	Normal stress in z direction
τ_{xz}	Shear stress in Cartesian coordinate system
$\tau_{\theta z}$	Shear stress in cylindrical coordinate system

Confidence in the long-term reliability of composite materials must precede full exploitation of their high specific strengths and stiffnesses. To gain this confidence, the fatigue behavior needs to be understood. Understanding composite fatigue behavior is a formidable task because of the diverse kinds of fibers, matrix materials, fiber orientations, and stacking sequences, and the concomitant variety of fatigue damage processes. Fatigue data in the form of stress versus cycles to failure for one laminate are not generally applicable to other laminates with different stacking sequences and certainly not those with different fiber orientations. To avoid testing all conceivable laminates, a generic fatigue analysis that predicts the fatigue behavior for any laminate from limited basic fatigue data is desirable.

The first step toward developing a generic fatigue analysis is to understand the basic fatigue damage processes in specific composite laminates. Tests (for example, Refs 1–4) and analyses (for example, Refs 5–7) reveal much about fatigue damage morphology and stress distributions in composites. Broken fibers, disbonded fibers, transthickness cracks, and intralaminar cracks have been shown to be primary forms of fatigue damage in boron/epoxy [*1*].[2] For graphite/epoxy laminates, Foye [*8*] has shown that delamination is an important mode of damage propagation. Interlaminar stresses at straight free edges have been shown to be responsible for the initiation of delamination in graphite/epoxy laminates.

The objective of this paper is to systematically examine fatigue damage in several notched graphite/epoxy laminates and to compare the damage with stress distributions. First, specimens subjected to fatigue loading were examined for damage type and location with ultrasonic C-scans, X-radiography, scanning electron microscopy (SEM), and light microscopy (of specimen sections). Next, stress distributions were calculated using three-dimensional finite element analysis and were compared with the distribution of delaminations. Finally, the effects of the fatigue cycling on residual strength and stiffness were measured for some of the specimens of each laminate type.

[2]The italic numbers in brackets refer to the list of references appended to this paper.

Experimental Procedure and Apparatus

Specimens and Loading

The notched specimen configuration is shown in Fig. 1. Ultrasonically drilled holes in the specimens represented structural discontinuities that cause stress concentration. Four laminates were considered. Two were orthotropic: $(0/\pm45/0)_s$ and $(45/0/-45/0)_s$. The other two were quasi-isotropic: $(90/\pm45/0)_s$ and $(45/90/-45/0)_s$. The specimens were made from T300/5209 tape. Average laminate thickness was about 0.96 mm; fiber volume was 67 percent. Approximately 48 specimens were used in the study.

The specimens were fatigue tested on servo-controlled hydraulic test machines under constant-amplitude, load-controlled, sinusoidal axial loading at a frequency of 10 Hz. For each stacking sequence of the two types of laminates, several specimens were tested with tension-tension ($R = 0.05$) loading. Several others were tested with compression-compression ($R = 20$) loading; guide plates (Fig. 2) prevented buckling during compression loading. The maximum absolute gross-section stress applied during cyclic loading was 254 MPa for the orthotropic laminates and 195 MPa for the quasi-isotropic laminates. These stresses, which were approximately two thirds of the ultimate tensile strength of the notched specimens, initially produced nominal absolute axial strains of 0.0033 and 0.0038 for the orthotropic

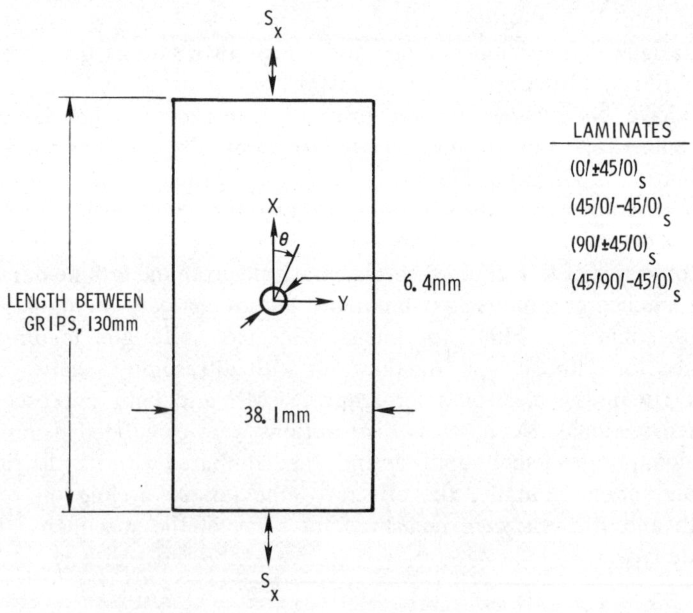

FIG 1—*Specimen configuration.*

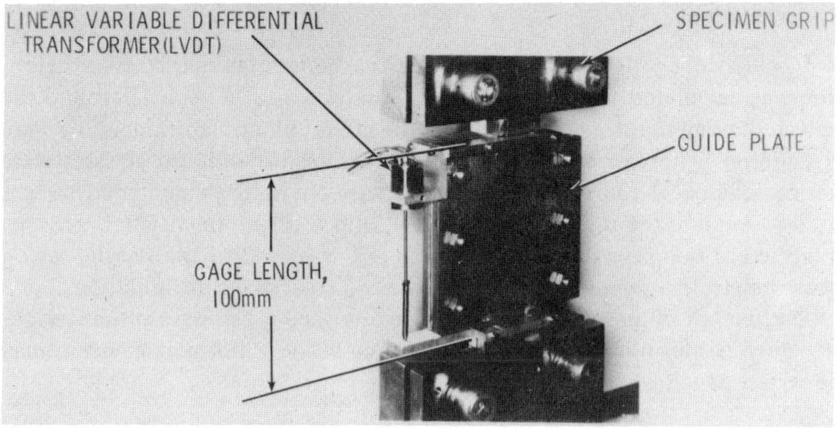

FIG. 2—*Apparatus for stiffness measurement in tension and compression and lateral guide plates for compression tests.*

and quasi-isotropic laminates, respectively. A few tensile fatigue tests were conducted at maximum cyclic stresses of approximately 80 percent of the static ultimate tensile strength. All measurements were taken in U.S. Customary Units.

Monitoring Fatigue Damage

Various techniques were used to locate damage. Ultrasonic C-scan records were used to locate delaminations over the length and width of the specimen. Some specimens were then sectioned with a low-speed diamond circular saw. Cutting-induced damage was minimized by sandwiching the graphite/epoxy between aluminum sheets. The sections were then polished for examination by light microscopy. These section studies were useful both for locating the particular ply interface at which delamination occurred and to locate damage, such as ply cracking, not detected by the C-scan. Radiographs, enhanced with tetrabromoethane, were used to determine the direction of ply-crack propagation. A scanning electron microscope (SEM) was used to examine the fracture surfaces of the specimens tested for residual strength. The SEM was used to detect fiber disbonding, as evidenced by fiber pullout. Also, the matrix in a few unbroken fatigue specimens was burned away [*9*] to separate the plies for individual examination for fiber breakage; this procedure is referred to as deplying.

The effects of fatigue damage on stiffness and strength were measured. Gross stiffness was calculated from elongation measurements made over a 100-mm gage length with a linear variable differential transformer (Fig. 2). Residual tensile strengths were measured after 10^7 tensile or compressive load cycles.

Stress Analysis

To help understand the fatigue behavior, the interlaminar stress distribution was calculated, with finite element analysis, for the regions around the hole and along the straight edge. Only the effects of mechanical loading were considered. Thermal residual stresses and moisture-induced stresses were not considered. A conventional three-dimensional finite element analysis was used to analyze the region near the hole, and a three-dimensional analysis, modified to impose uniform axial strain [5], was used to analyze the region near the straight edge. Both analyses were displacement formulations.

A schematic of an unnotched specimen analyzed for stresses at the straight free edge is shown in Fig. 3*a*. The requirement for uniform axial strain may be stated as follows:

$$u = e_0x + U(y,z)$$

$$v = V(y,z)$$

$$w = W(y,z)$$

where u, v, and w are the displacements in the x, y, and z directions, respectively. In the analysis, e_0 is the specified axial strain. Because the three

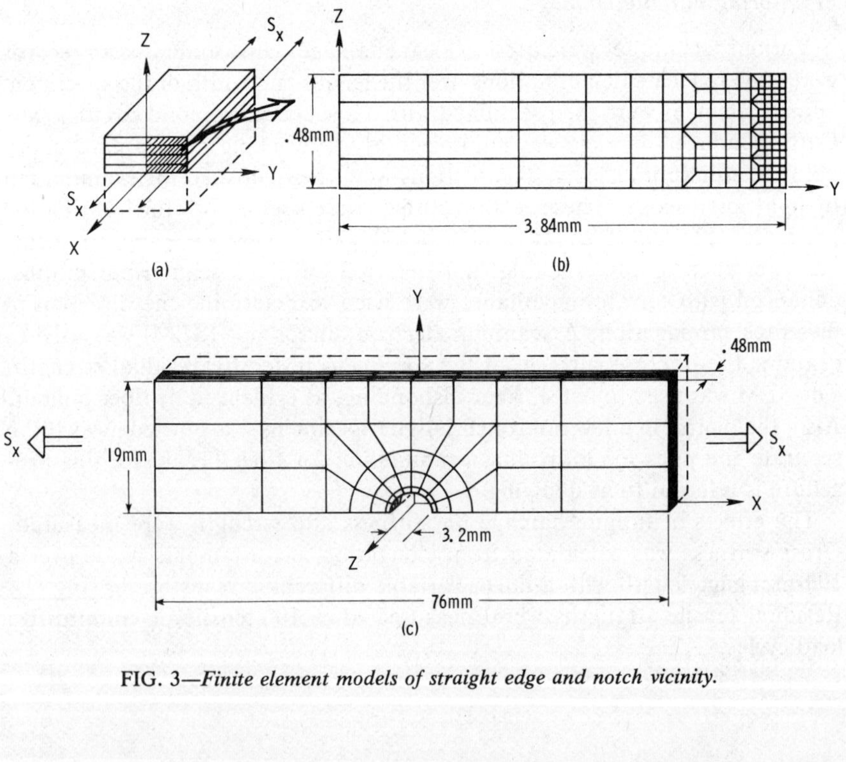

FIG. 3—*Finite element models of straight edge and notch vicinity.*

unknowns U, V, and W are functions of only y and z, the cross section of the specimen can be modeled by two-dimensional elements with 3 deg of freedom per node. Symmetry conditions permit solution of the problem by analysis of only one fourth of the cross section. The finite element model used in the edge analysis is shown in Fig. 3*b*. Eight-node isoparametric quadrilateral elements were used. Boundary conditions were, at $y = 0$ in the plane $x = 0$: $u = 0$, $v = 0$; and at $z = 0$: $w = 0$.

The region around the hole was analyzed with a conventional three-dimensional finite element analysis. The analysis used a 20-node isoparametric element. Exploiting the polar and midplane symmetries [*10*], only one fourth of the specimen was modeled, as shown in Fig. 3*c*. The total number of degrees of freedom in the model was 3198. Unit axial displacements were specified at $x = \pm 38$ mm. Other boundary conditions were

$$\text{at } y = 0 \qquad u(x, 0, z) = -u(-x, 0, z)$$

$$v(x, 0, z) = -v(-x, 0, z)$$

$$w(x, 0, z) = w(-x, 0, z)$$

$$\text{at } z = 0 \qquad w(x, y, 0) = 0$$

Material properties for a 0-deg ply were taken as follows [*5*]

$$E_x = 140 \text{ GPa}$$

$$E_y = E_z = 14 \text{ GPa}$$

$$G_{xy} = G_{xz} = G_{yz} = 5.9 \text{ GPa}$$

$$\nu_{xy} = \nu_{xz} = \nu_{yz} = 0.21$$

Properties for angle plies were obtained by appropriate coordinate transformations.

Results and Discussion

First, the types of fatigue damage observed are discussed. Next, the initiation sites and propagation paths of fatigue damage are described. Then delamination locations are compared with the calculated interlaminar stress distributions. Finally, the effects of the fatigue damage on strength and stiffness are discussed.

Damage Type

The specimens were examined for delaminations, ply cracks, fiber disbonds, and fiber breaks. Typical delaminations and ply cracks are shown in Fig. 4. Radiographs of the specimens revealed that the ply cracks grew parallel to the fibers.

A few specimens that had been fatigue loaded at up to 80 percent of the notched tensile strength were deplied. Figure 5 shows a micrograph of a deplied lamina. Very few broken fibers were found. Apparently, fiber breakage was not a significant fatigue degradation mode for the laminates examined. In contrast, progressive fiber breakage has been observed in boron/epoxy laminates [*1*].

Fracture surfaces showed that fibers pulled out only a very short distance (Fig. 6). The fiber pullout is indicative of fiber disbonding. However, fiber pullout was about the same for specimens with and without fatigue cycling, suggesting that the fatigue loading caused little fiber disbonding.

The various observations revealed that delamination and ply cracking were the primary mechanisms of fatigue degradation.

Location of Fatigue Damage

Generally, the plies cracked in the same region as they delaminated, but the density of ply cracks did not correlate with the extent of the delaminations. The 0-deg plies cracked axially along tangents to the hole in all of the laminates. However, as shown in Fig. 7*a*, axial cracks were sometimes confined to the 0-deg plies. Sometimes the axial cracks grew through the entire laminate thickness by linking of ply cracks with delaminations. In general,

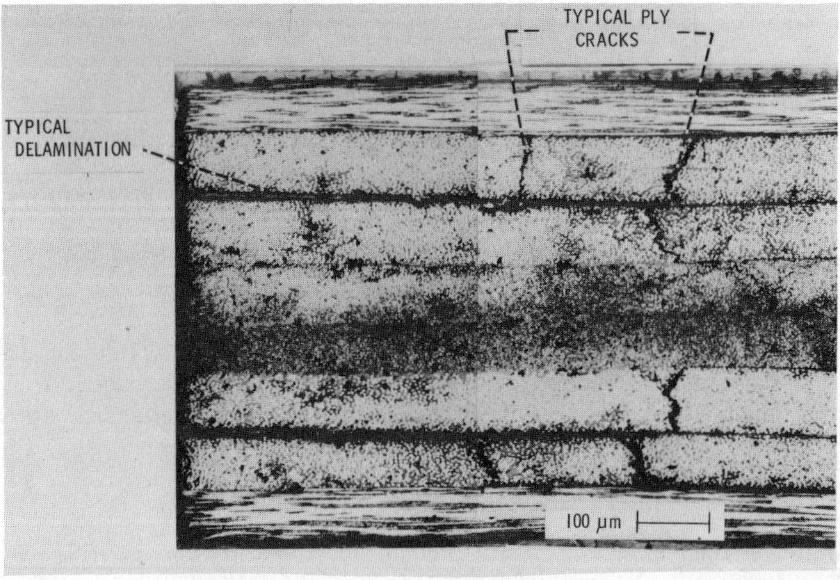

FIG. 4—*Transverse section showing typical delaminations and ply cracks [(90/±45/0)$_s$ laminate].*

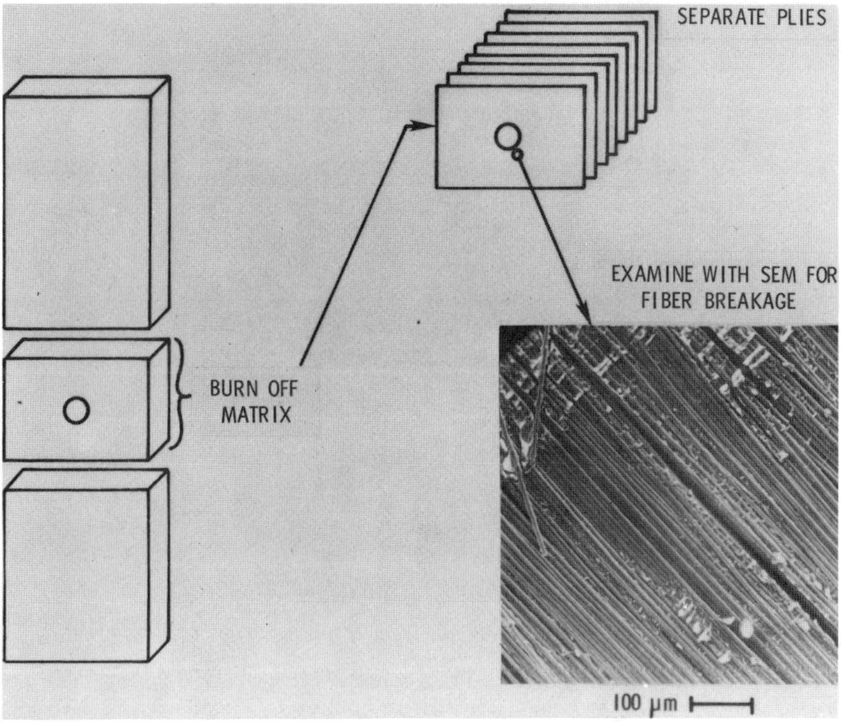

FIG. 5—*Deply procedure.*

ply cracks did not propagate into adjacent plies of differing fiber orientation. In contrast, axial cracks in boron/epoxy specimens grow relatively straight through the thickness without linking by delaminations, as shown in Fig. 7*b*.

The difference between the behaviors of the graphite/epoxy and boron/epoxy specimens may be a result of their relative tendencies to delaminate; the graphite/epoxy delaminated, but the boron/epoxy did not. Ply cracks are deterred from propagating straight through the thickness from one ply to the next by delaminations, which blunt the crack tip. The remainder of this section will concentrate on the location of delaminations.

Figure 8 shows typical delamination locations for the specimen after 10^7 tension or compression fatigue loads. Comparison of Fig. 8*a* and 8*b* reveals that stacking sequence affected delamination growth. The $(0/\pm45/0)_s$ specimen delamination above and below the hole, but the $(45/0/-45/0)_s$ specimen delaminated uniformly around the hole. The sign of the loading also affected delamination growth. For example, the $(45/0/-45/0)_s$ specimen delaminated much more extensively under compression (Fig. 8*d*) than tension (Fig. 8*b*). Fiber orientation also affected delamination growth; delamination growth from the hole was more closely aligned with the load

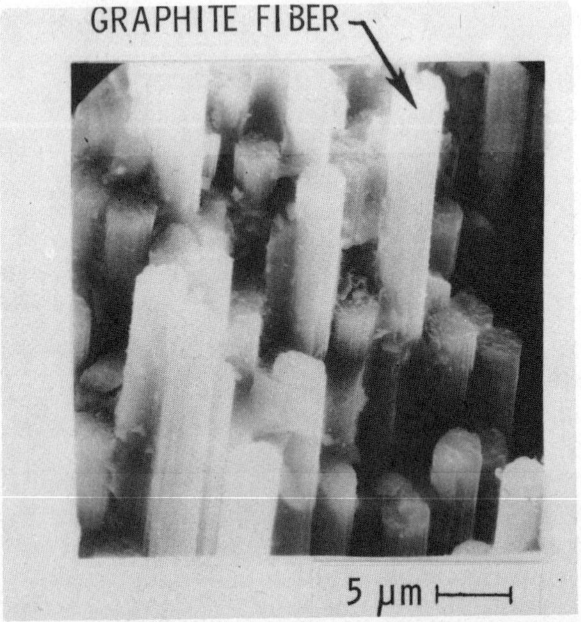

FIG. 6—*Fracture surface of graphite/epoxy fatigue specimen showing fiber pullout.*

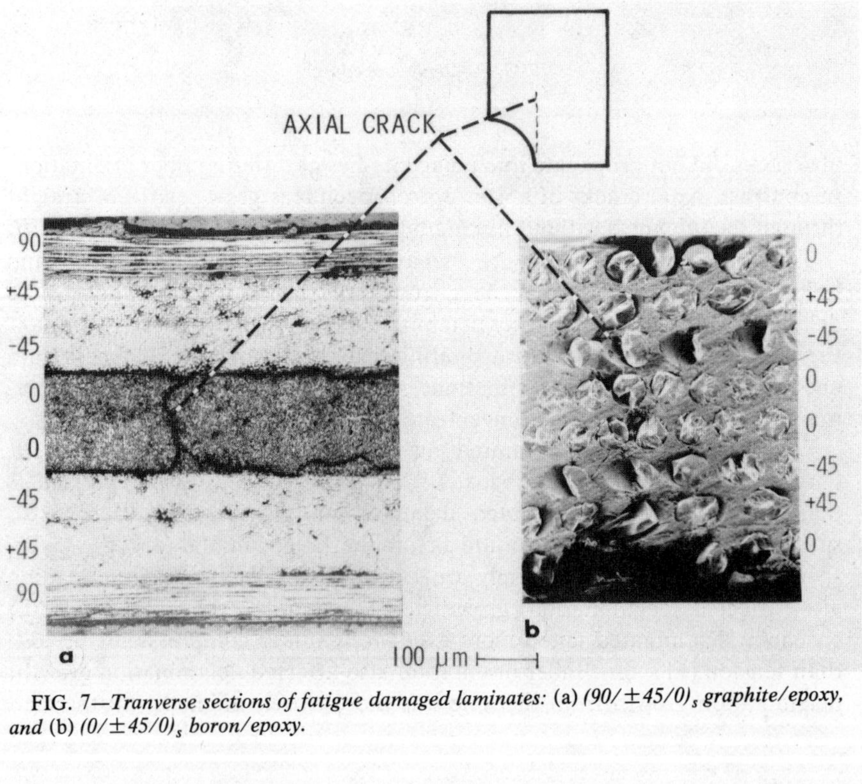

FIG. 7—*Tranverse sections of fatigue damaged laminates:* (a) *$(90/\pm45/0)_s$ graphite/epoxy, and* (b) *$(0/\pm45/0)_s$ boron/epoxy.*

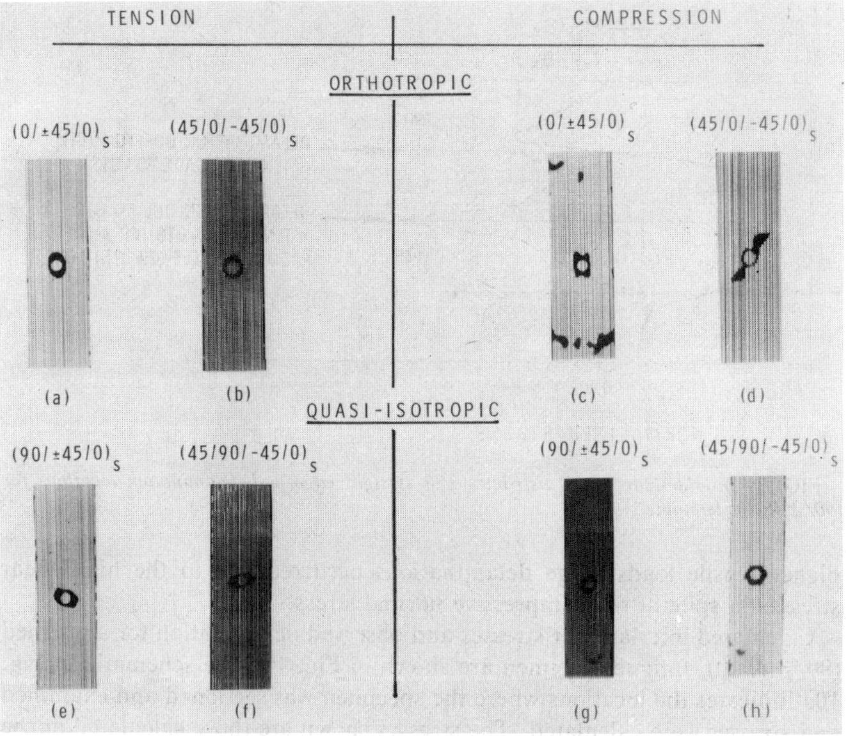

FIG. 8—*C-scan records of various notched laminates after* 10^7 *tensile or compressive fatigue cycles.*

direction for the orthotropic specimens than for the quasi-isotropic specimens.

The delamination locations compare well with the stress distributions determined by a finite element stress analysis. Delaminations were more likely in areas where both the interlaminar shear and tensile peel stresses were high. However, some delaminations were found in areas of high interlaminar shear but where the analysis indicated the peel stresses were compressive. Figures 9, 10, and 11 show typical results.

Figure 9 shows the stress distribution at the straight edge of a $(90/\pm45/0)_s$ specimen. The stresses are normalized with respect to the gross axial laminate stress. The highest interlaminar stresses occurred between the $\pm$45-deg and $-$45-deg plies. The peel stress, σ_z, at this interface was compressive for tensile loading. Thus, based on the sign of σ_z, delamination should be more likely under compression than tension loading. Note that the C-scans shown in Fig. 8*e* and 8*g* corroborate this prediction; under tension fatigue, straight-edge delamination did not occur; whereas under compression fatigue, straight-edge delamination did occur. In other tests under

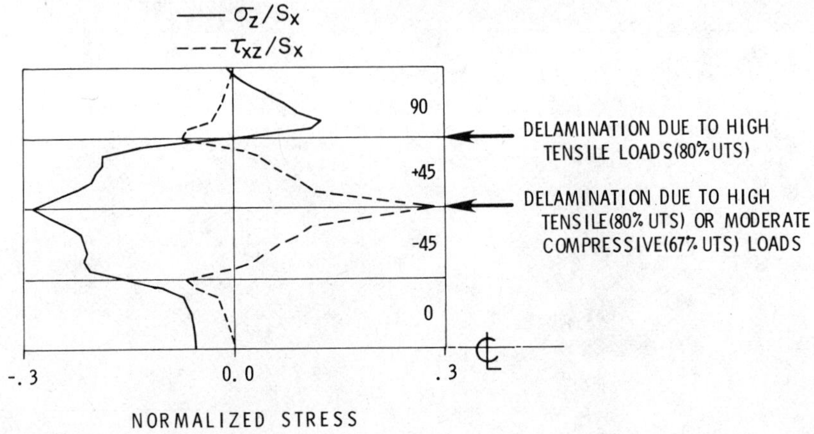

FIG. 9—*Interlaminar stress distribution at straight edge and delamination locations for (90/±45/0)$_s$ laminate.*

higher tensile loads, edge delaminations occurred due to the high shear stresses in spite of the compressive normal stress.

Calculated interlaminar stresses and observed delamination for a notched (90/±45/0)$_s$ fatigue specimen are shown in Fig. 10. The schematic in Fig. 10*a* indicates the locations where the specimen was sectioned and examined and stresses were calculated. The stresses shown are those calculated at the edge of the hole. In this case, the stresses are normalized with respect to the absolute value of gross axial stress. At 90 deg, the delaminations were associated with coincidental peaks in the shear and peel stresses between the 45-deg plies. At 120 deg, the shear ($\tau_{\theta z}$) stress peaks between the 0-deg and 45-deg plies caused delamination. At 160 deg, the tensile peel stress between the 0-deg plies appears to have governed the location of delamination.

Figure 11 shows results for a notched (45/90/−45/0)$_s$ fatigue specimen. At 90 deg, the delamination was associated with a small tensile peel stress and high shear stress. At 120 deg, the delamination was driven by high shear stresses, as it was in the (90/±45/0)$_s$ specimen at the same relative location. At about 175 deg, no delamination occurred at the edge of the hole. The delamination away from the edge was associated with shear-out of the 0-deg plies.

Delamination away from the edge of the hole cannot be predicted by an analysis based on an undamaged specimen because, after fatigue damage develops, stress distributions change. The change in stress distribution around the hole can even alter the direction of damage propagation, as shown in Fig. 12. For most of the test, the damage propagated at about 60 deg to the load direction. Later in the test, damage propagated axially. These results point up the need for a stress analysis capable of calculating stresses after fatigue damage occurs.

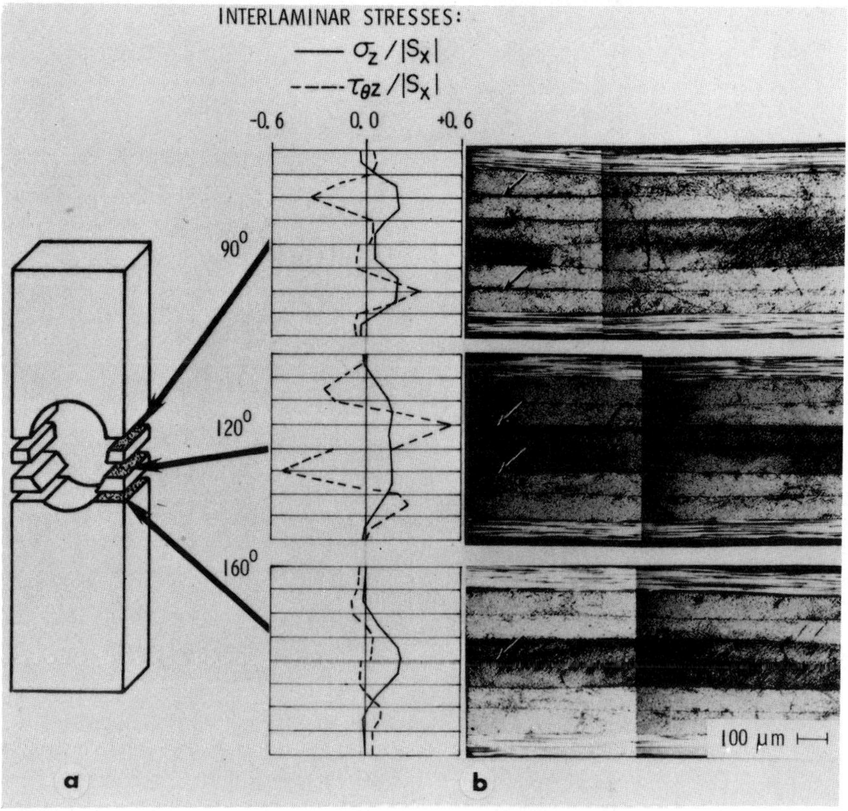

FIG. 10—*Interlaminar stress distribution at edge of hole and delamination location for $(90/\pm45/0)_s$ specimen subjected to compression fatigue.*

Residual Stiffness and Tensile Strength

Stiffness was monitored during the fatigue tests, and residual strength was determined for some of the specimens after fatigue loading.

Results of the stiffness measurements are presented in Table 1. The stiffness changes were generally quite small. Since the fatigue damage was restricted to the vicinity of the hole and the straight free edges, the lack of large changes was not surprising.

The strength measurements are shown in Fig. 13. In all cases, the residual strength after 10^7 cycles was greater than or equal to the virgin strength. The increase in strength with fatigue probably is due to stress redistribution that accompanies fatigue damage. Interestingly, the fatigue damage responsible for the stress redistribution may not become part of the fracture surface. For example, during tension fatigue of a $(0/\pm45/0)_s$ specimen, delaminations grew predominantly axially above and below the hole (see Fig. 8*a*). However, as shown in Fig. 14, the fracture was transverse.

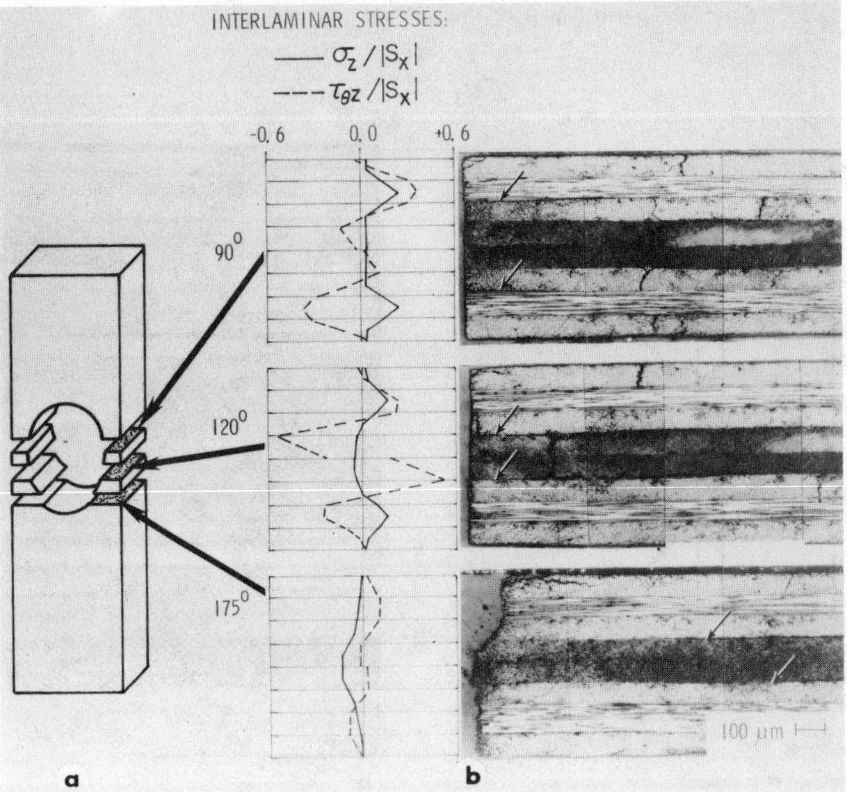

FIG. 11—*Interlaminar stress distribution at edge of hole and delamination location for $(45/90/-45/0)_s$ specimen subjected to tension fatigue.*

Concluding Remarks

The fatigue behavior of four notched graphite/epoxy laminates was studied. Two of the laminates were orthotropic: $(0/\pm45/0)_s$ and $(45/0/-45/0)_s$. The other two were quasi-isotropic: $(90/\pm45/0)_s$ and $(45/90/-45/0)_s$. The specimens were tested at relatively high constant-amplitude tensile or compressive loads. Specimens were examined for fatigue damage type and location using ultrasonic C-scans, X-radiography, scanning electron microscopy, and light microscopy.

Fatigue damage was primarily delamination and ply cracking parallel to the fibers. In general, ply cracks did not propagate into adjacent plies of differing fiber orientation. In all cases, the fatigue cycling of notched specimens resulted in residual strengths greater than or equal to the virgin strengths. Fatigue loading generally caused only small stiffness changes.

The location of delamination was sensitive to the fiber orientations, stacking sequence, and sign of the loading. Finite element stress analysis indicated

TABLE 1—*Effect of fatigue cycling on stiffness.*

Laminate	Specimen Number	Loading[a]	Change in Stiffness, %
$(0/\pm45/0)_s$	2A6	T	0
$(0/\pm45/0)_s$	2A7	T	0
$(0/\pm45/0)_s$	2A13	T	0
$(0/\pm45/0)_s$	2A15	T	+5
$(0/\pm45/0)_s$	2A16	C	−7
$(0/\pm45/0)_s$	2A19	C	−3
$(45/0/-45/0)_s$	1A6	T	0
$(45/0/-45/0)_s$	1A20	T	+5
$(45/0/-45/0)_s$	1A13	C	−7
$(45/0/-45/0)_s$	1A14	C	−7
$(90/\pm45/0)_s$	3A6	T	0
$(90/\pm45/0)_s$	3A8	T	0
$(90/\pm45/0)_s$	3A14	T	+4
$(90/\pm45/0)_s$	3A21	T	−6
$(90/\pm45/0)_s$	3A24	T	−4
$(90/\pm45/0)_s$	3A7	C	0
$(90/\pm45/0)_s$	3A19	C	0
$(90/\pm45/0)_s$	3A20	C	−4
$(45/90/-45/0)_s$	4A6	T	0
$(45/90/-45/0)_s$	4A13	T	0
$(45/90/-45/0)_s$	4A14	T	+4
$(45/90/-45/0)_s$	4A17	T	0
$(45/90/-45/0)_s$	4A1	C	5
$(45/90/-45/0)_s$	4A8	C	0
$(45/90/-45/0)_s$	4A20	C	−10

[a] 10^7 cycles at $|S|_{max}/S_{ult} = 67\%$; $T \equiv$ tension and $C \equiv$ compression.

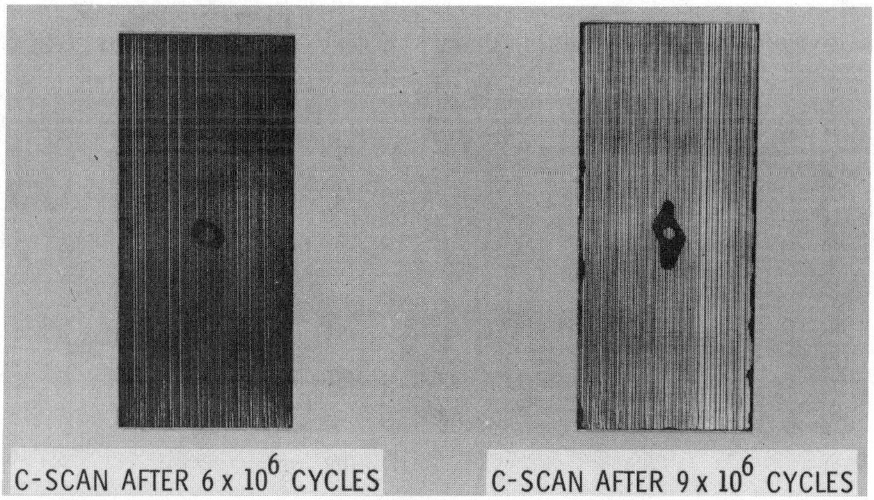

FIG. 12—*Delamination zones at two points in fatigue life of $(90/\pm45/0)_s$ laminate (S_{max}/S_{ult} = 80 percent).*

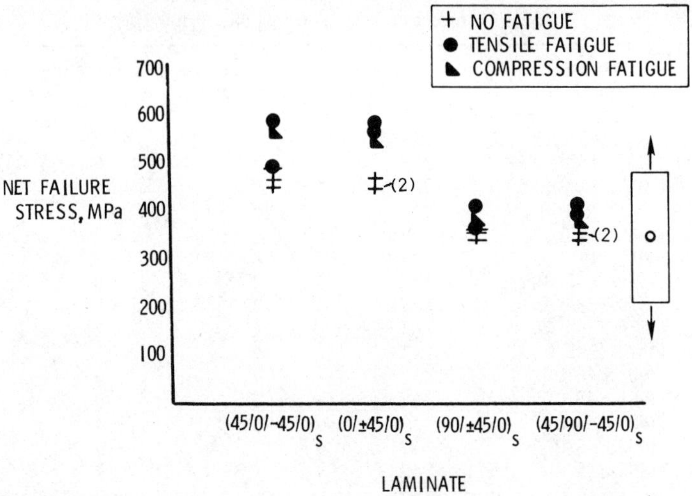

FIG. 13—*Tensile strength of notched laminates after 10^7 cycles* ($|S|_{max}/S_{ult}$ = *67 percent).*

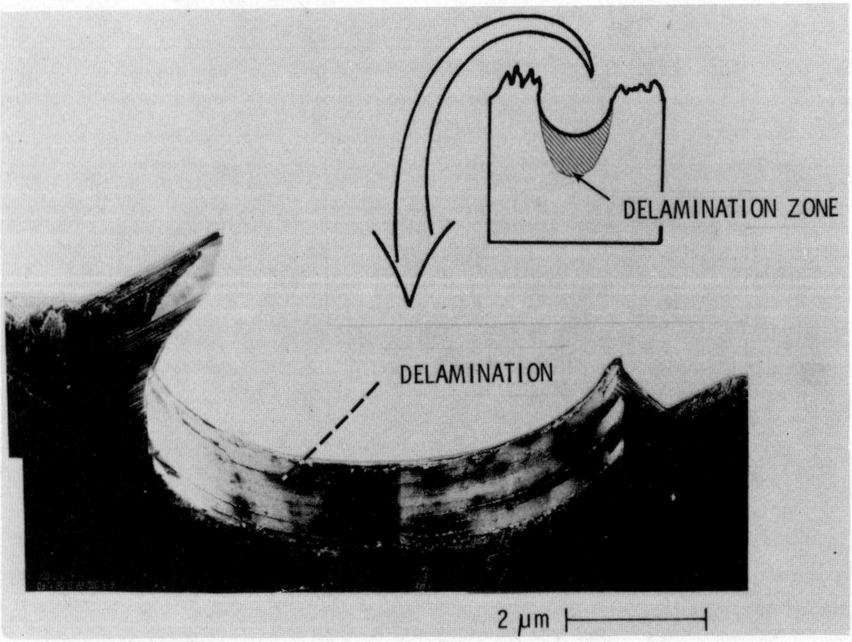

FIG. 14—*SEM photograph of fractured (0/±45/0)$_s$ tensile fatigue specimen.*

that both interlaminar normal stress and shear stress must be considered to explain the observed delamination. Furthermore, the altered stress distribution, concomitant with fatigue damage growth, can change the direction of delamination propagation.

Acknowledgments

The laboratory work was performed by J. C. Woolsey, a Langley Research Center engineering technician.

The computer codes used in this study were developed by I. S. Raju, a post-doctoral research fellow of the Joint Institute for Advancement of Flight Sciences, George Washington University, at Langley Research Center.

Specimens were deplied by Samuel Freeman of Lockheed-Georgia Company.

References

[*1*] Roderick, G. L. and Whitcomb, J. D. in *Fatigue of Filamentary Composite Materials, ASTM STP 636,* American Society for Testing and Materials, 1977, pp. 73–88.

[*2*] Daniel, I. M., Rowlands, R. E., and Whiteside, J. B., *Journal of Experimental Mechanics,* Jan. 1974, pp. 1–9.

[*3*] Ramani, S. V. and Williams, D. P. in *Fatigue of Filamentary Composite Materials, ASTM STP 636,* American Society for Testing and Materials, 1977, pp. 27–46.

[*4*] Sturgeon, J. B., *Journal of Composites*, Oct. 1977, pp. 221–226.

[*5*] Wang, A. S. D. and Crossman, Frank W., *Journal of Composite Materials*, Vol. 11, Jan. 1977, pp. 92–106.

[*6*] Rybicki, E. F. and Schmuesser, D. W., "Three-Dimensional Finite Element Stress Analysis of Laminated Plates Containing a Circular Hole, " AFML-TR-76-92, Air Force Materials Laboratroy, 1976.

[*7*] Ramkumar, R. L., Kulkarni, S. V., and Pipes, R. B., "Evaluation and Expansion of an Analytical Model for Fatigue of Notched Composite Laminates," NASA CR-145308, National Aeronautic and Space Administration 1978.

[*8*] Foye, R. L. and Baker, D. J., "Design of Orthotropic Laminates," presented at the AIAA/ASME 11th Conference on Structures, Structural Dynamics, and Materials, Denver, Colo., April 1970.

[*9*] Bailey, C. D., Freeman, S. M., and Hamilton, J. M., "Detection and Evaluation of Impact Damage in Graphite/Epoxy Composites," Proceedings of the Ninth National Technical Conference of the Society for the Advancement of Material and Process Engineering (Materials and Processes-In-Service Performance), Atlanta, Ga, Oct. 1977.

[*10*] Raju, I. S. and Crews, J. H., Jr., "Polar Symmetry in Three-Dimensional Analysis of Laminates With Angle Plies," NASA TM-81852, National Aeronautic and Space Administration, July 1980.

[*11*] Stinchcomb, W. W. and Reifsnider, K. L. in *Fatigue Mechanisms, ASTM STP 675,* American Society for Testing and Materials, 1979, pp. 762–787.

W. W. Stinchcomb,[1] *K. L. Reifsnider,*[1] *P. Yeung,*[1] *and J. Masters*[1]

Effect of Ply Constraint on Fatigue Damage Development in Composite Material Laminates

REFERENCE: Stinchcomb, W. W., Reifsnider, K. L., Yeung, P., and Masters, J., **"Effect of Ply Constraint on Fatigue Damage Development in Composite Material Laminates,"** *Fatigue of Fibrous Composite Materials, ASTM STP 723,* American Society for Testing and Materials, 1981, pp. 64–84.

ABSTRACT: Laminae of composite materials are bonded together to form a laminate for the purpose of achieving strength and stiffness in certain desired directions. Although a direct use of laminae strength and stiffness properties can be used to determine the properties of an undamaged laminate, it is well known that the manner in which local damage initiates, grows, and combines is not the same in an unconstrained lamina as it is when the same lamina is constrained by other laminae in a laminate. This paper addresses the question of constraint effects directly and discusses the consequences of such effects on the development of fatigue damage in graphite epoxy laminates. Results of monotonic tension tests, tension-tension fatigue tests, and nondestructive investigations are reported for several constraint situations and are compared with the results of several analyses.

KEY WORDS: composite materials, fatigue (materials), damage, constraint, nondestructive testing.

Engineering materials, including metals, polymers, ceramics, and composites, contain defects. The relationship between defects and the response of materials must be understood if materials are to be used in efficient and reliable engineering structures. When a crack grows in a self-similar manner in a metallic material, the response can often be predicted adequately by fracture mechanics. However, the growth of a particular damage mode in a composite laminate is controlled by the orientation of the plies containing the damage relative to the load directions and by the constraints imposed on the damaged plies by adjacent undamaged plies. Figure 1*a* shows the damage

[1]Associate professor, professor, graduate research assistant, and graduate assistant, respectively, Department of Engineering Science and Mechanics, Virginia Polytechnic Institute and State University, Blacksburg, Va. 24061.

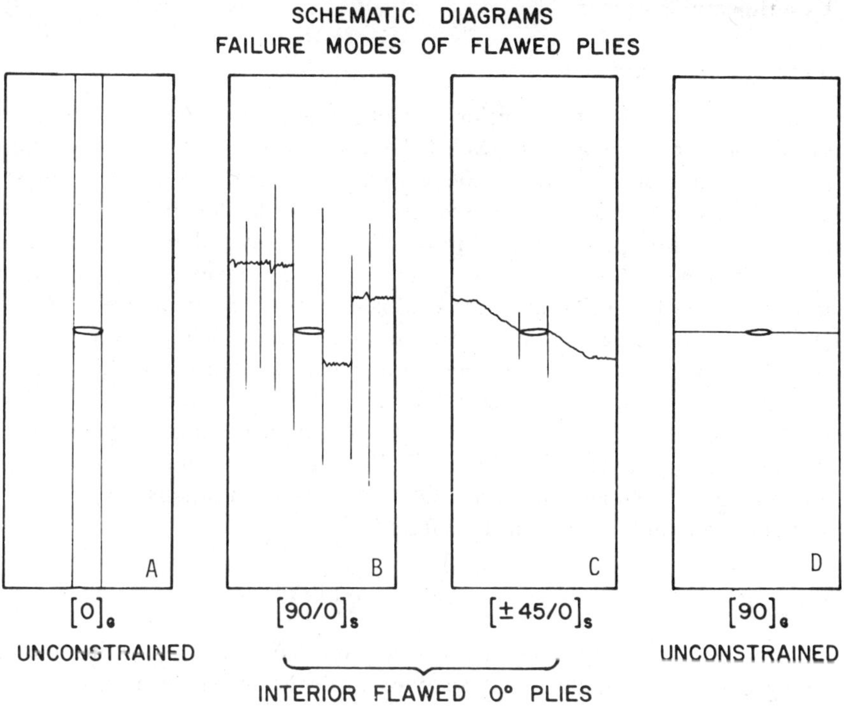

FIG. 1—*Tension-tension fatigue damage modes for several constraint situations.*

that develops in an unconstrained notched 0-deg specimen subjected to tension-tension cyclic loading. Longitudinal cracks initiate at the notch tips and grow parallel to the fibers to produce two ligaments. When a similarly notched unidirectional lamina is bonded between 90 or ±45-deg plies, the constraints imposed by the undamaged plies change the manner in which the damage develops as shown in Fig. 1*b* and *c*.

The nature of this constraint effect is poorly understood, and the consequences of constraint are infrequently used directly in strength and stiffness determinations beyond the usual lamination theory level. Although the in-plane stresses in the interior of a particular ply of symmetric laminates are insensitive to variations in the stacking sequence, the out-of-plane stresses and the state of damage in the laminates are changed [*1,2*].[2] There is an increasing amount of data that shows that variations of stacking sequence in quasi-isotropic laminates change the damage state as determined by the orientation of constraining plies, and affect tensile strength [*3*], shear strength [*4*], and residual tensile strength after cyclic loading [*5*].

[2]The italic numbers in brackets refer to the list of references appended to this paper.

Experimental Program

Materials and Specimens

Static tension tests and tension-tension fatigue tests were conducted on quasi-isotropic laminates of AS/3501 graphite epoxy and on $[90/0]_s$ and $[\pm45/0]_s$ laminates of T300/5208 graphite epoxy. The unnotched quasi-isotropic laminates had stacking sequences of $[0/\pm45/90]_s$ and $[0/90/\pm45]_s$ and were cut into specimens 25.4 mm (1.0 in.) wide by 178 mm (7.0 in.) long with a 76 mm (3.0 in.) test section. The $[90/0]_s$ specimens were 38 mm (1.5 in.) wide by 178 mm (7.0 in.) long and the $[\pm45/0]_s$ specimens were 38 mm (1.5 in.) wide by 254 mm (10.0 in.) long. The $[90/0]_s$ and $[\pm45/0]_s$ specimens were tested in unnotched and notched conditions. The notches were made by milling a 6.4 mm (0.25 in.) slit perpendicular to the fibers in two 0-deg pre-preg plies, filling the notch with liquid silicone rubber, adding the constraining 90-deg (Fig. 2) or ±45-deg (Fig. 3) plies, and curing to form a laminate with an embedded flaw. Other constraint situations have been investigated and are reported in Refs *6* and *7*.

Mechanical Tests

Specimens of each type were subjected to tension tests to determine tensile strength, elastic modulus, and fracture strain as shown in Table 1. The

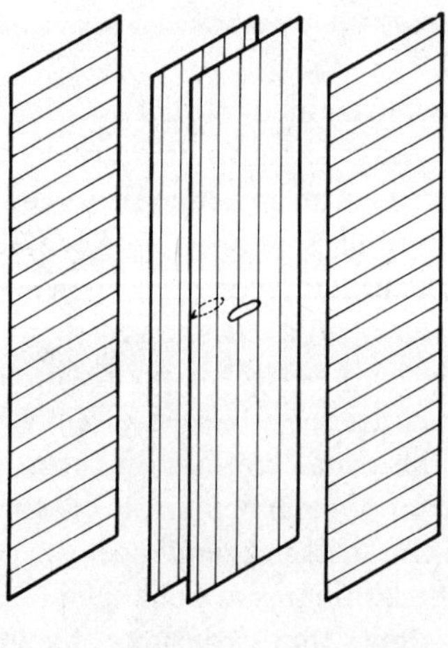

$[90,0]_s$ with Transverse Flaw in Interior 0° Plies

FIG. 2—*A $[90/0]_s$ laminate with embedded flaw in the 0-deg plies.*

FIG. 3—*A [±45/0]$_s$ laminate with embedded flaw in the 0-deg plies.*

TABLE 1—*Mean tensile properties of graphite epoxy specimens.*

Specimen	Tensile Strength, ksi	E_{xx} Elastic Modulus, msi	Fracture Strain %
[0/90/±45]$_s$	70.4	6.5	1.14
[0/±45/90]$_s$	67.3	6.2	1.04
[90/0]$_s$ unnotched	100.7	10.1	0.97
[90/0]$_s$ notched	82.0	...	...
[±45/0]$_s$ unnotched	85.6	8.0	1.16
[±45/0]$_s$ notched	51.5	...	...

values of static tensile strength were used to select the maximum stresses imposed during cyclic loading at a stress ratio of R = 0.1. The cyclic load tests were run at a frequency of 15 Hz on the quasi-isotropic laminates and 10 Hz on the [90/0]$_s$ and [±45/0]$_s$ laminates.

Damage Detection

A number of investigative methods were employed before, during, and after the mechanical tests to determine the damage state in the specimens and measure the response of the material. The methods include microscopy, ultrasonic C-scan, ultrasonic attenuation, acoustic emission, replication, X-radiography, stiffness, thermography, and sectioning. During the static and fatigue tests, attempts were made to follow the initiation and development of damage using the nondestructive methods. When possible, these investigations were carried out in real time or with only short interruptions in the loading history. The techniques were used in concert to monitor damage development and assist in the interpretation of results. A short description of each method is given in Ref *7*, and some results relevant to constraint effects are reported in Ref *8*.

Observations

Figure 4 shows the decrease in crack spacing (increase in the number of cracks) in the 90-deg plies of $[0/\pm45/90]_s$ and $[0/90/\pm45]_s$ laminates during cyclic loading at several maximum stress levels. Although both laminates are quasi-isotropic and the interior in-plane stresses are equal in plies of the same orientation, the damage state in the two laminates is different and is a characteristic of the stacking sequence. In the $[0/\pm45/90]_s$ laminate, the weak 90-deg plies are joined together and are constrained by 45-deg angle

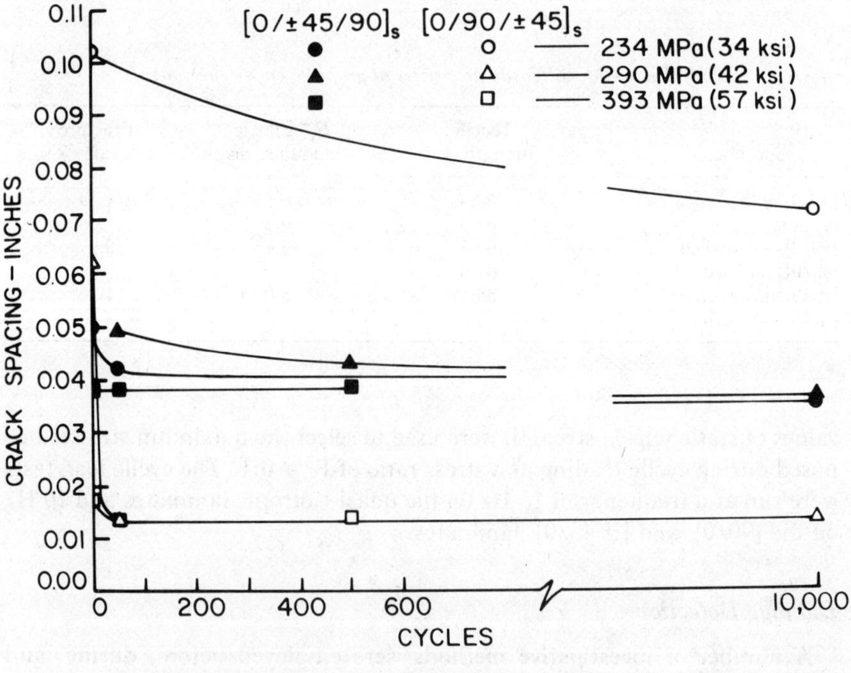

FIG. 4—*Change in spacing of cracks in 90-deg plies for two different quasi-isotropic laminates during cyclic loading.*

plies; however, the 90-deg plies in the $[0/90/\pm 45]_s$ laminate are separated and each is constrained by 0-deg and 45-deg plies. The transverse cracks that develop in the 90-deg plies form a regular pattern with a spacing governed by the constraint imposed on the 90-deg plies.

The number of cracks in the off-axis plies increases (the crack spacing decreases) sharply during the initial portion of the cyclic loading and then levels off as the crack spacing reaches a stable or equilibrium value. This characteristic damage state is determined by material properties, ply orientation, and stacking sequence. For the stresses shown in Fig. 4, the crack spacing in the 90-deg plies of the $[0/\pm 45/90]_s$ laminates stabilize early in the fatigue life. At applied stresses greater than one half of the tensile strength, an equilibrium spacing of cracks in the off-axis plies is usually reached before one million cycles [*9*]. For a particular type of laminate, the equilibrium crack spacings for different maximum cyclic stresses and quasi-static loading are the same. Thus, the characteristic damage state is a well-defined property of the laminate and is independent of loading history.

When an unconstrained, transversely notched, unidirectional specimen is cyclically loaded in the fiber direction, matrix cracks initiate at the notch tip and propagate parallel to the fibers to form an "H" pattern. The rate and extent of propagation are functions of the cyclic stress level. The response of a constrained, transversely notched lamina is dependent on the orientation of the constraining plies. Figure 5 compares the damage revealed by ultrasonic C-scans of $[90/0]_s$ and $[\pm 45/0]_s$ specimens at several stages during cyclic loading. In both cases, the maximum cyclic stress on the specimen is such that the nominal maximum stress in the 0-deg plies is equal to the tensile strength of the unconstrained notched material.

Although the $[\pm 45/0]_s$ notched laminates have a lower tensile strength than the similarly notched $[90/0]_s$ laminates, the ± 45-deg constraining plies are more effective in preventing the spread of delaminations during cyclic loading. Sectioning studies show that the damage indicated in the C-scans is delamination of ± 45 and $-45/0$ ply interfaces in $[\pm 45/0]_s$ laminates and 90/0 interfaces in $[90/0]_s$ laminates [*8*]. The sections also show axial cracks in the constrained 0-deg plies that are similar to the cracks described previously for the unconstrained, notched situation. Again, the ± 45-deg constraining plies are more effective in restricting the growth of axial cracks in the constrained 0-deg plies. After one million cycles, the cracks do not extend beyond the delaminated zone in the $[\pm 45/0]_s$ laminate shown in Fig. 5. However, the cracks in the 0-deg plies of the $[90/0]_s$ laminate extend beyond the delamination zone toward the gripped region. Figure 6 shows an X-ray and a C-scan of a $[90/0]_s$ laminate before and after 100 000 cycles of loading at a nominal 0-deg ply stress equal to the tensile strength of the unconstrained notched specimen. After 100 000 cycles, the delamination of the 90/0 interface appears as shown. The X-ray, which was made without an image enhancement agent, reveals the axial cracks in the constrained 0-deg

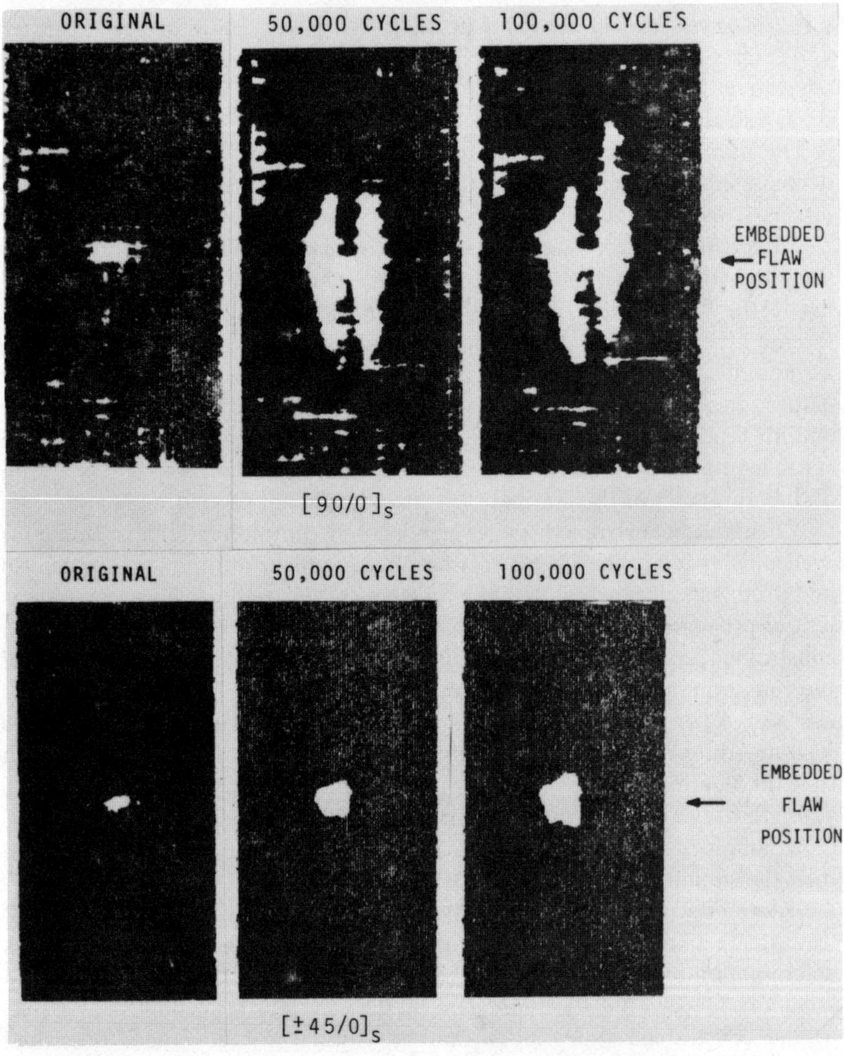

FIG. 5—*C-scans of a* $[90/0]_s$ *laminate and a* $[\pm 45/0]_s$ *laminate during cyclic loading at a maximum nominal stress in the 0-deg plies equal to the tensile strength of the notched, unconstrained, 0-deg lamina.*

plies. The arrows in the diagram indicate the tips of the axial cracks as detected by radiography. In all cases, the axial cracks extend beyond the delaminated region shown in the C-scan.

Discussion

Constraint effects can be classified in two categories, in-plane effects and through-the-thickness effects. The manner in which the two types of con-

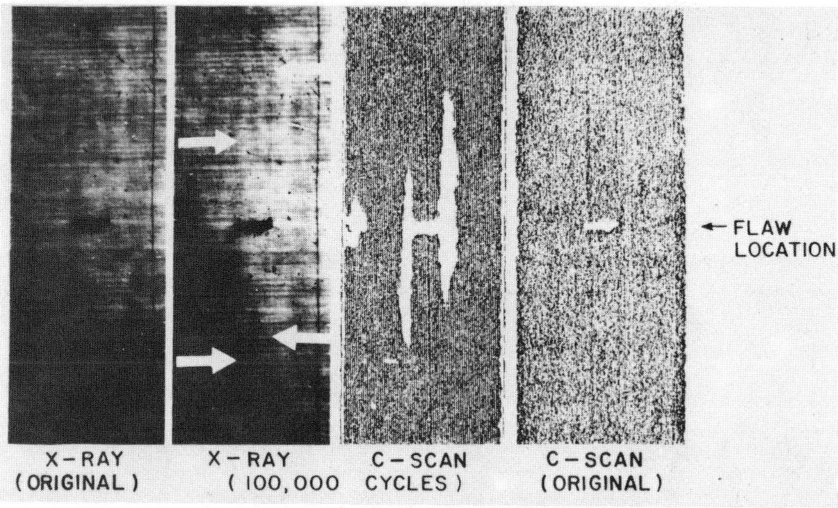

FIG. 6—*Initial and 100 000 cycle radiographs and C-scans of a* $[90/0]_s$ *specimen with an embedded flaw. Radiograph shows axial cracks in 0-deg plies and C-scan shows delamination of 90/0 interfaces due to cyclic loading.*

straint influence response is distinct, and the mechanics involved in the two cases is different. The in-plane response and the in-plane constraint effects are the principal contributors to notched strength and changes in notched strength during quasi-static loading. The major effect of through-the-thickness constraint stresses is on fatigue response as affected by delamination, longitudinal cracking, and the growth and coupling of transverse cracks in off-axis plies.

Before addressing the in-plane problem, a more precise definition of constraint effects is given. If a lamina or laminae of identical type and orientation are unconstrained, unidirectional behavior is being discussed. If plies of some other orientation are bonded to the unidirectional ply or plies, they constrain the response of the unidirectional material accordingly. That situation defines constraint effects. For the unnotched in-plane problem, the constraint process is represented by Fig. 7. The cross section of a laminate consisting of 0, 45, and 90-deg plies is shown at the top of the figure. If an axial (out of the page) force is applied to the laminate, the Poisson effect causes the transverse dimensions (horizontal in the figure) to change in the proportions shown in Fig. *7b*. (The thickness changes, which will be uniform, have been ignored.) Of course the laminae are bonded together so that the transverse strains shown in Fig. *7b* are not allowed. Rather, each of the plies applies a transverse stress to its neighbor(s) so that a common transverse strain occurs as shown in Fig. *7c*. The proportional magnitude of those transverse (constraint) stresses is also shown in Fig. *7c*. For the present case, the 0-deg plies are nearly "unconstrained" from the standpoint of the transverse elastic stress state, while the 45 and 90-deg plies experience large

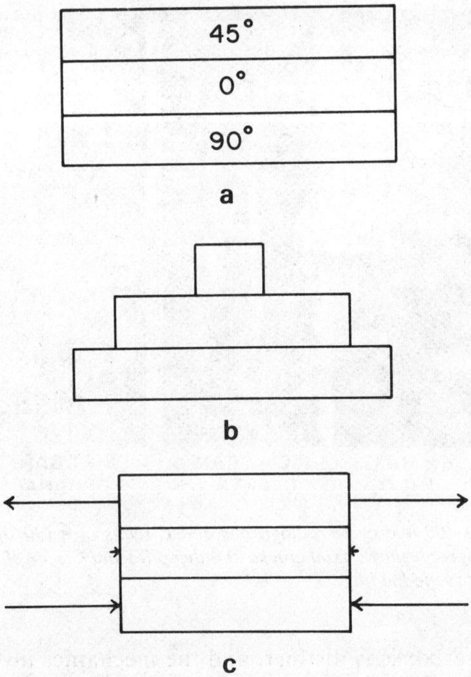

FIG. 7—*Cross section of (a) unloaded, (b) loaded but unconstrained, and (c) loaded and constrained transverse (Poisson) displacement of a quasi-isotropic laminate under axial loading.*

constraint stresses. The nature of the constraint is, of course, controlled by the orientation of the plies in a given laminate.

A list of unidirectional and laminate Poisson's ratios is shown in Table 2. The calculated values were determined from a laminate analysis. The Poisson contraction controls the magnitude of the constraint stresses and can be used to calculate the transverse (σ_y) component. While using engineering modulii, which are determined from unidirectional stress tests, is not a recommended general procedure for the determination of lamina stresses, the approximate value of the transverse stress (σ_y) can be quickly and easily obtained by a very simple procedure if Poisson's ratios are known. The most important aspect of the scheme is that it provides a conceptually simple rationale for understanding and mentally estimating the transverse normal stresses that are created by constraint alone. To apply the scheme, one first calculates a "differential Poisson's ratio" defined as the absolute difference between the Poisson's ratio of the ply in question when that ply is unconstrained and the laminate Poisson's ratio when the subject ply is bonded into the laminate of interest. Analytically, the expression is stated as

$$\Delta\nu = |\nu\,(\text{constrained}) - \nu\,(\text{unconstrained})| \tag{1}$$

For any given axial loading, represented by an axial strain of $\epsilon_x{}^\circ$, the "constraint effect" in the transverse direction is then given by

$$\Delta\nu\epsilon_x{}^\circ \tag{2}$$

The transverse normal stress in the subject ply is then approximated by

$$\sigma_y{}^i = E_y{}^i \, \Delta\nu\epsilon_x{}^\circ \tag{3}$$

where $E_y{}^i$ is the engineering Young's modulus of the subject ply in the transverse (y) direction. The accuracy of such a simplified scheme is indicated by the comparisons shown in Table 3. The "constraint stresses" shown there are σ_y stresses, and the "exact" values are those stresses calculated by a standard laminate analysis. The only noticeable difference in approximate and exact values is seen to occur for the $[\pm 45/0]_s$ laminate. The Poisson's ratio for a 45-deg (or ±45-deg) lamina is not a well-established number and may have contributed to the difference. In the context of the discussion, it is especially important to notice the large difference in the constraint environment of the 0-deg plies in the two cases examined in Table 3. In one case, the σ_y constraint stress is positive and, in the other case, it is negative. (In both cases an axial strain of 1000 $\mu\epsilon$ was applied to the laminate.) The transverse stress is significant in magnitude in both cases. It would be difficult to imagine that any damage that develops in the 0-deg plies would be unaffected by such a stress or that the nature of damage development in the 0-deg plies would be identical when the σ_y stress changes from a

TABLE 2—*Poisson Ratios.*

Laminate	Measured Values	Calculated Values
$[0]_n$	0.29	...
$[90]_n$	0.031	...
$[\pm 45_n]_s$	0.78[a]	...
$[0, 90]_n$	0.051	0.039
$[\pm 45, 0]_s$	0.68	0.688
$[0, 90, \pm 45]_s$	...	0.299

[a]From literature.

TABLE 3—*Constraint Stress.*

Constrained Ply	Exact	Approximate
0-deg ply in $[0, 90]_s$	+1418	+1413
90-deg ply in $[0, 90]_s$	−1418	−1399
0-deg ply in $[\pm 45, 0]_s$	−1720	−1719
±45-deg plies in $[\pm 45, 0]_s$	+860	+779 (ν = 0.8)

large positive to a large negative value. In fact, the results show that these stresses do have a significant influence on damage development. While some occasional suggestion of that fact has been made elsewhere from time to time, no damage growth model that we are aware of explicitly includes the influence of σ_y constraint stresses. In the case of unnotched response, damage in the 0-deg plies consists, primarily, of longitudinal splitting along the 0-deg fibers in the unconstrained case. While final fracture requires transverse fiber breakage, the transverse fracture is nearly always discontinuous with regions separated by longitudinal splitting. The effect of transverse σ_y constraint stresses on this damage mode is easily noticed. For the $[90/0]_s$ laminate, the transverse σ_y stress in the 0-deg ply is positive, as shown in Table 3. One might expect such a stress to contribute to an increase in longitudinal splitting, compared to the unconstrained behavior. It is, in fact, so. The results of nondestructive investigations and the sectioning studies show a definite increase in the number of longitudinal cracks that form in the 0-deg plies for a given load or for a given number of cycles at a given stress amplitude when the 90-deg plies are added to the unidirectional 0-deg plies. At the same time, adding the 45-deg plies to form the $[\pm 45/0]_s$ laminate, so that the transverse (σ_y) stress in the 0-deg plies is compressive, decreases the longitudinal splitting essentially to an unobservable level.

The second general category of constraint effects is through-the-thickness effects. These effects are due to interlaminar stresses that result from the different response of each lamina in the laminate. Two basic consequences of these stresses are most significant from the standpoint of influencing response. First, independent of the action of an (in-plane) notch, the through-the-thickness constraint controls the pattern of transverse cracks that form in the off-axis plies of a laminate, which, in turn, controls the state of stress and state of strength of the laminate in the unnotched case. Second, the through-the-thickness stresses in the neighborhood of a notch (or transverse crack in a given lamina) control the tendency for delamination and crack growth along laminae boundaries.

The first of these effects has been reported earlier [*9*]. Perhaps the most important effect of constraint in unnotched laminates is the creation of a characteristic damage state (CDS). This state consists of regular arrays of cracks in the off-axis plies of an angle-ply laminate. The arrays form in exactly the same way under quasi-static or cyclic loading, independent of load history. In fact, the crack spacing in these arrays is controlled by the constraint of the laminate on the cracked plies. The CDS is a laminate property, controlled by the properties of each lamina and the stacking sequence of the laminae in the laminate. The CDS is a stable condition that is well defined, and always developes prior to fracture of the laminate. In that sense, it is quite similar to the single crack situation in a homogeneous material in that it suggests a well-defined mechanics representation of a damage state that controls the state of stress and state of strength prior to failure. An example of the crack patterns in the CDS is shown in Fig. 8.

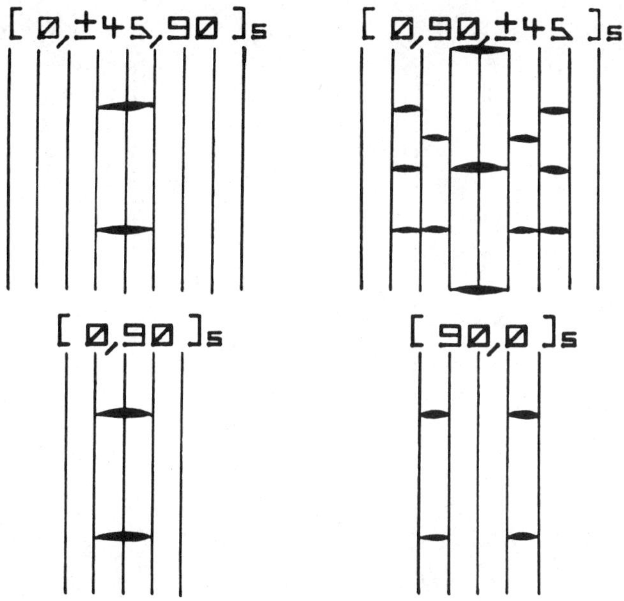

FIG. 8—*Examples of crack patterns that form the characteristic damage state (CDS) of the laminates.*

The manner in which the through-the-thickness constraint controls formation of the CDS is explained by a simple modeling operation. The modeling method chosen was an equilibrium element approach similar to that which has been used by other authors for other fracture problems in composite materials [*10,11*]. A schematic of the idealized damage situation is shown in Fig. 9. In that figure, a crack is assumed to have formed in two plies having a total width a. The schematic represents an edge view of the cracked laminate. The constraint layers next to the crack have an orientation of α degrees to the load axis, and the disturbance caused by the broken layer does not extend beyond the first constraining (unbroken) ply on either side of the broken ply. Finally, on the basis of experimental observation, any gradients in response from ply to ply occur over a distance, b, which extends for a distance of one twentieth of the ply width into each ply, or about the thickness of one tow or bundle on either side of the ply interface in the case of graphite epoxy, for example. The analysis was used to predict the characteristic saturation spacing of cracks by assuming that the constraining layers transferred stress back into the broken ply on either side of the cracks until the stress that created the first crack was reached again. At that distance away from the first crack, a new crack will then form.

Some results of the analysis are shown in Table 4, along with the spacing values measured experimentally. The experimentally determined values are expressed as the range of values included within one standard deviation

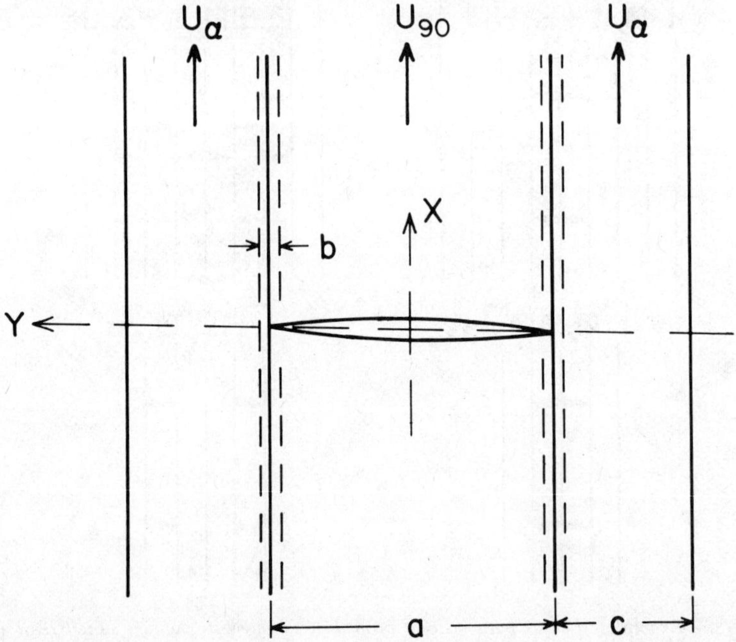

FIG. 9—*Schematic diagram of a crack in two 90-deg plies at the edge of a laminate.*

above and below the average measured value. Both static and fatigue data are shown in one case.

The agreement between predicted and observed crack spacings is quite satisfactory. It is especially interesting to notice the large difference in crack spacing caused by only a change in stacking sequence, a clear indication of the importance of constraint effects. For example, the number of cracks in the 90-deg plies of $[0/90]_s$ laminates is nearly twice that in $[90/0]_s$ laminates, and the $[0/90]_s$ laminates also have the lower tensile strength. In each case, the constraint of each ply on its neighbors and the constraint of the laminate response on the individual plies determine the equilibrium crack spacing. Figure 10 shows a typical example of this equilibrium damage state for a $[0/90/\pm 45]_s$ laminate. (The observed state is on the top and the predicted one is on the bottom.) Figure 11 shows the observed and predicted CDS when the stacking sequence is changed to $[0/\pm 45/90]_s$. These characteristic damage states are created at moderate quasi-static loads or by cyclic loading. The plies that form cracks are determined by the magnitude of the load level, but the total number of cracks in the equilibrium characteristic damage state in a given ply is a laminate property, independent of load history or geometry. This characteristic damage state controls the state of stress and state of strength in the laminate until the final failure events begin, quite close to fracture, and is a direct result of through-the-thickness constraint

TABLE 4—*Results of analysis.*

Specimen Type	Predicted Crack Spacing, mm		Observed Crack Spacing, mm
Cracks in two center 90-deg plies of $[0, 90]_s$ laminate	0.882		1.087 to 0.532
Cracks in outside 90-deg plies of $[90, 0]_s$ laminate	1.66		1.73 to 0.775
Cracks in center two 90-deg plies of $[0, \pm45, 90]_s$ laminate	0.76 0.76	static fatigue	1.51 to 0.62 1.44 to 0.47
Cracks in two center 45-deg plies of $[0, 90, \pm45]_s$ laminate	1.21		1.25 to 0.995
Cracks in single 90-deg plies of $[0, 90, \pm45]_s$ laminate	0.411		0.423 to 0.241
Cracks in +45-deg plies of $[0, 90, \pm45]_s$ laminate	0.875		0.909 to 0.524
Cracks in outside 90-deg plies of $[90_2, 0, \pm45]_s$ laminate	1.12		1.40 to 1.07
Cracks in −45-deg plies of $[0, \pm45, 90]_s$ laminate	0.879		0.960 to 0.889
Cracks in 90-deg plies of $[0, 90_2, \pm45]_s$ laminate	0.701		0.600 to 0.460
Cracks in +45-deg plies of $[0, 90_2, \pm45]_s$ laminate	0.890		1.05 to 0.650
Cracks in −45-deg plies of $[0, 90_2, \pm45]_s$ laminate	1.223		1.38 to 0.963

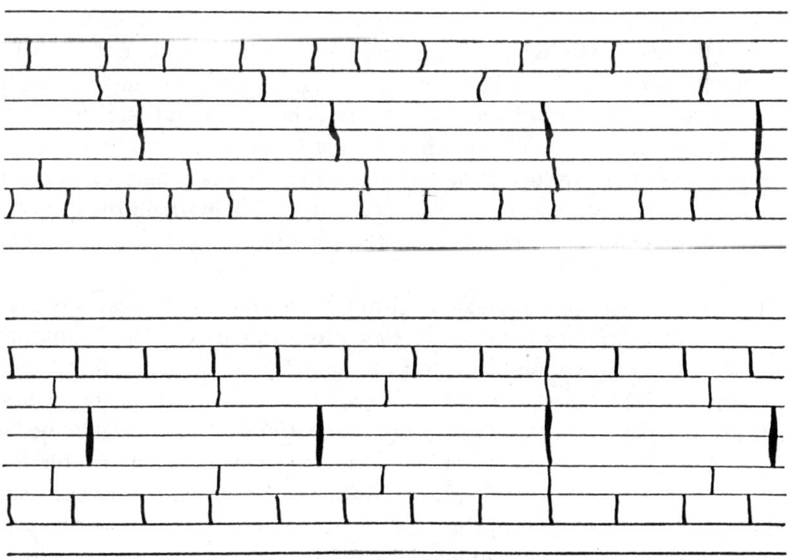

FIG. 10—*Observed* (top) *and predicted* (bottom) *characteristic damage state (CDS) for a [0, 90, $\pm45]_s$ graphite epoxy laminate.*

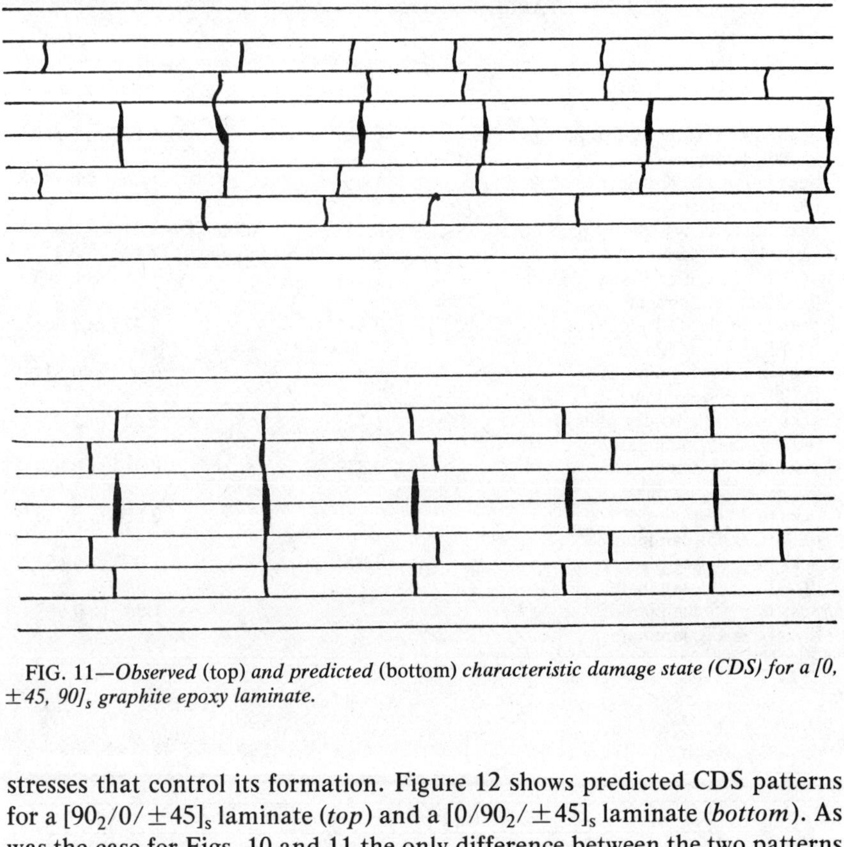

FIG. 11—*Observed* (top) *and predicted* (bottom) *characteristic damage state (CDS) for a [0,* $\pm$*45, 90]$_s$ graphite epoxy laminate.*

stresses that control its formation. Figure 12 shows predicted CDS patterns for a $[90_2/0/\pm45]_s$ laminate (*top*) and a $[0/90_2/\pm45]_s$ laminate (*bottom*). As was the case for Figs. 10 and 11 the only difference between the two patterns is the stacking sequence of the plies. Hence, the substantially different patterns change because of interlaminar (constraint) stresses. These different CDS patterns also produce different strengths. The laminate in Fig. 10 is stronger than its counterpart in Fig. 11, and the laminate on the bottom of Fig. 12 is stronger than the one on the top that differs only in stacking sequence.

The second basic consequence of through-the-thickness stresses is the influence of those stresses on notch or flaw growth along laminae boundaries. This effect can also be modeled, but the model must be accurate enough to give fairly reliable three-dimensional (complete) field stress information. Such an analysis, using a three-dimensional finite difference scheme, was used to calculate complete stress fields for the cracked lamina problem [*9*].

Although the one-dimensional analysis discussed earlier is very convenient and instructive, the details of the mechanics involved in crack formation and especially crack growth cannot be extracted from such an analysis. The precise nature of the stress field in the neighborhood of a crack in an off-axis ply will determine how that crack grows into the next ply, along the ply inter-

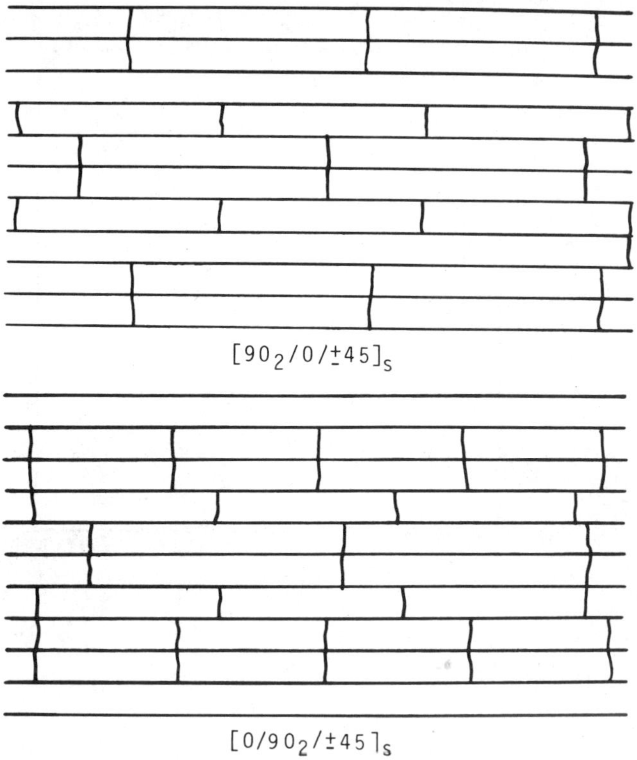

FIG. 12—*Predicted characteristic damage states (CDS) for a [90$_2$, 0, $\pm$45]$_s$* (top) *and a [0, 90$_2$, +45]$_s$* (bottom) *laminate.*

faces, couples with other cracks, or participates in a laminate fracture event. The mechanics of these important aspects must be investigated by means of a more general and more precise analysis scheme. The problem was set in a manner similar to the one-dimensional treatment so that the complete stress field around a flaw in one ply of an angle-ply laminate could be analyzed. The equations of equilibrium were solved by a finite difference scheme under the assumption that all stresses were independent of the (in-plane) specimen width. The equilibrium equations were also satisfied at boundary nodes for increased accuracy. Cracks were simulated by stress-free surfaces.

Some of the stress distributions in damaged laminates obtained by the use of this analysis are presented in the following paragraphs. The applied strain values for all examples discussed here are $\epsilon_x = 1000\mu\epsilon$ ($\epsilon_x = 10^{-3}$ in./in.) and $\epsilon_y = -200\mu\epsilon$ ($\epsilon_y = -2 \times 10^{-4}$ in./in.)

Figure 13 shows the axial normal stress through the laminate thickness for three laminates when the interior plies always consist of a pair of 0-deg laminae with a center notch. The ordinate of the figure is actually the stress

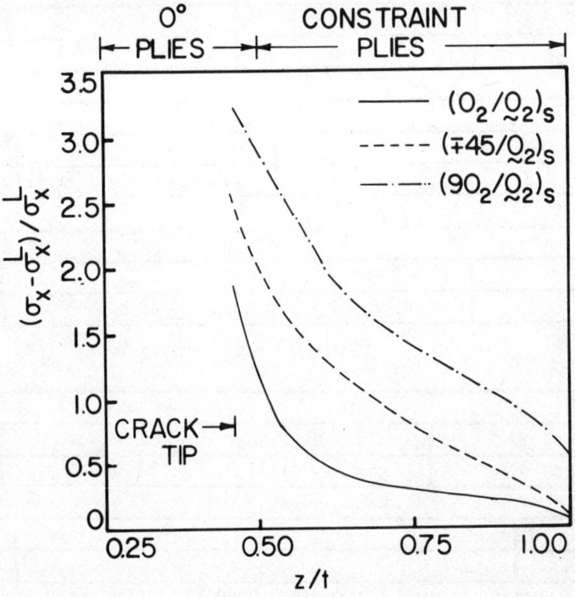

FIG. 13—*Normalized stress increase across specimen thickness for a notched 0-deg ply constrained by three different types of constraining plies.*

elevation due to the notch for three different constraint layers. Interestingly enough, the lowest modulus (and strength) constraint layer has the highest stress elevation due to a notch. Figures 14 and 15 show a stress that would be completely absent in an uncracked laminate. The through-the-thickness shear stress, τ_{xz}, comes about because the broken lamina tries to "slide out" in a manner similar to our earlier discussion for the in-plane problem. From the standpoint of damage growth this stress contributes to delamination growth between the plies, that is, a spread of damage above and below the notch or ply crack. The magnitude of the calculated stresses is quite high, high enough to participate in damage growth. The distribution of shear stress for the two laminates is quite similar suggesting that the 0-deg plies control that distribution.

Figure 16 shows another stress component that is entirely caused by constraint around the internal crack or notch. It is a through-the-thickness normal stress, σ_z, that develops at the interface between the broken and unbroken plies. The three constraint situations discussed earlier are shown. Again, the $[90/0]_s$ case shows the greatest σ_z stress. The σ_z component of stress would be expected to produce delamination modes of damage, and the stress levels calculated are significant. Experimental results showed that the $[90/0]_s$ laminates delaminated extensively in the neighborhood of the notch,

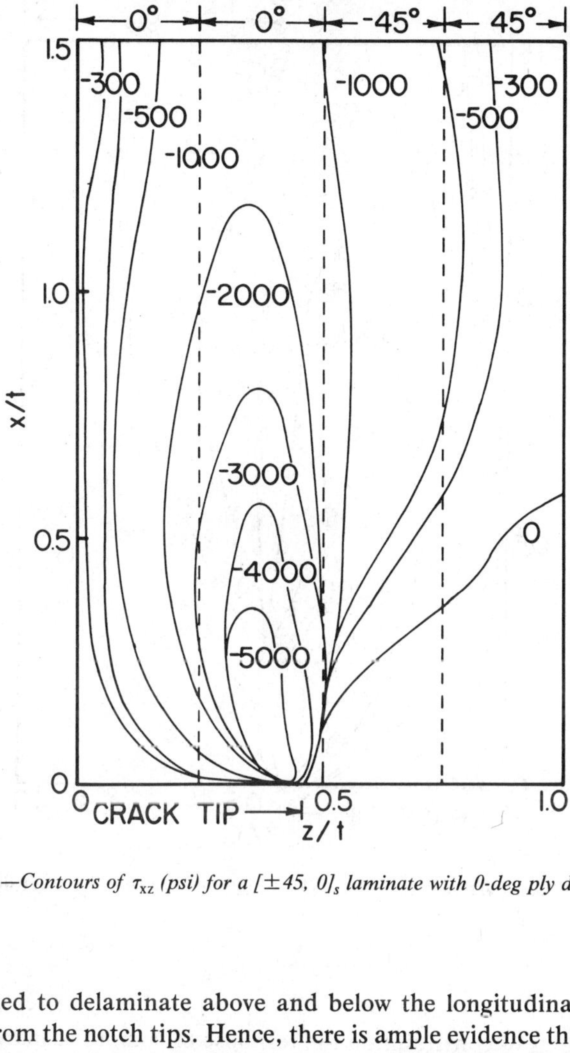

FIG. 14—*Contours of* τ_{xz} *(psi) for a* $[\pm 45, 0]_s$ *laminate with 0-deg ply damage.*

and continued to delaminate above and below the longitudinal crack that grew away from the notch tips. Hence, there is ample evidence that the model is helping to anticipate the details of constraint-stress-controlled flaw growth. Such a modeling scheme can also be applied to failure-theory analysis to predict flaw growth through the thickness as well as along ply boundaries. A preliminary example of such a scheme is provided in Fig. 17 that shows the axial normal stress in the unbroken 0-deg ply of a $[0/90/\pm 45]_s$ and a $[0/\pm 45/90]_s$ laminate when all other plies are broken. The stress is different in the two cases because the compliance of the free crack faces is different causing a different constraint on the crack opening. It also happens that the laminate that has the higher stress in the 0-deg ply, the $[0/\pm 45/90]_s$ laminate, also has the lower strength.

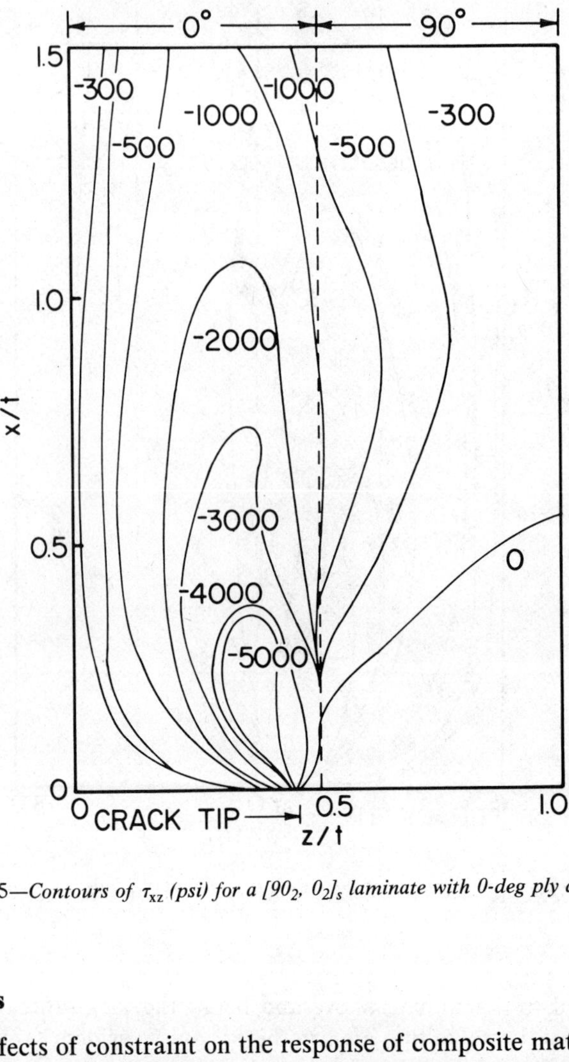

FIG. 15—*Contours of* τ_{xz} *(psi) for a* $[90_2, 0_2]_s$ *laminate with 0-deg ply damage.*

Conclusions

1. The effects of constraint on the response of composite materials can be classified in two categories: (*a*) in-plane effects and (*b*) through-the-thickness effects.

2. In-plane constraint is the principal contributor to notched strength and changes in notched strength under quasi-static loading.

3. Constraint situations that produce greatest static strength do not minimize the extent of damage that develops during static or cyclic loading. Likewise, minimum damage situations do not correspond to maximum strength cases.

4. Through-the-thickness constraint controls the pattern and spacing of transverse cracks in the characteristic damage state that determines the state of stress and state of strength in unnotched laminates.

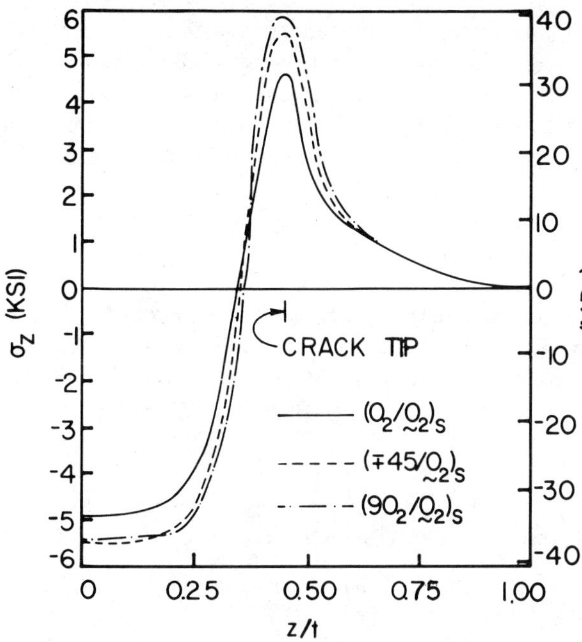

FIG. 16—*Constraint effect on the through-the-thickness normal stress distribution.*

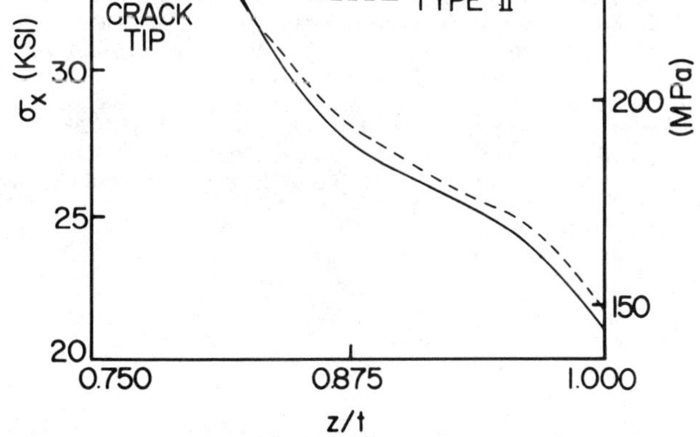

FIG. 17—*Through-the-thickness variation of σ_x ahead of the crack tip for complete off-axis ply failure as the crack extends into the 0-deg ply.*

5. Out-of-plane stresses produced by constraints are influential on the growth of damage along ply interfaces, especially during cyclic loading.

6. The mode of damage and the extent of damage in constrained notched and unnotched plies is governed by the stress state in those plies, as determined by the constraining plies, and the relationship of the stress state to the strength state.

Acknowledgments

The authors appreciate the support of the Air Force Materials Laboratory, Contract No. F33615-77-C-5044, monitored by N. J. Pagano and NASA Langley Research Center, Grant NSG 1364, monitored by J. D. Whitcomb. They also extend their thanks to Frances Carter for typing the manuscript.

References

[*1*] Pagano, N. J. and Pipes, R. B., *Journal of Composite Materials,* Vol. 5, Jan. 1971, pp. 50–57.

[*2*] Pagano, N. J. and Pipes, R. B., *International Journal of Mechanical Sciences,* Vol. 15, 1973, pp. 679–688.

[*3*] Bjeletich, J. G., Crossman, F. W., and Warren, W. J., "The Influence of Stacking Sequence on Failure Modes in Quasi-Isotropic Graphite Epoxy Laminates," Failure Modes in Composites IV, American Institute of Mining, Metallurgical and Petroleum Engineers, 1978.

[*4*] Duggan, M. F., "An Experimental Evaluation of the Slotted Tension Test for Composite Materials," presented at the 1979 SESA Spring Meeting, San Francisco, Calif. May 1979.

[*5*] Reifsnider, K. L., Henneke, E. G., and Stinchcomb, W. W., "Defect-Property Relationships in Composite Materials," AFML-TR-76-81, Part III, June 1978.

[*6*] Stinchcomb, W. W., Reifsnider, K. L., and Yeung, P., "Investigation and Characterization of Constraint Effects on Flaw Growth During Fatigue Loading of Composite Materials," Third Semi-Annual Status Report, NASA Grant NSG-1364, June 1978.

[*7*] Stinchcomb, W. W., Reifsnider, K. L., Yeung, P., and Gibbins, M. N., "Investigation and Characterization of Constraint Effects on Flaw Growth During Fatigue Loading of Composite Materials," Final Report, NASA Grant NSG-1364, June 1979.

[*8*] Yeung, P. C., Stinchcomb, W. W., and Reifsnider, K. L., *Nondestructive Evaluation and Flaw Criticality for Composite Materials, ASTM STP 696,* American Society for Testing and Materials, 1979, pp. 316–338.

[*9*] Reifsnider, K. L., Henneke, E. G., and Stinchcomb, W. W., "Defect-Property Relationships in Composite Materials," AFML-TR-76-81, Part IV, June 1979.

[*10*] Zweben, C., *Journal of Mechanics and Physics of Solids,* Vol. 19, 1971, pp. 103–116.

[*11*] Kulkarni, S. V., McLaughlin, P. V., and Pipes, R. B., "Fatigue of Notched Fiber Composite Laminates," NASA CR-145039, April 1976.

C. T. Robinson[1] *and P. H. Francis*[2]

Damage Initiation in a Three-Dimensional Carbon-Carbon Composite Material

REFERENCE: Robinson, C. T. and Francis, P. H., "**Damage Initiation in a Three-Dimensional Carbon-Carbon Composite Material,**" *Fatigue of Fibrous Composite Materials, ASTM STP 723,* American Society for Testing and Materials, 1981, pp. 85-100.

ABSTRACT: Small coupons of a three-dimensionally woven, carbon-carbon composite were axially loaded in a specially-designed SEM load stage. Notched and unnotched specimens were fatigue cycled at room temperature. Crack initiation was found to be in a sliding mode, most often near a fiber-matrix interface. The cracks would immediately turn 90 degrees to an opening mode and propagate through a matrix pocket until encountering a longitudinal fiber bundle. In most cases, the crack would then propagate along the fiber bundles, again in a sliding mode. Yarn bundle failures were not limited to the immediate region of an approaching crack, leading to the conclusion that this material is not notch sensitive. Acoustic emission was found very useful in characterizing the different stages of microstructural damage.

KEY WORDS: carbon-carbon composite, damage initiation, graphite, crack propagation, scanning electron microscopy, acoustic emission, fatigue (materials), composite materials

Carbonaceous materials reinforced with a three-dimensional network of graphite fibers have been found to be very useful in high-temperature applications such as rocket exhaust nozzles, missile nose tips [*1*],[3] and high-performance brake shoes. These carbon-carbon materials have good strength retention at high temperatures (~3000°C), excellent ablative resistance [*2*], and good environmental tolerance when compared to other high-temperature materials such as pyrolitic graphite and refractory metals. The unique thermomechanical properties of these carbon-carbon composites have effected substantial improvements in high-temperature reliability and efficiency. However, the unique structure poses problems

[1]Research engineer, Southwest Research Institute, San Antonio, Tex. 78284.

[2]Chairman, Department of Mechanics and Aerospace Engineering, Illinois Institute of Technology, Chicago, Ill. 60616.

[3]The italic numbers in brackets refer to the list of references appended to this paper.

in adequately characterizing the thermal, physical, and mechanical properties of these materials.

Limited work has been done attempting to correlate the microstructure with impact damage, cracking, and failure of carbon-carbon composites [*3,4*]. Ablative resistance, as it relates to microstructure, has also been studied experimentally [*5*]. Possibly the most important parameters affecting the performance of these composites are the processing parameters of temperature and pressure [*6–8*]. Along these lines, direct studies of the resulting microstructures have been compared to the processing variables [*9*] and the resulting damage mechanisms [*10*].

Analytical methods have included a nonlinear modeling technique to account for the temperature-dependent material behavior [*11*] and an approximate method in which an "equivalent" matrix is introduced, which approximates the actual material and for which the mechanical properties are easily calculated [*12*]. Most of the analytical work, however, has been related to, or is a development leading to a system that is compatible with various computer codes. One such code affects a micromechanical and macromechanical analysis in which it is possible to account for the effects of voids, cracks, and other nonlinear features within the material [*13*]. Others include an approximate technique enabling the use of two-dimensional axisymmetric finite element codes for analysis of three-dimensional orthotropic materials with axisymmetric shapes and with an axisymmetric loading geometry [*14*] and a characterization of filament-wound carbon-carbon materials that is compatible with either two-dimensional plate and shell codes or with three-dimensional orthotropic finite element codes [*15*]. In other work, experimental results were found to have reasonable correlation with a continuum model for the response to thermostructural loading [*16*]. This model was applied through a finite element code. Other experimental programs have included measurement of lateral strain in order to verify predicted Poisson ratio values over a temperature range [*17*] and standard fracture mechanics tests to ascertain the applicability of fracture mechanics parameters to carbon-carbon materials [*18*].

One major difficulty that arises in attempting to test a carbon-carbon material is its weak shear strength. A tension test is an example of a standard mechanical test that is very difficult to perform because of low shear strength. Alternate test methods include flexure tests [*19*] and testing of ring-shaped specimens [*20*]. Ring specimens are especially convenient in the mechanical testing of cylindrically woven composites, as the weave geometry matches the test specimen geometry. One type of measurement that has proven very useful in characterizing composite materials is monitoring of acoustic emission [*21*]. The fracture of fibers is readily noticeable, in relation to general yielding of matrix material, due to the very high amplitude (approximately four times as great) of the associated acoustic signal.

In summary, very little work has been done in the area of mechanical properties characterization and identification of damage mechanisms in carbon-carbon materials [*22*]. The purpose of the present work was to attempt to characterize the initiation of damage and flaw growth in carbon-carbon materials by post-test analyses and by real-time observation through electron microscopy. It is hoped that characterization of damage mechanisms will provide insights that will facilitate analytical modeling of the very complex internal structure of this material. An understanding of load carrying mechanisms and an ability to model carbon-carbon composites will enhance the ability to predict thermomechanical performance and reliability of these composites.

Material System

The material system investigated was a three-dimensional, orthogonally-woven, carbon-carbon composite. By convention, this system is designated as a 2,2,5 material, indicating the number of fiber ends per site in the X, Y, and Z directions, respectively. A unit cell of this cartesian material is shown schematically in Fig. 1. The precursor for the fibers was HM-3000, a polyacrylonitrile (PAN) yarn manufactured via a high-modulus process. The matrix precursor was Ashland 240, a petroleum pitch. The manufacturing process was a proprietary process of Fiber Materials, Inc. (FMI) that was very similar to the sequence used for rocket nozzles. Details of the material system are given in Ref *22*.

Static Load Tests

Static load tests were conducted to characterize the material for its tensile, compressive, and shear properties. These tests, with the aid of acoustic emission analysis and post-failure fractographic examination, provided some gross information regarding failure mechanisms. Details of specimen geometries and test procedures are given in Ref *22*.

A typical stress versus strain curve for a tensile test is shown in Fig. 2. Most notable is the brittle behavior of the material as indicated by the extremely linear curves. There is, however, a phenomenon resembling a small pop-in prior to final failure (at an approximate stress of 70 MPa) that was typical of all of the tension tests. From all indications, this seems to be related to a fracture of a small number of fibers shortly before failure of the entire specimen. The other obvious feature of the typical stress-strain curve is the negligible transverse strain. The compressive stress-strain curves were identical in form to those for the tension tests. Once again the transverse strains were essentially zero. It seems that the structure of this inhomogeneous material masks the Poisson strains in the

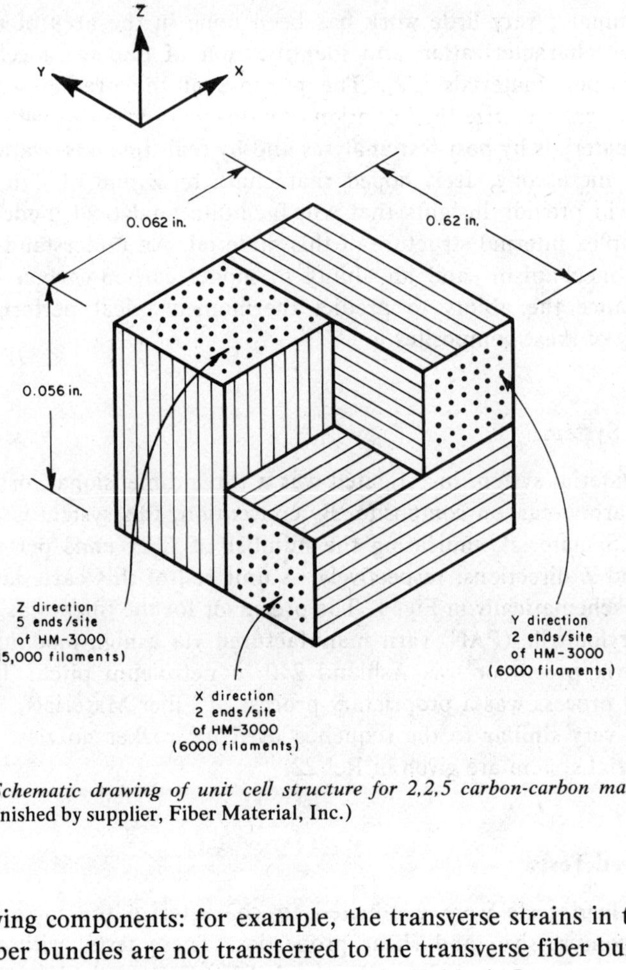

FIG. 1—*Schematic drawing of unit cell structure for 2,2,5 carbon-carbon material.* (Dimensions furnished by supplier, Fiber Material, Inc.)

load-carrying components: for example, the transverse strains in the longitudinal fiber bundles are not transferred to the transverse fiber bundles.

Results of the mechanical tests are given in Table 1. In most cases there were three valid tests, in which case the standard deviation is given. These values are intended only to give an indication of the scatter in the data, as there was not a sufficient number of data points to give statistically significant data. In the cases where there were only two valid tests, both values are given. The "specimen orientation" indicates which fiber bundles are oriented along the loading direction of the specimen. It should be noted that the fiber volume in the X direction is the same as that in the Y direction, but the Z direction has on the order of 150 percent the fiber volume of the X and Y directions. Since the fibers are the main load-carrying component, it is expected (as seen in Table 1) that "Z direction" specimens will show higher load carrying capabilities.

One phenomenon that is noted from Table 1 is that the material tends to become slightly stronger as temperature is increased. Probably the most

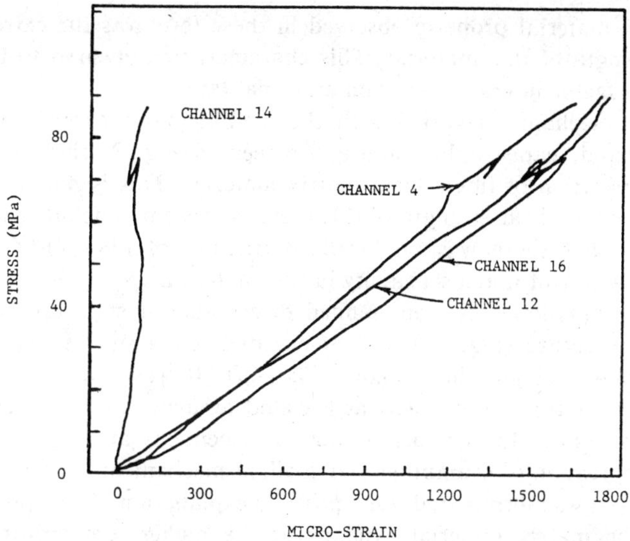

FIG. 2—*Typical stress-strain relationships for three longitudinal strain gages on a tension specimen. A strain reading from a transverse gage is shown for comparison. (Channel 14, Specimen T-7, Y direction).*

Table 1—*Mechanical test data.*

Loading Mode	Temperature, °C	Specimen Orientation	σ_{MAX}, MN/m^2	E, GN/m^2
Compression	21	Z	176 ± 18	163 ± 11
Compression	21	X	103 ± 3	66 ± 2
Compression	1370	Z	157, 168	...
Compression	1370	Y	118, 122	...
Compression	1950	Z	198, 209	...
Compression	1950	Y	143, 160	...
Tension	21	Z	185[a]	...
Tension	21	Z	297 ± 37[b]	140 ± 29
Tension	21	Y[c]	96 ± 3	48 ± 8
Tension	21	Y[d]	128 ± 13	63 ± 10
Shear	21	Z	15 ± 2	...
Shear	21	Y	11, 17	...

[a] Constant radius reduced test section: only one valid test.
[b] Specimens were half as thick as other specimens and had longer tabs for gripping.
[c] Z fiber direction oriented across width of specimen.
[d] Z fiber direction oriented across thickness of specimen.

NOTE: Values are given with standard deviation if there were tnree or four tests. If there were only two valid tests, both values are given.

important material property observed in these tests was the extremely low shear strength of this material. This characteristic appeared to be the determining factor in crack initiation and final failure.

Failure mechanisms varied with the loading mode. Tensile specimens showed rapid, catastrophic failure. As seen in Fig. 3, there was partial to full separation of fibers from matrix material. This is not surprising in light of the low shear strength of this material system. The other interesting feature is that there was no localized fracture of fiber bundles, which indicates a lack of notch sensitivity in this material.

The compression specimens tended to crumble or spall upon failure at room temperature (Fig. 4*a*), but at elevated temperatures the specimens failed along a single shear plane (Fig. 4*b*). It was noticed, during the tests, that failure of specimens at elevated temperatures was very sudden and catastrophic. In contrast, failure of specimens at room temperature was a progression of crumbling and spalling mechanisms, which continued until the test was terminated. One possible explanation of this phenomenon is that the matrix material expands at the higher temperatures, giving

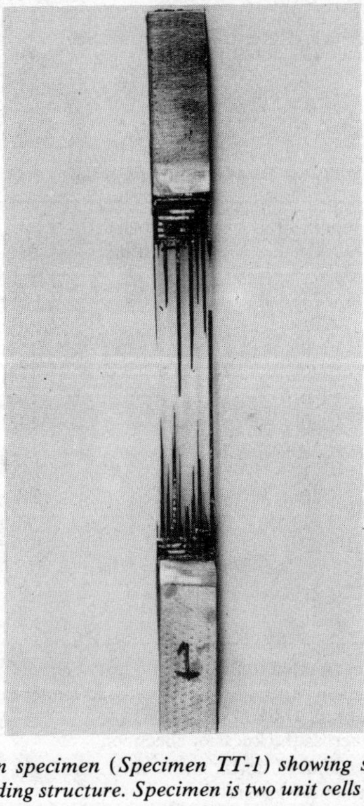

FIG. 3—*Failed tension specimen (Specimen TT-1) showing some fibers that pulled completely out of the surrounding structure. Specimen is two unit cells thick.*

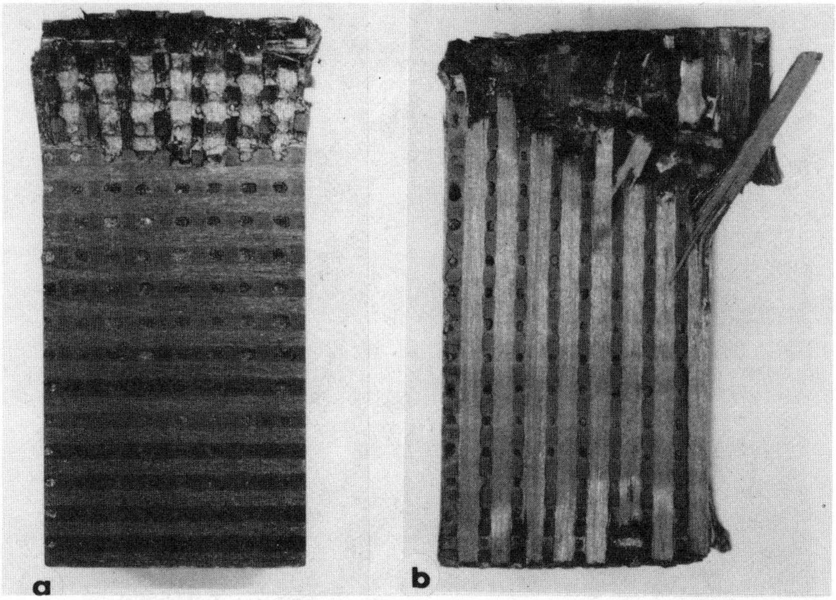

FIG. 4—*Failed compression specimens:* (a) *Specimen C-24 showing brooming phenomenon typical of room temperature failure, and* (b) *Specimen C-11 showing shear failure typical of elevated temperature (1930°C) tests.*

the load-carrying fiber "columns" more support [*23*]. Thus, the fibers do not buckle as soon, but the failure is very sudden and complete.

The most notable feature during the shear tests was the phenomenal amount of deformation (elastic and plastic) that the material could sustain without failing (Fig. 5). Though great amounts of strain were sustained, it required very low stress levels to deform the material in shear.

Valuable insights into the damage chronology were obtained by acoustic emission (AE) monitoring. The AE crystal was a 6.35 mm diameter by 2.35 mm thick PZT-5 crystal with a resonant frequency of approximately 200 KHz. Data for the tests are displayed in bar graph format: the horizontal axis represents the magnitude of the signal and the vertical axis is a logarithmic scale representing the number of counts at a given amplitude level. It was discovered during the testing that the higher-amplitude signals also had a longer duration.

Figure 6 shows a typical test sequence for a tension test. Early in the test there were numerous low-level signals as the matrix material and fibers began to strain (Fig. 6*a*). When the first load drop occurred (as just mentioned) prior to failure, there was a small number of markedly higher signals (Fig. 6*b*). Finally, at the end of the test, there were considerably more lower-level signals and several very high-amplitude signals (Fig. 6*c*), resulting from final failure.

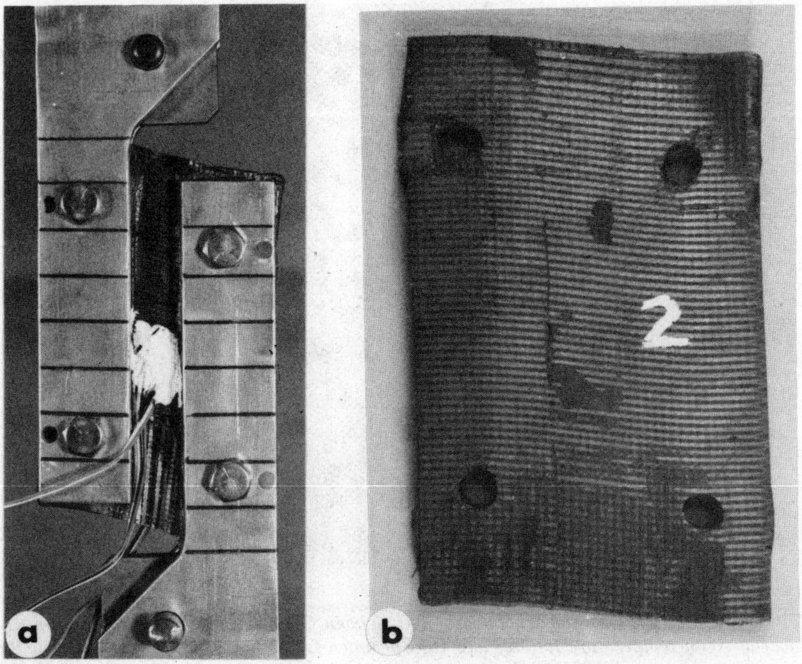

FIG. 5—*Rail shear Specimen S-2:* (a) *at maximum load (17.2 MPa), and* (b) *plastic deformation after test was terminated.*

The conclusions drawn from these observations are that the lower-level signals were due to matrix crumbling (or shearing) or to some relative slip between fiber bundles and matrix regions, and the small number of isolated, higher-level signals were due to sudden fiber failure.

SEM Observation of Fatigue Loading

Observation of specimens, under load, in a scanning electron microscope (SEM) provided the most valuable insights into damage initiation in this material. Southwest Research Institute's ETEC Autoscan SEM has been fitted with a load stage with a tensile load capacity of 4900 N (1100 lb) (Fig. 7). Cycling frequency used for these tests was one cycle per 2.5 s (0.40 Hz) [*24*].

The load carrying mechanism for this facility consists of two pins at the edges of each end of the specimen. In order to sustain this shear load across the specimen, aluminum tabs were bonded to the specimens. The specimens were machined with a constant-radius reduced cross section in order to localize the damage initiation so it could be more easily observed in the limited area covered by the electron beam. Some specimens

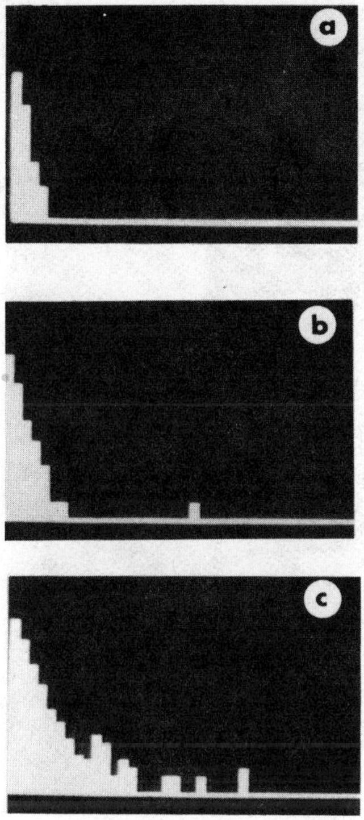

FIG. 6—*AE count distribution graph for Specimen T-3. Horizontal axis is relative amplitude of the signal and vertical axis is a logarithmic scale indicating the number of counts at the given amplitude:* (a) *53 MPa (7.8 ksi),* (b) *80 MPa (11.6 ksi), and* (c) *138 MPa (19.4 ksi).*

were notched as well. The net cross-sectional area of the specimens in the reduced region was approximately 40 to 65 mm^2 (0.06 to 0.10 in.^{2}). This area is roughly four unit cells thick and four to six unit cells in width. A picture of this geometry is shown in Ref *22*. The R-ratio for these tests was 0.05 to 0.10. The small minimum load was necessary to keep from damaging the load stage. Maximum stresses ranged from 60 to 100 MPa (8.7 to 14.5 ksi). No test exceeded 10 000 cycles.

Specimens of all three orientations were tested, and the failure mechanisms were found to be the same in all cases. Despite the specimen preparation, many of the specimens developed cracks at locations other than in the reduced area. The most common location for initiation of damage was in the shoulder of the reduced region, where the fiber direction was at an oblique angle to the specimen surface. In this orientation, there is considerable shear strain between the matrix region and fiber bundle due to the compliance difference between the two components. Figure 8 shows a crack that started in such a manner. The fiber-surface geom-

FIG. 7—*Loading stage for ETEC Autoscan SEM. Carbon-carbon specimen is shown loaded in stage.*

etry, the loading direction, and the extremely low shear strength of the material, as well as the previously-mentioned compliance difference, are all expected to favor this mode of initiation. Figure 9 shows a similar crack that started in the shoulder, and is seen to extend completely across the top edge of the specimen, and down into the matrix region.

As can be noted in Figs. 8 and 9, the cracks turned from an orientation parallel to the loading direction to an orientation perpendicular to the loading direction within the matrix region. The cracks invariably propagated in this mode until encountering a longitudinal fiber bundle. The cracks then turned 90 deg again and propagated along the loading direction, very near to the matrix-fiber interface, but within the matrix region. It was thus concluded that, in this material, the bond between the fiber bundles and the matrix regions was stronger than the matrix shear strength.

In a special case in which a notch root was located in the middle of a longitudinal fiber bundle (that is, half the bundle was cut and half of the fibers remained intact), the first sign of damage was a crack that initiated within the bundle and propagated, parallel to the applied load, down the entire viewable length of the specimen. This would indicate

FIG. 8—*SEM photograph (Specimen SEM-14, X direction) showing crack that initiated in shear mode* (arrow), *propagated in opening mode, and finally turned 90 deg (to shear mode again) and ran along fiber bundle;* ×*120, 73 MPa (10.5 ksi).*

that the shear strength between the individual fibers, within the bundles, is also quite low.

The cracks appeared to have no difficulty in shearing through transverse fiber bundles as propagation continued along longitudinal fiber bundles. The only geometric anomalies that seemed to influence crack paths were rather large voids. Cracks would invariably alter their course to pass through a void that was a small distance from the fiber-matrix interface. Immediately upon passing through a void, a crack would turn and again propagate into the region near the interface. Figure 10 shows a crack that altered course for two consecutive voids. Multiple cracking within the matrix regions was very common. Figure 11 shows a rather extensive crack network emerging from a notch root.

Final fracture occurred very suddenly. As a result, there is no record of critical damage to fiber bundles prior to failure. The insights into this process were gained through post-test inspection. In many areas of the specimen fiber, bundles were found to have been sheared completely or partially out of the matrix so that fiber failure was not at the same location as apparent matrix failure. Figure 12 shows a notched specimen in

FIG. 9—*SEM photograph (Specimen SEM-14,* × *direction) showing crack running across top edge and down front face of specimen;* ×*60, 76 MPa (11.0 ksi).*

which the fiber failure was spread over a rather large region in the area under the notch. The apparent matrix failure was in the shoulder of the specimen. However, the separation of fiber bundles from matrix material (which could be considered a type of matrix failure) occurs in the same region as the fiber failures that, in this case, are under the notch.

The conclusion drawn from these observations is that fiber breakage is of a rather random nature. The location of fiber fracture seems to depend on weaknesses of the individual fiber bundles rather than stress concentrations due to gross geometric features such as notches or voids. In this respect, the material could be considered to be insensitive to notches. However, notches do introduce geometries that facilitate differential shear strains, thus leading to initiation of cracks.

FIG. 10—*SEM photograph of crack running in the proximity of the interface between the fiber bundle and matrix* (arrows). *Crack alters course to pass through sizeable voids.*

Conclusions

1. Compressive strength seems to increase slightly as temperature increases from room temperature to approximately 2000°C. At elevated temperatures, the final failure was much more catastrophic than at room temperature. Poisson's ratio was not measurable.

2. Tensile strength at room temperature is comparable to the compressive strength at room temperature. Both are found to depend on fiber volume ratio in the test direction.

3. The material is weak and highly ductile in shear.

4. Acoustic emission is effective in identifying damage mechanisms: low-amplitude signals indicate matrix crumbling and shear failures; and the high-level signals indicate sudden, total failure of fibers.

5. Cracks appear to originate at geometric locations that lend themselves to differential shear strains between fiber and matrix constituents. Cracks would then turn 90 deg and propagate in an opening mode until encountering a longitudinal fiber bundle. They would then turn 90 deg

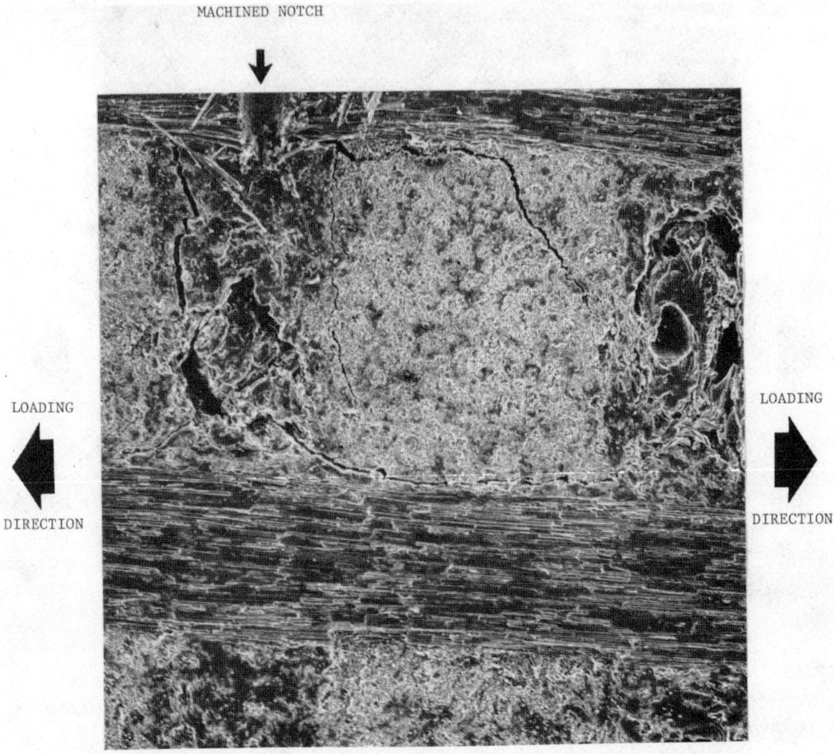

FIG. 11—*Crack network emanating from notch root. Specimen loaded to 66 MPa (9.6 ksi), ×40.*

again and propagate through the matrix material, generally parallel to and in the immediate vicinity of the fiber-matrix interface.

6. Fiber bundle failure was randomly distributed over a material volume, indicative of the high nonhomogeneity of the microstructure and the redundant nature of the material system. Random fiber failures also indicate the material is not sensitive to notches or other geometric anomalies except in that they affect differential shear strains, which in turn lead to crack initiation.

Acknowledgments

The research reported herein was conducted under the sponsorship of the Office of Naval Research, Contract N00014-77-C-0510.

FIG. 12—*Failed SEM specimen (Specimen SEM-21) at* ×*17. Specimen was notched* (arrow) *in an effort to localize damage.*

References

[*1*] Stetson, J. R. and Schultzer, J. C., "Evaluation of Carbon-Carbon Composite Nosetip Materials," AMMRC CTR 76-34, Army Materials and Mechanics Research Center, Oct. 1976.

[*2*] Ossin, A. and Kendall, P., "Evaluation of CONAP Concept for Advanced ABM Nosetips," AMMRC CTR 76-38, Army Materials and Mechanics Research Center, Nov. 1976.

[*3*] Zimmer, J. E., White, J. L., Evangelides, J. S., and Meyer, R. A., "Reentry Vehicle Materials Technology (REVMAT) Program: Carbon Nose-Tip Materials Technology," Aerospace Corporation, El Segundo, Calif., Sept. 1975.

[*4*] Adams, D. F., *Materials Science and Engineering*, Vol. 23, April 1976, pp. 55–68.

[*5*] Evangelides, J. S., Rubin, L., White, J. L., and Zimmer, J. E., "Reentry Vehicle Materials Technology (REVMAT) Program: Carbon Nose-Tip Materials Technology," Aerospace Corporation, El Segundo, Calif., Sept. 1976.

[*6*] Evangelides, J. S., Zimmer, J. E., Rubin, L., White, J. L., and Buechler, M., "Strategic Missiles Materials Technology (SMMT) Program: Carbon Materials Analyses," Aerospace Corporation, El Segundo, Calif., Dec. 1976.

[7] Evangelides, J. S., White, J. L., Buechler, M., Zimmer, J. E., and Rubin, L., "Strategic Missiles Materials Technology (SMMT) Program: Carbon Materials Analyses," Aerospace Corporation, El Segundo, Calif., March 1977.
[8] Zimmer, J. E., White, J. L., Evangelides, J. S., Rubin, L., and Buechler, M., "Strategic Missiles Materials Technology (SMMT) Program: Carbon Materials Analyses," Aerospace Corporation, El Segundo, Calif., June 1977.
[9] McSwain, R. H. and Bates, C. E., "Microstructural Relationships in 2-2-3 Carbon Reinforced Carbon Composites," AFML-TR-75-117, Air Force Materials Laboratory, Wright-Patterson Air Force Base, Ohio, Sept. 1975.
[10] Evangelides, J. S., Sines, G. H., and Batdorf, S. B., "Damage Mechanisms and Modeling of Carbon-Carbon Composites," University of California at Los Angeles, May 1978.
[11] Jones, R. M., "Nonlinear Multiaxial Modeling of Graphitic and Carbon-Carbon Materials," AFML-TR-76-215, Air Force Materials Laboratory, Wright-Patterson Air Force Base, Ohio, Dec. 1976.
[12] Tarnopol'skii, Y. M., Polyakov, V. A., and Zhigun, I. G., *Polymer Mechanics*, Vol. 9, No. 5, May 1975, pp. 754–159.
[13] Ross, A. L., *Mechanical Engineering*, April 1975, pp. 32–37.
[14] Pardoen, G. C., *AIAA Journal*, Vol. 13, No. 6, June 1975, pp. 756–761.
[15] Reuter, R. C., Jr., and Priddy, T. G., "Analytical Characterization of Filament-Wound, Carbon-Carbon Material Behavior with Experimental Correlation," SC-RR 720888, Sandia Laboratories, Albuquerque, N.M., Jan. 1973.
[16] Starrett, H. S., Weiler, F. C., and Pears, C. D., "Thermostructural Response of Carbon-Carbon Materials Under High Heat Flux Environments," AFML-TR-74-232, Air Force Materials Laboratory, Wright-Patterson Air Force Base, Ohio, June 1975.
[17] Starrett, H. S. and Pears, C. D., "Elastic Compliances for ATJ-S Graphite and Mod 3a Carbon-Carbon," AFML-TR-74-271, Air Force Materials Laboratory, Wright-Patterson Air Force Base, Ohio, Feb. 1975.
[18] Cull, A. D. and Starrett, H. S., "Fracture Mechanics of Thermal Protective Materials," AFML-TR-77-19, Air Force Materials Laboratory, Wright-Patterson Air Force Base, Ohio, March 1977.
[19] Whitney, J. M. and Hussman, G. E., *Experimental Mechanics*, Vol. 18, No. 5, May 1978, pp. 185–190.
[20] Bush, A. L., Legg, J. K., and Pears, C. D., "The Development and Demonstration of a High Temperature Ring Test Facility," AFML-TR-74-254, Air Force Materials Laboratory, Wright-Patterson Air Force Base, Ohio, Jan. 1975.
[21] Carlyle, J. M., *Experimental Mechanics*, Vol. 18, No. 5, May 1978, pp. 191–195.
[22] Francis, P. H. and Robinson, C. T., "Damage Mechanisms and Failure of 3-D Carbon-Carbon Composites," Southwest Research Institute, San Antonio, Tex., Feb. 1979.
[23] Personal conversation Dave Walrath and Professor D. F. Adams, University of Wyoming.
[24] Davidson, D. L. and Nagy, A., *Journal of Physics E;* Scientific Instruments, Vol. 11, 1978, pp. 207–210.

M. Gouda,[1] *K. M. Prewo,*[2] *and A. J. McEvily*[3]

Mechanism of Fatigue in Boron-Aluminum Composites

REFERENCE: Gouda, M., Prewo, K. M., and McEvily, A. J., **"Mechanism of Fatigue in Boron-Aluminum Composites,"** *Fatigue of Fibrous Composite Materials, ASTM STP 723,* American Society for Testing and Materials, 1981, pp. 101–115.

ABSTRACT: A fatigue model for unidirectionally reinforced composites that are loaded parallel to the fibers is proposed that is based upon the growth of interfacial fatigue cracks. Certain predictions of the model are verified by experimental results, namely, that the rate of crack growth is independent of crack length and that the fatigue life at $>10^6$ cycles is relatively insensitive to the mean stress. Depending upon whether the matrix is elastic or elastic-plastic, a different dependency of the rate of crack growth on stress results, but the data available are insufficient to make a precise distinction between the two cases. In addition, at higher stress amplitudes the tensile rupture mode of separation becomes increasingly important, a fact dealt with only empirically by the model. It is concluded that for loading parallel to the fibers the most fatigue resistant composites will consist of strong, uniform, and defect-free fibers in a relatively soft matrix.

KEY WORDS: fatigue (materials), fatigue crack growth, composite materials, mechanical properties, boron-aluminum composites

The high strength-to-weight ratio of composite materials such as boron-aluminum is certainly an attractive structural characteristic, particularly for aerospace applications. However, in many circumstances consideration must often also be given to the fatigue properties, and a number of investigations have indeed dealt with this topic [*1–9*]. In the particular case of the boron-aluminum system, prior studies have shown that fatigue cracks are initiated early in life at fiber breaks [*1*] that are present prior to testing or that occur at isolated locations at the start of cyclic loading as a result of the wide variation in the strength properties of individual fibers as indicated in Fig. 1 [*10*]. The boron fibers themselves are quite resistant to fatigue, and it is considered by some that fibers that fail after the first loading do so only because

[1]Formerly, graduate assistant, now at Department of Materials Science, North Carolina State University, Raleigh, N.C. 27607.
[2]Principal scientist, United Technologies Research Laboratories, E. Hartford, Conn. 06108.
[3]Professor, Institute of Materials Science, University of Connecticut, Storrs, Conn. 06268.
[4]The italic numbers in brackets refer to the list of references appended to this paper.

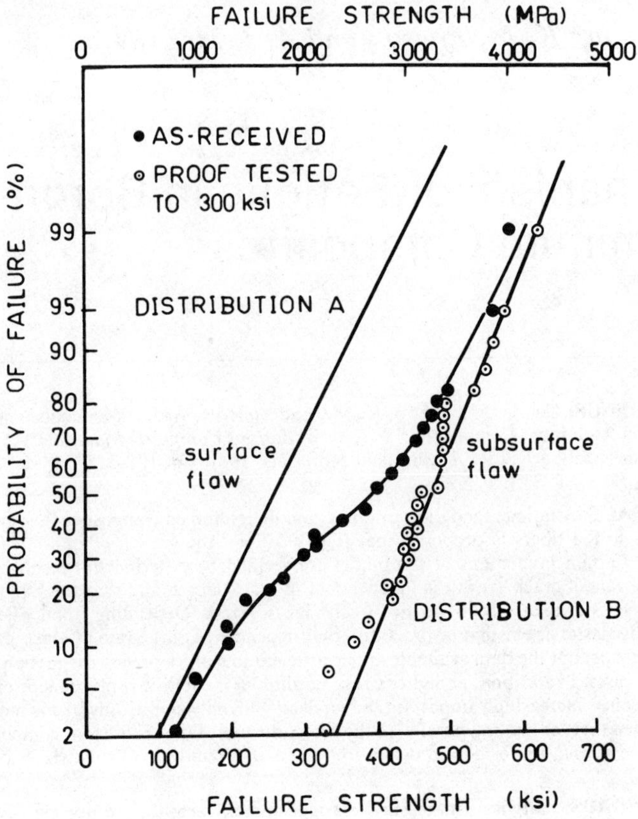

FIG. 1—*Distribution of fracture strength of boron fibers.* (After Metcalfe as cited by Tetelman [*10*].

an approaching crack results in a local stress in excess of the static strength of the fiber [*11*]. Final failure occurs as the result of the loss of cross-sectional area as a number of cracks link up.

On the basis of this background, it is clear that one could develop a model of fatigue based upon crack growth processes alone. However, the nature of the crack path will have a strong influence on the overall fatigue behavior. For cyclic loading parallel to the fibers, in cases where the interface between fiber and matrix is relatively weak, a crack will tend to grow along an interface between fiber and matrix (axial crack) and will be damaging only when a weak site in the fiber is reached. On the other hand, when the interface is relatively strong, the cracks will tend to grow across (transverse crack) rather than along the fibers. Drastically different fatigue behavior will result depending upon which mode of growth is dominant.

The intent of this paper is to present a simplified analysis of fatigue of boron-aluminum composites based upon such crack growth considerations and to compare the predictions with new experimental results as well as those taken from the literature.

The Model

We consider that fiber breaks are created on initial loading, and, that for cyclic stresses applied parallel to the fibers, crack growth occurs in the matrix either transverse to (Type A) or parallel to (Type B) the fibers as in Figs. 2 and 3. Type A growth implies that the interface is relatively strong, Type B implies that it is relatively weak. Crack growth of Type A is similar to the usual Mode I type of growth observed in alloys in which the rate of crack

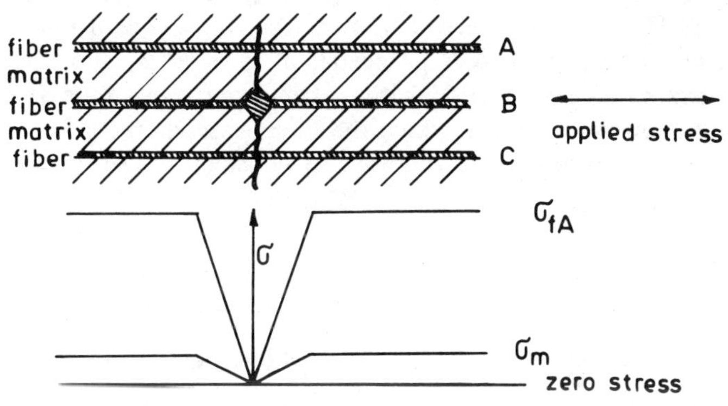

FIG. 2—*Schematic representation of the variation of fiber and matrix stresses in vicinity of transverse crack.* σ_m *is the matrix stress;* σ_{fA} *is the fiber stress in the A fiber (Type A).*

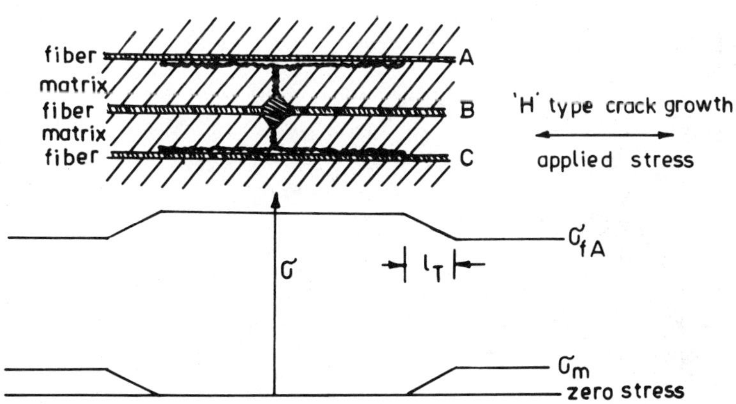

FIG. 3—*Schematic representation of the variation of fiber and matrix stresses in vicinity of H-type crack.* l_t *is the transfer length.* σ_m *is the matrix stress;* σ_{fA} *is the fiber stress in the A fiber (Type B).*

growth as a function of ΔK, the range of the stress intensity factor, is sigmoidal between the limits of ΔK_{th}, the threshold level for crack growth, and K_c, the fracture toughness. An *S-N* curve computed from such crack growth rate data will lead to the usual form of the *S-N* curve.

For crack growth of Type B, however, the shape of the *S-N* curve may be quite different. For example, if as the axial crack grows along the interface no weak sites in the fiber are encountered to allow some transverse growth, then this crack will not contribute to final failure. In fact, if all the fibers had the same ideal tensile strength except for a few isolated weak points, then fatigue failure would not occur, and the *S-N* curve would be a horizontal line at the tensile strength of the composite. Figure 4 indicates the two extremes for either completely Type A (transverse) or Type B (axial) crack growth behavior. The actual fatigue behavior of a composite will fall between these two extremes with one or the other limit being approached as the relative strengths of fiber and interface vary. Note that the shape of the *S-N* curve does not depend upon the *S-N* properties of the matrix, and, in fact, the same matrix can be associated with either Type A or Type B behavior depending upon the ease with which the fibers can be cracked.

We next attempt to develop this view by considering two limiting matrix

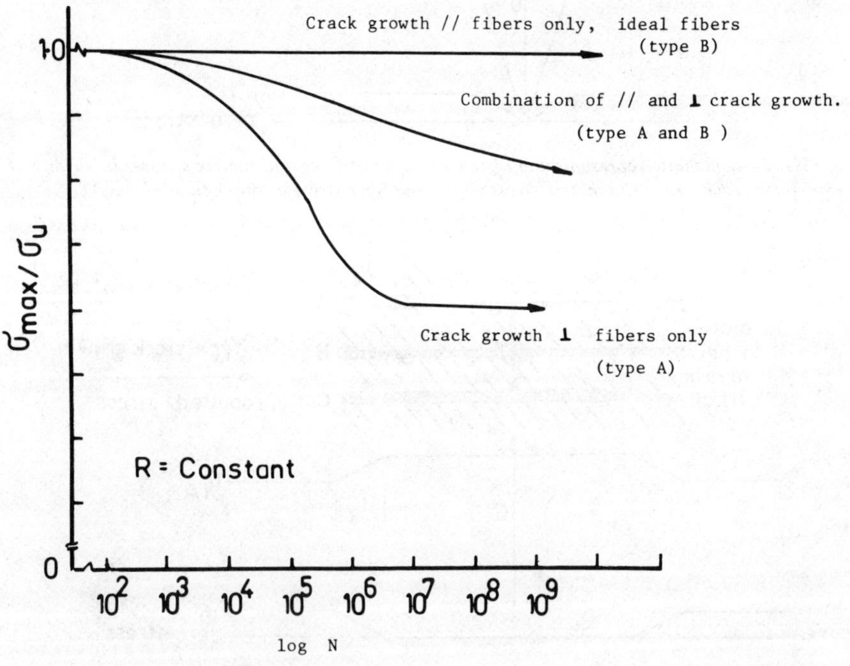

FIG. 4—*Shapes of* S-N *curves in unidirectional composites as influenced by crack path. Alternating stress applied parallel to fibers.*

classifications and Type B (axial) crack growth behavior. In Case 1, the matrix is treated as elastic, whereas in Case 2 the matrix is treated as elastic-plastic of yield stress σ_y. For each case, it is postulated that the fatigue crack growth process along the interface occurs as the result of debonding along the interface, and that the extent of growth per cycle is proportional to the relative displacement, δ, of fiber and matrix at the crack tip. In addition, the tension-tension loading usually used in fatigue testing of composites will result in compression-compression cycling in the lateral direction of the specimen due to the Poisson's ratio effect. This type of cyclic lateral straining may also contribute to the growth of cracks in the matrix, but is not treated explicitly herein.

In both cases, it is assumed that the strains in fiber and matrix are the same at distances greater than at l_t, the transfer length, ahead of the crack tip, or

$$\frac{\sigma_f}{E_f} = \frac{\sigma_m}{E_m} = \frac{\sigma_c}{E_c} \tag{1}$$

where the subscripts f, m, and c refer to the fiber, matrix, and composite, respectively, σ is the stress in either fiber, matrix, or average stress in composite, E is Young's modulus, and E_c is given by the rule of mixtures, that is, $E_c = E_f V_f + E_m V_m$, where V is the volume fraction. Over the region of load transfer, l_t, (Fig. 3), in progressing from the uncracked into the cracked region, the stresses in the fiber rise and those in the matrix drop. (It is this rise in fiber stress at the crack tip that can trigger failure of the fiber as the crack advances and weak sites are encountered.) We approximate these stress changes for Case 1 as follows.

At the point where load transfer to the fiber begins, that is, at a plus l_t, where a is the crack length and l_t is the transfer length, the stresses in fiber and matrix are given as

$$\sigma_f = \frac{E_f}{E_c} \sigma_c \tag{2}$$

$$\sigma_m = \frac{E_m}{E_c} \sigma_c \tag{3}$$

At the crack tip, σ_m is equal to zero, and the corresponding increase in σ_f is given as

$$\Delta\sigma_f = A\sigma_m \frac{V_m}{V_f} = A \frac{E_m}{E_c} \sigma_c \frac{V_m}{V_f} \tag{4}$$

where A indicates the fraction of the load transferred from the matrix to a particular fiber. For the simple case shown in Fig. 3, A would be of the order of unity. The increase in fiber stress can also be expressed as

$$\Delta\sigma_f = \frac{4l_t\tau_i}{d} \tag{5}$$

where τ_i is the interfacial shear stress and d is the fiber diameter. The expression for the transfer length, l_t, obtained by combining Eqs 4 and 5, is as follows

$$l_t = A\,\frac{\sigma_c}{\tau_i}\,\frac{E_m}{E_c}\,\frac{V_m}{V_f}\,\frac{d}{4} \tag{6}$$

Since the matrix stress goes to zero over the transfer length, whereas the stress in the fiber increases, the strains in both fiber and matrix are no longer the same, and a displacement differential between fiber and matrix will develop. This displacement differential, δ, is given as

$$\delta = \left(\frac{1}{2}\,\frac{\sigma_c}{E_f}\,\frac{E_m}{E_c}\,\frac{V_m}{V_f} + \frac{\sigma_f}{E_f} - \frac{1}{2}\,\frac{\sigma_m}{E_m}\right) l_t \tag{7}$$

$$\delta = \left(\frac{1}{2}\,\frac{\sigma_c}{E_f}\,\frac{E_m}{E_c}\,\frac{V_m}{V_f} + \frac{\sigma_f}{E_f} - \frac{1}{2}\,\frac{\sigma_m}{E_m}\right) A\,\frac{\sigma_c}{\tau_i}\,\frac{E_m}{E_c}\,\frac{V_m}{V_f}\,\frac{d}{4} \tag{8}$$

where $V_m \approx V_f$ and $E_c \gg E_m$, and with A taken as unity, Eq 8 can be approximated as

$$\delta = \frac{\sigma_f}{E_f}\cdot\frac{\sigma_c}{\tau_i}\,\frac{E_m}{E_c}\cdot\frac{V_m}{V_f}\,\frac{d}{4} \tag{9}$$

$$\delta = \frac{\sigma_c^{\,2}}{E_c^{\,2}}\cdot\frac{E_m}{\tau_i}\,\frac{V_m}{V_f}\,\frac{d}{4} \tag{10}$$

It is assumed that the rate of fatigue crack growth is proportional to δ, and therefore, if the matrix is elastic, to depend upon the square of the stress amplitude and to decrease with decrease in fiber diameter and increase in V_f. It is also noted that the model indicates that the rate of crack growth should be independent of crack length and mean stress (at least for tension-tension loading), since the crack tip displacements depend only upon the shear transfer in a direction parallel to the applied tensile stress over the distance l_t.

For Case 2, where the matrix is elastic-plastic, it is assumed that at distances remote from fiber and matrix cracks, that the stress in the matrix is equal to the yield stress, σ_{ym}. The transfer length, l_t, is the length over which the matrix stress rises from zero to σ_{ym}, and is given by

$$l_t = \frac{\Delta\sigma_f}{4\tau_i} d \qquad (11)$$

The increase in fiber stress over the length l_t is

$$\Delta\sigma_f = A\sigma_{ym} \cdot \frac{V_m}{V_f} \qquad (12)$$

where A is defined as before. The expression for l_t becomes

$$l_t = A \frac{\sigma_{ym}}{\tau_i} \frac{V_m}{V_f} \frac{d}{4} \qquad (13)$$

The expression for the strain differential that develops between fiber and matrix in the transfer region is

$$\delta = \left(\frac{A}{2} \frac{\sigma_{ym}}{E_f} \frac{V_m}{V_f} + \frac{\sigma_f}{E_f} - \frac{1}{2} \frac{\sigma_{ym}}{E_m}\right)\left(A \frac{\sigma_{ym}}{\tau_i} \frac{V_m}{V_f} \frac{d}{4}\right) \qquad (14)$$

As before, the expression for δ can be approximated as

$$\delta = \frac{\sigma_f}{E_f} \frac{\sigma_{ym}}{\tau_i} \frac{V_m}{V_f} \frac{d}{4} \qquad (15)$$

and with $\sigma_f = \sigma_c/V_f$ ($E_m \approx 0$). Then

$$\delta = \frac{\sigma_c}{E_f} \frac{\sigma_{ym}}{\tau_i} \frac{V_m}{V_f^2} \frac{d}{4} \qquad (16)$$

In this case, the rate of crack growth should be directly proportional to the stress range, rather than to its square as in Case 1, but again mean stress and crack length are not factors.

In both cases, if τ_i reflects the strength of the matrix then an increase in matrix strength will lead to a decrease in crack growth rate. It is also noted that if σ is expressed as a fraction of σ_u, the composite tensile strength, then the rate of crack growth decreases with V_f. Therefore, it is expected that at any given fraction of the tensile strength, the fatigue lifetime will increase as the volume fraction of fibers increases.

Experimental Results

In order to assess the validity of predictions made in the previous section, we make use of experimental results obtained for a 1.75 mm (0.07 in.) thick unidirectional six-ply laminate of a boron (57 percent by volume 0.14 mm (0.0056 in.) in diameter) 6061-0 aluminum alloy composite. The specimens were carefully cut from this laminate and the grip ends were reinforced by fiberglass doublers cemented to the laminate. Similar test specimens were used for both tension and fatigue (R = 0.05, 30 Hz) testing and measured 7.5 mm (0.3 in.) in width and 100 mm (4 in.) in length. No relaxation of mean stress during cycling prior to failure was noted, even at the highest stress amplitude employed.

Figure 5 shows the *S-N* curve as well as tensile strengths obtained. In these tests, fatigue failure occurred either in the test section or in the grips. At in-

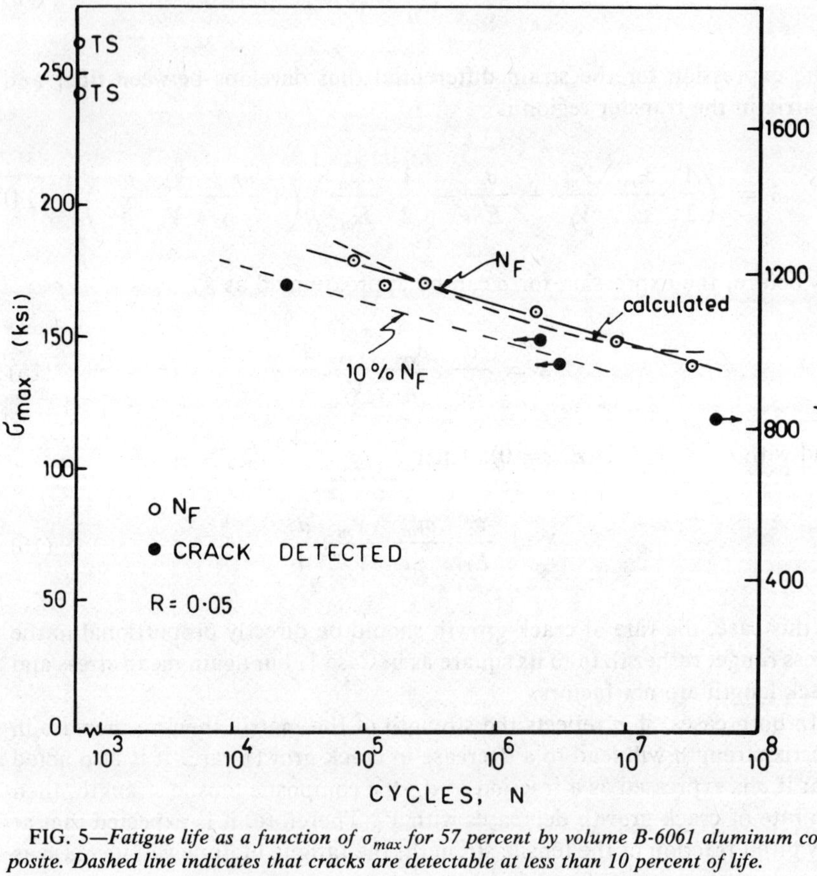

FIG. 5—*Fatigue life as a function of* σ_{max} *for 57 percent by volume B-6061 aluminum composite. Dashed line indicates that cracks are detectable at less than 10 percent of life.*

tervals during the cyclic tests, the fatigue tests were interrupted and replicas were made of the surface to determine crack initiation sites and the progress of crack growth. These replicas were subsequently examined in a scanning electron microscope at ×60 magnification. From these replicas, it was possible to detect cracks at an early stage in life as indicated in Fig. 5. In accordance with other results [*1*], cracks in these composites can be observed prior to 10 percent of life even in the high cycle range. Such early crack development contrasts strongly with the behavior of metals and alloys, wherein cracks are usually observed in the high cycle range only at more than 90 percent of the lifetime. Radiographic examination of selected specimens was also carried out at intervals during the cycling process to determine if fiber breaks could be observed. However, in contrast to other work [*1*], no fiber breaks were found by this method, an indication that the fibers in our tests may have been of superior quality to those previously tested.

A series of fatigue tests was also carried out on specimens that had the aluminum matrix in the mid-region of the specimen leached away in dilute hydrochloric acid. It was thought that the fiber bundles, if they survived the first cycle, might never fail in fatigue since there were no matrix cracks available to link up weak sites. Nevertheless, the bundles did fail, and with about the same lifetime as for the composite specimens; however, it was realized this was not an ideal experiment since cracks may have started in the grip sections. In addition, the rubbing of fibers may also have contributed to failure. Nevertheless, the results are shown for completeness in Fig. 6.

Comparison with the Model

An example of the type of crack that was observed by the replica technique is shown in Fig. 7. In these tests, a characteristic "H" type of surface crack often developed, with the bar of the H located at a fiber break and the legs of the H being two parallel cracks growing along fiber interfaces adjacent to the fiber break as in Fig. 3. By polishing away the surface layer, it was possible to ascertain that indeed cracking had occurred along the interface. In addition, after fracture had occurred fiber pull-outs, uncoated by aluminum, could be observed; a further indication of a weak interface.

The crack length as a function of the number of cycles for several of these cracks could be determined from the replica information and is plotted in Fig. 8. In agreement with either model, the rate of crack growth is independent of crack length. In order to determine the stress dependency of the rate of crack growth, the rates are plotted as a function of σ_{max} in Fig. 9. As indicated in Fig. 9, the results appear to favor a σ^2 dependency (Case 1) rather than a linear dependency (Case 2), but more data would be needed to establish the dependency on stress with greater certainty. It is also apparent from Fig. 9 that a lower threshold limit, designated as σ_0, must be exceeded

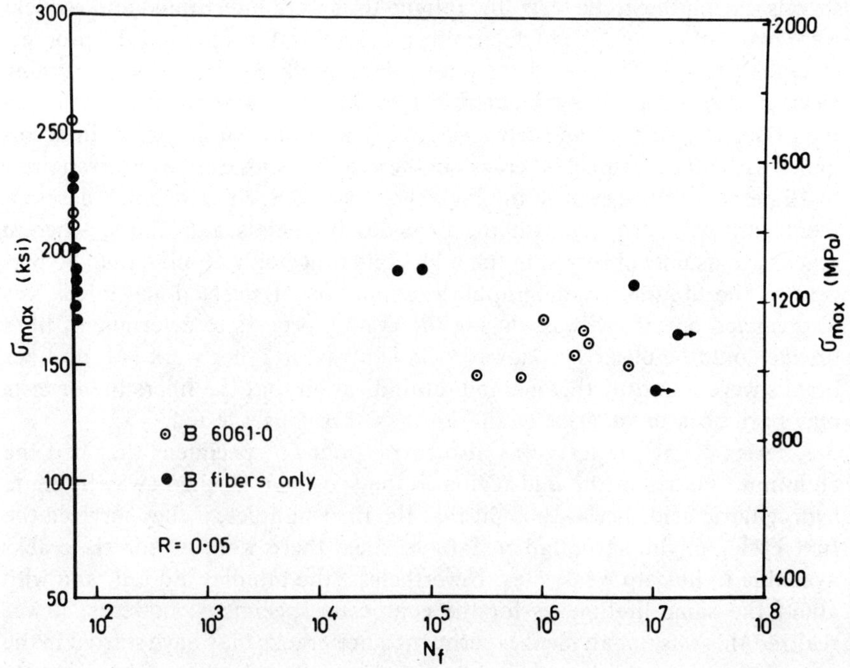

FIG. 6—*Comparison of fatigue behavior of boron-aluminum composite and boron fiber bundle. Stresses computed on basis of same cross-sectional area.*

in order for crack propagation to occur if we assume the crack growth below 3×10^{-8} mm/cycle (10^{-9} in./cycle) is unlikely. The rate of crack propagation for the present results can be expressed as

$$\frac{\Delta a}{\Delta N} = 0.8 \times 10^{-10} [\sigma - \sigma_0]^2 \quad (17)$$

where $\sigma_0 = 952$ MPa and $\Delta a/\Delta N$ is in mm/cycle.

Most of the fatigue lifetime is spent in propagating these interfacial cracks until they link up, and if failure is assumed to occur shortly after first link up, an estimate of the fatigue lifetime can be obtained in integrating the prior rate equation to obtain

$$N = \frac{a_f}{0.8 (\sigma - \sigma_0)^2} \times 10^{10} \quad (18)$$

where a_f represents the final interfacial crack length at the time of link-up. The value of a_f should decrease with increase in stress, and if, to provide

FIG. 7—*Appearance of H type crack*, $\sigma_{max} = 966$ *MPa*, N = *2.8* × *10*^{7} *cycles*, × *60.*

agreement with the data of Fig. 5, a_f is taken to be equal to 0.016×10^{-11} $(\sigma_u - \sigma)^{4.67}$ mm, then N, the number of cycles to failure is given by

$$N = 0.002 \frac{(\sigma_u - \sigma)^{4.67}}{(\sigma - \sigma_0)^2} \text{ (MPa units)} \tag{19}$$

This expression is plotted as a function of σ in Fig. 5 (calculated).

A further prediction of the model in the high cycle range is that the fatigue life does not depend upon mean stress. In order to assess this prediction, results from the literature for B-6061-0 as well as the present results are plotted in the Goodman diagram, Fig. 10, for lives in excess of 10^6 cycles. Indeed, there appears to be little influence of mean stress for this class of material indicating that interfacial crack growth is a dominant factor in determining the fatigue life, but this conclusion needs to be more thoroughly substantiated. Further, in systems where fibers are weaker or matrices stronger than for B-6061-0, Type A (transverse) crack growth will occur, and in such cases the allowable stress amplitude should decrease with in-

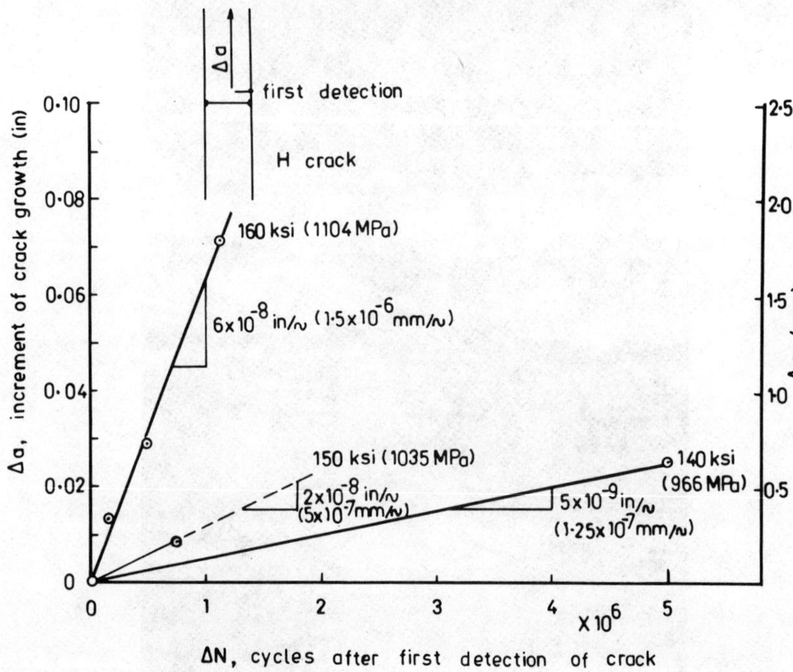

FIG. 8—*The increment of crack length as a function of cycles past the point of first crack detection at three stress levels.*

crease in mean stress level. It is noted that at $R = 0.4$, the σ_{max} values at 10^6 cycles or more are at the level of σ_u, perhaps a surprising result. However, such a result is reasonable if cyclic loading creates Type B cracks that minimize the severity of initial defects [*8*]. Also, with respect to the model, it is noted in Fig. 10 that a higher V_f results in a higher value of σ_a/σ_u, as expected.

Finally in Fig. 11, we plot the *S-N* curves for a variety of boron-metal matrix composites to emphasize that as the strength of the matrix goes up (for example, a Ti-6A1-4V matrix) or that of the fibers goes down (for example, in the case of the 40 V_f [*1*] B-6061-0 extensive fiber cracking was detected), the resultant *S-N* curve will shift from that reflecting interfacial growth (Type B) to that reflecting transverse growth (Type A).

Conclusions

As a result of this study, it is concluded that:

1. In unidirectional composites, cracks are initiated early in life at defects in fibers. Growth of these cracks along the fiber-matrix interface can account for a major portion of the fatigue lifetime in systems in which the fiber-matrix strength ratio is high. In systems wherein the fiber-matrix strength

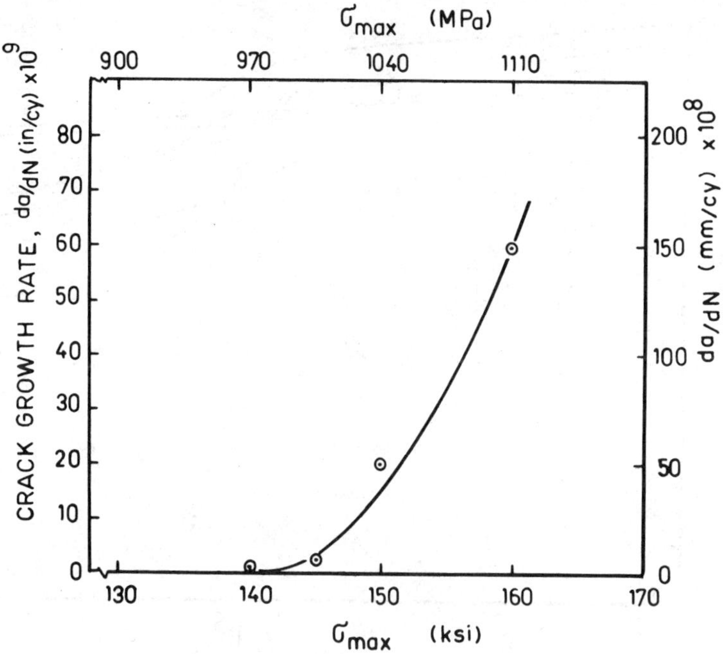

FIG. 9—*Comparison of observed crack growth rates with parabolic rate dependency* (σ_0 = *966 MPa*).

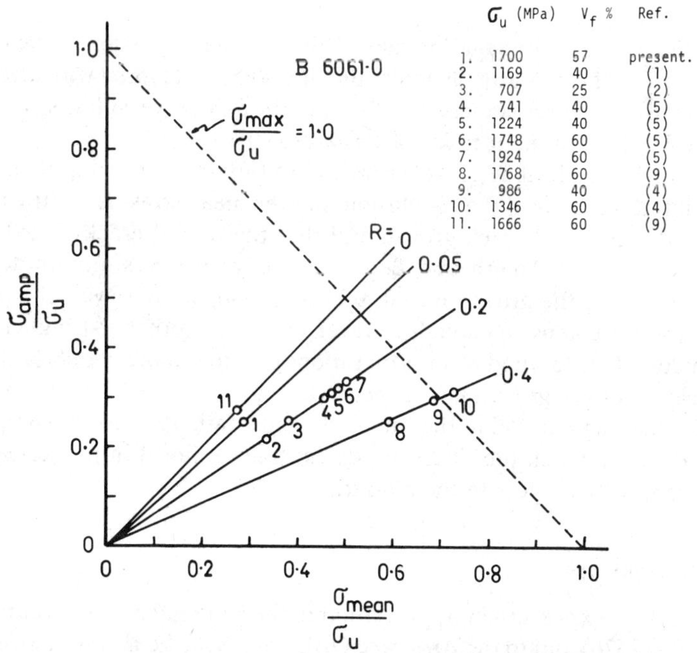

FIG. 10—*Dependence of* σ_{amp} on σ_{mean} *for fatigue lives* ≥ 10^6 *cycles.*

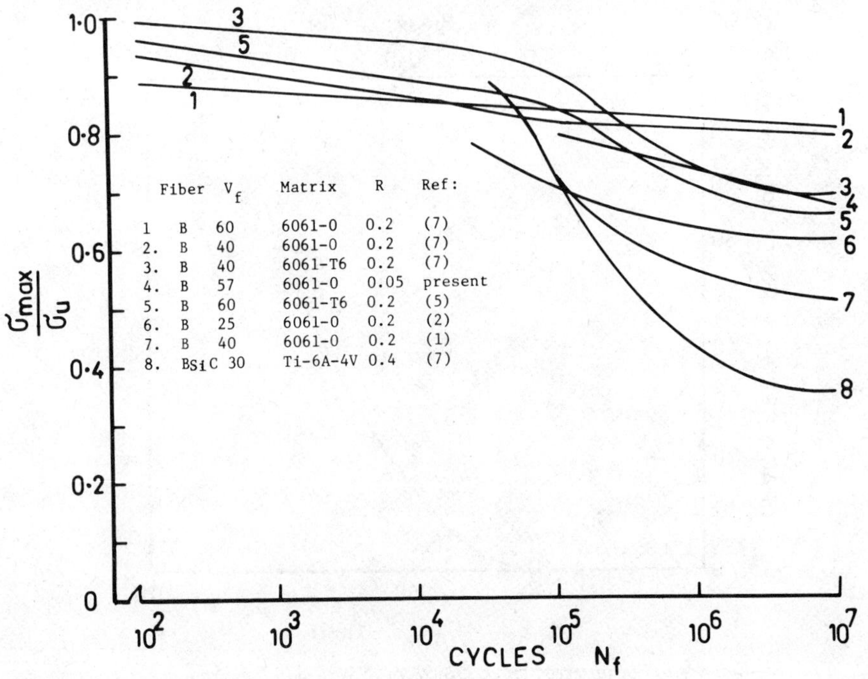

FIG. 11—*A comparison of S-N curves of boron composites.*

ratio is lower, crack propagation may again be a major portion of the fatigue life but a path of crack growth across the fibers will be favored with attendant lessening of fatigue resistance. The shape of the *S-N* curve will be strongly influenced by the dominant mode of crack growth.

2. A model for fatigue crack growth based upon debonding at the fiber matrix interface leads to the conclusion that the mean stress is of little importance in this mode of crack growth and that the rate of crack growth is independent of crack length and dependent only on stress amplitude. Two limiting forms of the growth law have been developed, however, the present results are not extensive enough to favor one or the other. At higher stress amplitudes, tensile modes of separation become more important, and modification of the growth laws is needed.

3. For loading parallel to the fibers, the most fatigue resistant composites will be those in which the fibers are defect free, uniform in properties, and quite strong with respect to the matrix.

Acknowledgments

The authors express their appreciation to the National Science Foundation (DMR 77-12526) and to the Air Force Office of Scientific Research (Contract No. F49620-79-C-004) for support of this investigation.

References

[1] Hancock, J. R. in *Composite Materials*, Broutman and R. H. Knock, Eds., Vol. 5, Academic Press, N.Y., 1974, p. 370.

[2] Toth, I. J. in *Composite Materials: Testing and Design, ASTM STP 460,* American Society for Testing and Materials, 1969, p. 236.

[3] Toth, I. J. in *Composite Materials: Testing and Design* (*Third Conference*), *ASTM STP 546,* American Society for Testing and Materials, 1974, p. 542.

[4] Menke, G. D. and Toth, I. J., "The Time Dependent Mechanical Behavior of Composite Materials," AFML-TR-70-174, Air Force Materials Laboratory, June 1970.

[5] Toth, I. J. and Shimmin, K. D., "Fatigue and Fracture of Metal Matrix Composites," *Proceedings,* Air Force Conference on Fatigue and Fracture of Aircraft Structural Materials, AFFDL-TR-70-144, 1970, p. 343.

[6] Hancock, J. R., "Mechanisms of Fatigue in Filament-Reinforced Metals," *Proceedings,* Air Force Conference on Fatigue and Fracture of Aircraft Structural Materials, AFFDL-TR-70-144, 1970, p. 295.

[7] Shimmin, K. D. and Toth, I. J. in *Failure Modes in Composites,* American Institute of Mining, Metallurgical, and Petroleum Engineers, 1973, p. 357.

[8] Prewo, K. M., and Kreider, K. G. in *Failure Modes in Composites,* American Institute of Mining, Metallurgical, Petroleum Engineers, 1973, p. 395.

[9] Menke, G. D. and Toth, I. J., "The Time Dependent Mechanical Behavior of Composite Materials," AFML-TR-71-102, Air Force Materials Laboratory, 1971.

[10] Metcalfe, A. G., cited by A. S. Tetelman in Fracture Processes in Fiber Composite Materials, *Composite Materials: Testing and Design, ASTM STP 460,* American Society for Testing and Materials, 1969, p. 473.

[11] Shimmin, K. D. unpublished result as referred to in Refs *4* and *9*.

A. S. D. Wang,[1] P. C. Chou,[1] and J. Alper[1]

Effects of Proof Test on the Strength and Fatigue Life of a Unidirectional Composite*

REFERENCE: Wang, A. S. D., Chou, P. C., and Alper, J., **"Effects of Proof Test on the Strength and Fatigue Life of a Unidirectional Composite,"** *Fatigue of Fibrous Composite Materials, ASTM STP 723,* American Society for Testing and Materials, 1981, pp. 116-132.

ABSTRACT: The effects of a static proof testing on the statistical distribution of the static strength and fatigue life of a unidirectional graphite-epoxy laminate (AS-3501-05) are investigated experimentally. Loading mode for both the static and fatigue tests is restricted to uniaxial tension in the fiber direction. Six-ply tension coupons with dimensions of 22.9 cm by 1.9 cm are used; and all tests are conducted using a closed-loop Instron tester, under room temperature (~21°C) and ambient humidity (~60% percent relative humidity) conditions. Test data are analyzed by means of a two-parameter Weibull distribution in conjunction with the maximum likelihood method for parameter estimation. Results show that proof testing can guarantee a minimum static strength, and to a lesser degree, the method can also assure a minimum fatigue life.

KEY WORDS: composite materials, fatigue (material), graphite-epoxy, proof test, statistical analysis, unidirectional laminates

Nomenclature

$\hat{F}$	Median rank
$F_X(x)$	Cumulative distribution function for random variable X
$F_{X,S}(x)$	Conditional cumulative distribution function for random variable X
j	Failure order number
k	Number of specimens following a suspended set
m	Sample size

*This work is supported by the Air Force Materials Laboratory, Wright-Patterson Air Force Base, Ohio.

[1] Professor of Applied Mechanics, billing professor of Mechanical Engineering, and graduate student, respectively, Drexel University, Philadelphia, Pa. 19104.

n Value of fatigue life
n_γ Value of guaranteed fatigue life
N Random variable of fatigue life
$P[X \leq x]$ Cumulative distribution function for random variable X
$P[X \leq x | X > S]$ Conditional cumulative distribution function for random variable X
S Maximum stress applied in fatigue cycling
S_m Sample mean
$\bar{x}$ Sample mean
x Value of static strength
x_γ Value of proof load
X Random variable of static strength
α Weibull shape parameter for static strength
α_1 Weibull shape parameter for fatigue life
β Weibull scale parameter for static strength
β_1 Weibull scale parameter for fatigue life
γ Percentage of surviving specimens
γ_a Percentage of specimens surviving a proof load among the total population
Δ_j Failure order number increment of all specimens following a suspended set

Recent years have seen increased interests in the fatigue damage states in composite laminates; particularly, the problems of post-fatigue residual strength and life of laminates have attracted the most attention. Fatigue test data usually exhibit large scatter, and statistical methods are employed in the analysis. For post-fatigue residual strength or life or both, several fatigue degradation models have been proposed in the past [*1–5*].[2] A comprehensive assessment on the various degradation models may be found in Ref [*5*]. However, in all of the existing analysis methods, a common assumption has been made either explicitly or implicitly in the mathematical modeling. Essentially, it is assumed that there exists a unique relationship between the static strength and the fatigue life of a laminate. Or, statistically, for a specimen in a given population, the assumption stipulates that a specimen's rank in the static strength distribution shall be the same as its rank in fatigue life distribution. This is known as the "equal-rank" assumption.

In an effort to verify experimentally this "equal-rank" assumption, Hahn and Kim [*2*] employed the concept of proof test and studied the effects on the creep behavior of unidirectional glass-epoxy laminates. Later, Awerbuch and Hahn [*6*] conducted proof tests on unidirectional graphite-epoxy laminates and examined the effects on the fatigue behavior under cyclic load. In both experiments, test data seemed to support the "equal-rank" assumption,

[2]The italic numbers in brackets refer to the list of references appended to this paper.

although the amount of their test data was considered insufficient for a definite conclusion.

The objective of this paper is to investigate experimentally the effects of proof load on both the post-proof static strength behavior and the post-proof fatigue life behavior of unidirectional graphite-epoxy laminates. The "equal-rank" assumption is evaluated more closely through an enlarged test data base. In addition, it is observed that proof loading does not change the essential features in the static strength distribution of the unidirectional laminates, and consequently, the procedure removes the weaker specimens from the population, thus guaranteeing a minimum strength for the specimens that survive the proof test. Moreover, proof loading is found to degrade only slightly the fatigue property of the specimens; hence, the procedure can still guarantee a minimum life with a rather high degree of confidence.

Experiment

In the present experiment, the AS-3501-05 graphite-epoxy system is used. All tests are conducted under room temperature and ambient humidity (about 60 percent relative humidity) conditions. The overall test program consists of eight different tests, and a total of 304 data points are obtained. Break down of the experimental program includes the following. (1) Baseline data generation: this consists of static tension to failure, fatigue life at a constant maximum load of 71 percent of the static mean strength (0.71 S_m), and fatigue life at 0.81 S_m maximum fatigue load, both in a tension-tension mode. (2) Post-proof static strength study: this consists of proof loading the specimens statically to 88 percent S_m and 95 percent S_m, followed by static tension to failure. (3) Post-proof fatigue life study: here, two tests are conducted for specimens proof loaded to 88 percent S_m and then followed by fatigue load to failure at 71 percent S_m and 81 percent S_m maximum loads, respectively; and one test for specimens proof loaded to 95 percent S_m and followed by fatigue at 81 percent S_m load. Additional details of these tests are discussed in the following subsections, and the number of specimens tested in each case is listed in Table 1.

Specimens

The AS-3501-05 material system used in the experiment had a nominal fiber content of 65 percent by volume. All test specimens were cut from panels supplied directly from the manufacturer.[3] The dimensions of the test specimens are 6-ply thick (0.084 cm), 1.9 cm wide and 22.9 cm long with glass-epoxy end tabs of 3.8 cm in length. Thus, the test gage length is 15.3 cm.

[3]Hercules, Inc., Magna, Utah.

TABLE 1—*Number of specimens tested.*

		Fatigue Life	
	Static Strength	Maximum Stress $= 0.71\ S_m{}^a$, 1.034 GPa	Maximum Stress $= 0.81\ S_m$, 1.179 GPa
Baseline Data (no proof load)	24 (A)	130 (B)	25 (C)
Proof load to 0.88 S_m, 1.29 GPa	25 (D)	25 (F)	25 (G)
Proof load to 0.95 S_m, 1.39 GPa	25 (E)	. . .	25 (H)

[a] S_m = sample mean of static strength.
[b] Letter in parenthesis gives section in Table 2 where details are given.

Static Tension Test

The static tension tests are conducted on a closed-loop Instron tester under room temperature (~21°C) and ambient humidity (~60 percent relative humidity) conditions. The loading rate selected is approximately 1800 kg/min (4000 lb/min). The selection of test specimens follows a random number schedule.

Fatigue Life Test

The fatigue tests are also conducted on the Instron tester under the same room temperature and ambient humidity conditions as in the static tests. But no effort is made to monitor any temperature change in the specimen during fatigue. The loading procedure is as follows: the specimen is first loaded statically in tension to a prescribed mean stress level; it is then followed by prescribed oscillatory loading with the maximum to minimum stress amplitudes ratio of $R = 0.1$. The running cycle frequency is 9.5 Hz. Most tests are carried to fatigue failure; some are suspended at some prescribed life for purposes of either residual strength measurement,[4] or the reduction of testing time.

Proof Test

1. Post-proof tensile strength—this part of the proof-test consists of loading the specimens statically to the prescribed proof load, releasing this load, and reloading the surviving specimens to failure.

2. Post-proof fatigue life—this part of the proof test consists of loading

[4] Specimens suspended after a prescribed life were tested for both tensile residual strength and compressive residual strength. Data obtained in these residual strength tests were analyzed in Ref [7].

specimens statically to the previously selected proof load, releasing this load, and then subjecting the surviving specimens to a fatigue life test at the prescribed fatigue load. There is no significant resting period between loading steps.

Strength and Life Distribution Equations

Let x be the value of static strength. The two-parameter Weibull probability function for the static strength is given by (see, for example, Ref *8*)

$$F_X(x) = P[X \leq x] = 1 - \exp\left[-\left(\frac{x}{\beta}\right)^{\alpha}\right] \quad (1)$$

To compare the static strength distribution with the fatigue life distribution obtained under the maximum fatigue stress, S, the static strength distribution for those specimens that have strength larger than S should be used. This is known as the conditional probability

$$\begin{aligned} F_{X,S}(x) &= P[X \leq x | X > S] \\ &= 1 - \exp\left[-\left(\frac{x}{\beta}\right)^{\alpha} + \left(\frac{S}{\beta}\right)^{\alpha}\right] \end{aligned} \quad (2)$$

Let n be the value of fatigue life, the two-parameter Weibull distribution function for fatigue life is similar to Eq 1

$$F_N(n) = P[N \leq n] = 1 - \exp\left[-\left(\frac{n}{\beta_1}\right)^{\alpha 1}\right] \quad (3)$$

The strength-life equal-rank assumption stipulates that, for a given specimen with static strength, x_γ, and fatigue life, n_γ, the following relationship exists

$$F_N(n_\gamma) = F_{X,S}(x_\gamma) = 1 - \gamma \quad (4)$$

where 100γ is the percentage of the specimens that have strength larger than x_γ among those that have strength above S. Similarly, there will be 100γ percent of specimens that have life greater than n_γ.

When a group of samples, randomly selected from the population, is proof loaded to a value of x_γ, the percentage of surviving specimens among those that have strength larger than S is

$$\gamma = \exp\left[-\left(\frac{x_\gamma}{\beta}\right)^{\alpha} + \left(\frac{S}{\beta}\right)^{\alpha}\right] \quad (5)$$

After a proof load, x_γ, the percentage of surviving specimens among the total population will be denoted by γ_a, which is given by

$$\gamma_a = \exp\left[-\left(\frac{x_\gamma}{\beta}\right)^\alpha\right] \tag{6}$$

The guaranteed life under the maximum fatigue stress, S, is obtained by replacing x and n by x_γ and n_γ, respectively, from Eqs 2, 3, and 4

$$n_\gamma = \beta_1 \left[+\left(\frac{x_\gamma}{\beta}\right)^\alpha - \left(\frac{S}{\beta}\right)^\alpha\right]^{1/\alpha_1} \tag{7}$$

Similarly, the post proof-load static strength distribution may be described by the "truncated" Weibull distribution subjected to a proper conditional probability due to the proof-load, x_γ

$$\begin{aligned} F_{X,x_\gamma}(x) &= P[X \leq x | X > x_\gamma] \\ &= 1 - \exp\left[-\left(\frac{x}{\beta}\right)^\alpha + \left(\frac{x_\gamma}{\beta}\right)^\alpha\right] \end{aligned} \tag{8}$$

The parameters α and β in Eq 8 are determined by post proof-test strength data. When the value of α and β for the virgin specimens are used in Eq 8, the resulting distribution represents the "top γ-percent" strength distribution of the virgin specimen population.

In graphical presentations of the experimental data of either static strength or fatigue life cycles, the median rank of each data point is used to calculate the cumulative distribution. The median rank, $\hat{F}$, can be approximated by the formula

$$\hat{F} = \frac{j - 0.3}{m + 0.4} \tag{9}$$

where j is the failure order number, and m is the sample size. When there are suspended or censored specimens, the order number increment of all specimens following the suspended sets is given by

$$\Delta j = \frac{(m + 1) - j}{1 + k} \tag{10}$$

where k is the number of specimens following a particular suspended set.

The maximum likelihood method [9] is used in estimating the Weibull parameters, including those of the "truncated" Weibull distribution, Eq 8.

Results and Analysis

All test data are tabulated in a summary as displayed in Table 2.

Static Strength Distribution

A total of 24 virgin samples are tested here (Table 2*A*); the cumulative distribution of strength is shown in Fig. 1. The two-parameter Weibull function fits well to the test data, with $\alpha = 10.2$ and $\beta = 1531.7$ Mpa. The sample mean is $\bar{x} = S_m = 1462$ Mpa.

From Fig. 1, it is possible to infer the percentage of surviving specimens (γ_a) that are proof loaded to x_γ. For example, in the proof tests, two proof

TABLE 2—*Summary of test results.*

A. Static strength of virgin specimens; 24 specimens failed. Unit in MPa.

1096	1339	1445	1534	1634
1221	1347	1471	1544	1689
1227	1416	1476	1573	1729
1287	1421	1481	1576	1760
1305	1429	1491	1584	

Sample mean strength $S_m = 1462$.

B. Fatigue life of virgin specimens. Cycles; maximum stress = 1034 MPa (71% S_m), $R = 0.1$, $f = 9.5$ Hz; 130 specimens, 98 suspended, and 32 failed.

1(2)[a]	2330	14260	95606	441030
29	8350	27300	96310	531170
450	9550	37770	96360	844080
844	10000[b](35)	57450	100000[b](35)	1000000[b](28)
860	10810	68517	222220	1049160
1770	12781	76890	327580	1874600
2315	13261	86580	398480	

[a] Failed before maximum load is reached; not included in life distribution calculation.
[b] Specimen suspended before fatigue failure.

C. Fatigue life of virgin specimens. Cycles; maximum stress = 1179 MPa (81% S_m), $R = 0.1$, $f = 9.5$ Hz; 25 specimens, 4 suspended, and 21 failed.

1[a]	286	5653	15520	149356
30	288	5984	15754	1000000[b]
69	380	8609	18995	1000000[b]
90	1570	11362	22570	1066620[b]
260	3269	12119	97009	3302720[b]

[a] Failed before maximum load is reached; not included in life distribution calculation.
[b] Specimen suspended before fatigue failure.

D. Static strength of proof-loaded specimens. Proof load = 1290 MPa (88% S_m), unit in MPa; 25 specimens, 6 failed during proof loading.

1041[a]	1289[a]	1407	1482	1551
1207[a]	1358	1441	1482	1558
1241[a]	1358	1462	1531	1593
1262[a]	1400	1469	1531	1600
1269[a]	1407	1476	1538	1744

[a] Failure during proof load.

TABLE 2—*Continued.*

E. Static strength of proof-loaded specimens. Proof load = 1393 MPa (95% S_m); unit in MPa; 25 specimens, 8 failed during proof loading.

1041[a]	1310[a]	1517	1545	1620
1151[a]	1338[a]	1524	1545	1655(2)
1227[a]	1365[a]	1524	1572	1669
1255[a]	1447	1538	1593	1682
1303[a]	1476	1545	1600	

[a] Failure during proof load.

F. Fatigue life of proof-loaded specimens. Cycles, proof load = 1290 MPa (88% S_m); maximum fatigue stress = 1034 MPa, (0.71 S_m); 25 specimens, 3 failed during proof load, 4 failed during fatigue load, and 18 suspended at 70 000 cycles.

0(1117 Mpa)[a]	20481
0(1172 Mpa)[a]	39000
0(1269 Mpa)[a]	39643
289[b]	70000(18)[c]

[a] Failure during proof load.
[b] Specimen showed severe damage after proof load.
[c] 18 specimens suspended at 70 000 cycles.

G. Fatigue life of proof-loaded specimens. Cycles; proof load = 1290 MPa (88% S_m); maximum fatigue stress = 1179 MPa (0.81 S_m); 25 specimens, 4 failed during proof load, 18 failed during fatigue load, 3 suspended at 1 000 000 cycles.

0(1241 Mpa)[a]	1060	22860
0(1241 Mpa)[a]	1200	29440
0(1248 Mpa)[a]	1870	68010
0(1261 Mpa)[a]	2510	368280
100	3210	782120
150	5430	964760
200	9050	1000000(3)[b]
960	16910	

[a] Failure during proof load.
[b] 3 specimens suspended at 10^6 cycles.

H. Fatigue life of proof-loaded specimens. Cycles; proof load = 1393 MPa (95% S_m); maximum fatigue stress = 1179 MPa (0.81 S_m); 25 specimens, 9 failed during proof load, 14 failed during fatigue load, 2 suspended at 1 000 000 cycles.

0(1255 Mpa)[a]	0(1379 Mpa)[a]	13510
0(1255 Mpa)[a]	50	13770
0(1289 Mpa)[a]	470	16770
0(1289 Mpa)[a]	2340	22980
0(1289 Mpa)[a]	2370	142870
0(1303 Mpa)[a]	4210	167760
0(1379 Mpa)[a]	7230	866070
0(1379 Mpa)[a]	8980	1000000(2)[b]

[a] Failure during proof load.
[b] 2 specimens suspended at 10^6 cycles.

loads are selected; they are 1290 Mpa and 1393 Mpa that correspond, respectively, to 0.88 S_m and 0.95 S_m. For these two proof loads, the probabilities of failure are 0.16 and 0.32, or γ_a equal to 0.84 and 0.68, respectively. This means that if proof tests are actually performed, the guaranteed post-proof

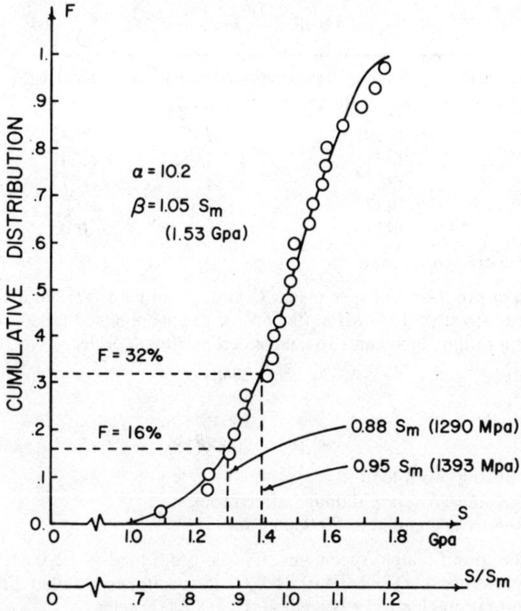

FIG. 1—*Cumulative distribution of static strength of virgin specimens.*

strengths would be greater than or equal to 1290 Mpa and 1393 Mpa, respectively.

Fatigue Life Distribution

Two fatigue tests are conducted using the virgin specimens, one at the maximum fatigue load of 0.71 S_m, or at 1034 Mpa; and one at 0.81 S_m, or 1179 Mpa. In the first case, a total of 130 samples are tested, with 32 failed, 35 suspended at 10^4 cycles, 35 suspended at 10^5 cycles, and 28 suspended at 10^6 cycles (Table 2*B*).[5] Parameter estimation using the maximum likelihood method yields a cumulative distribution function shown in Fig. 2, with $\alpha = 0.419$ and $\beta = 4.59 \times 10^6$ cycles. From Eq 5, with $S = 1034$ Mpa, the value of $(1 - \gamma)$ corresponding to the 0.88 S_m proof load is 0.145. Thus, if the equal-rank assumption is valid, the guaranteed post-proof life cycle would be equal to or greater than 55.4×10^3. This is shown in Fig. 2 and Table 3.

In the second fatigue case, a total of 25 samples are tested, of which 21 failed, 4 suspended at over 10^6 cycles, see Table 2*C*. The best fit to the Weibull distribution is depicted in Fig. 3. Here, $\alpha = 0.28$ and $\beta = 59.8$ kc. From Eq 5, with $S = 1179$ Mpa, the values of $(1 - \gamma)$ corresponding to the

[5]The suspended samples are subsequently tested for residual strength study. Results and analysis are contained in Ref [7].

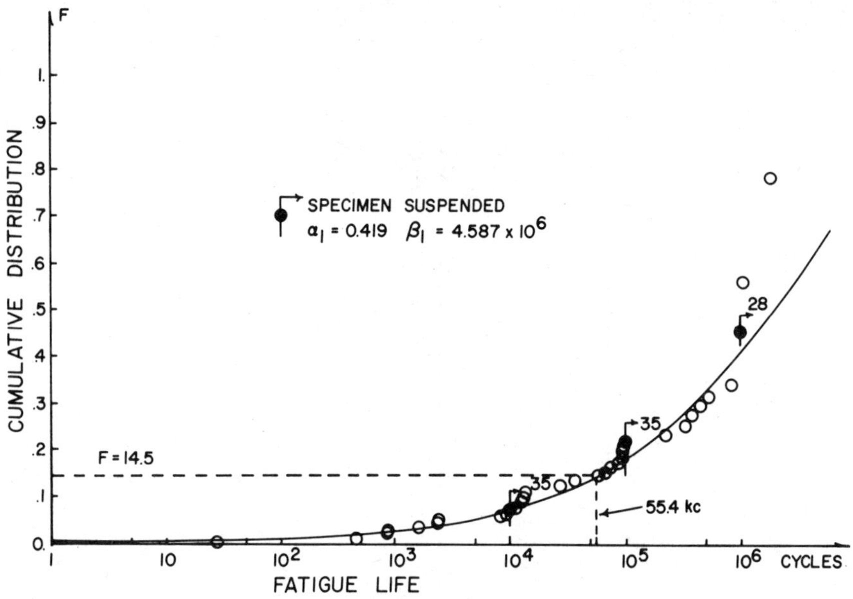

FIG. 2—*Cumulative distribution of fatigue life; virgin specimens; fatigue load* = *0.71* S_m *(1.034 GPa).*

two proof loads ($x_\gamma = 0.88\ S_m$ and $0.95\ S_m$) are 0.269 and 0.1, respectively, the corresponding guaranteed minimum life would be 914 cycles and 18 cycles, respectively. These results are shown in Fig. 3 and Table 3.

Post-Proof Static Strength Experiment

Test results for the post-proof static strength are tabulated in Table 2*D* and *E*. In the first case, the proof load is 1290 Mpa (0.88 S_m); here, 25 samples are tested with 6 samples failing during the proof loading while 19 survived. In subsequent reloading, all 19 samples have a strength larger than the proof load. The post-proof sample mean strength is 1489 Mpa. The modified Weibull distribution for the post-proof static strength, as expressed by Eq 8, is used to fit the experimented data, Fig. 4. Maximum likelihood estimation of parameter gives $\alpha = 12$ and $\beta = 1510$ Mpa. This distribution is compared with the distribution of the top 84 percent data of the virgin specimens (those larger than 1290 Mpa in Table 2*A*, sample mean is 1510 Mpa). These two distributions are practically identical, although a slight tendency of reduced scatter in the post-proof strength is indicated. This can be seen more clearly by comparing their respective density functions, Fig. 5.

In the second case the proof load is 1393 Mpa. A total of 25 samples are tested, of which 8 failed during proof loading and 17 survived. A Weibull fit

TABLE 3—*Parameters in guaranteed fatigue life after proof loading.*

		Fatigue Load, S					
		$0.71\ S_m$			$0.81\ S_m$		
Proof Load	Percent of Failure Among All Specimens	Percent of Failure Among $X \geq S$	Guaranteed Life, cycles	Experimental Percentage Exceeding Minimum Life	Percent of Failure Among $X \geq S$	Guaranteed Life, cycles	Experimental Percentage Exceeding Minimum Life
x_γ	$1 - \gamma_a$	$1 - \gamma$	n_γ	...	$1 - \gamma$	n_γ	...
$0.88\ S_m$ (1.290 GPa)	16	14.5	55350	85.7	10	17.8	100
$0.95\ S_m$ (1.393 GPa)	32	...	...	...	26.9	914	87.5

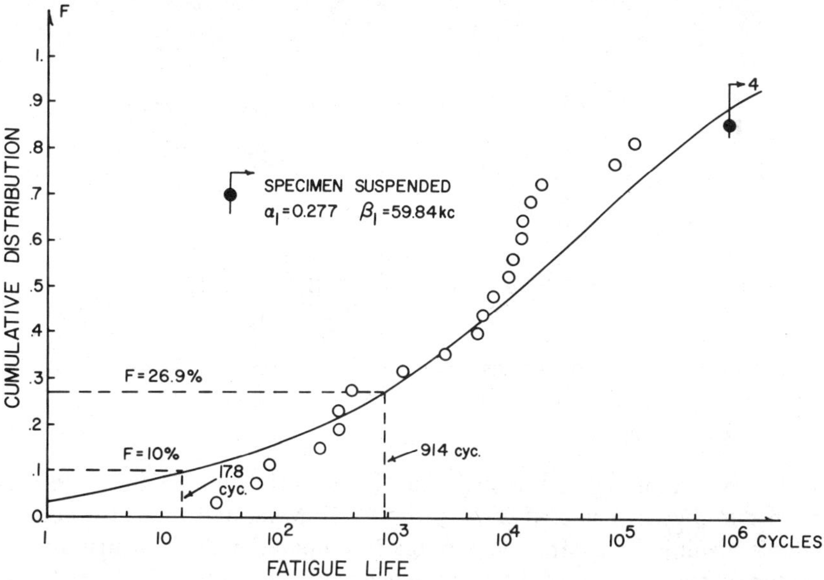

FIG. 3—*Cumulative distribution of fatigue life; virgin specimens; fatigue load* = *0.81* S_m *(1.179 GPa).*

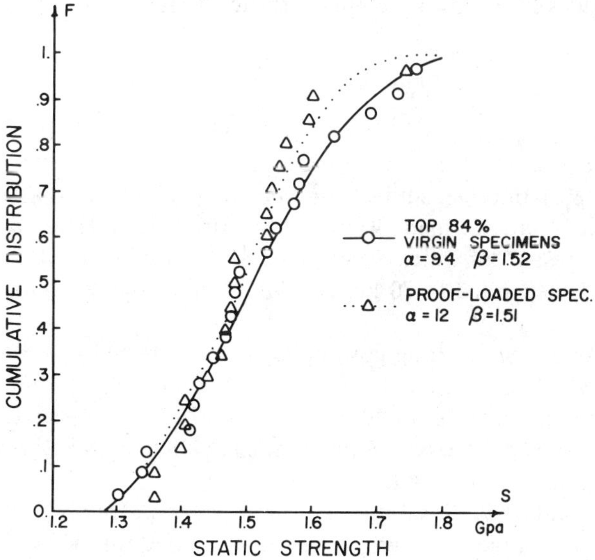

FIG. 4—*Cumulative distribution of post-proof static strength; proof laod* = *0.88* S_m *(1.29 GPa).*

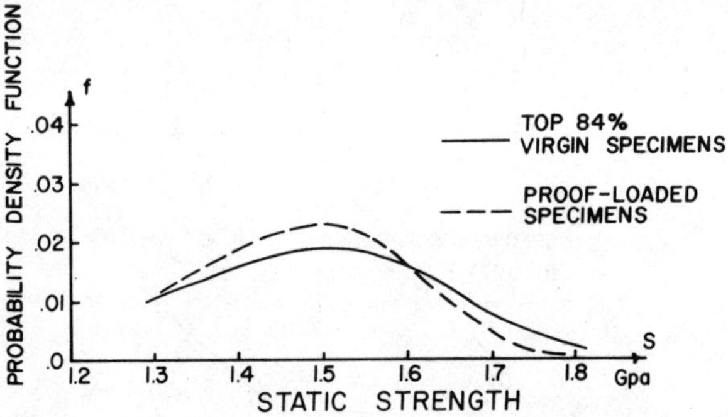

FIG. 5—*Weibull density function for post-proof static strength; proof load* = *0.88* S_m *(1.29 GPa).*

of the post-proof strength distribution is shown in Fig. 6. Here, the post-proof distribution has $\alpha = 23$ and $\beta = 1572$ Mpa, and the corresponding sample mean is 1572 Mpa. This is again compared with the distribution of the top 68 percent data of the virgin specimens (those larger than 1393 Mpa, sample mean is 1545 Mpa). Again, proof loading seems to reduce the scatter, while it does not affect the sample mean strength. The respective distribution density functions are shown in Fig. 7, where a more pronounced reduction in scatter for the post-proof samples is indicated.

Figure 8 gives a summary of the data points and various sample mean strength values for the two post-proof static strength cases.

Post-Proof Fatigue Life Experiment

Results of the post-proof fatigue life tests are tabulated in Table 2*F, G,* and *H*. The first case pertains to proof loading the specimens to 1290 Mpa (88 percent S_m) and then subjects the surviving specimens to a fatigue test at the maximum fatigue load of 1034 Mpa (0.71 S_m). Here, a total of 25 samples are tested, of which 4 failed the proof load, 3 failed during fatigue test, and 18 suspended at 70 kc. An *S-N* scan for the test data is shown in Fig. 9.

Earlier, in the virgin fatigue samples, it was inferred that the 1290 Mpa proof load would screen out 14.5 percent of the low life specimens, and the guaranteed minimum life would be 55.4 kc. In Fig. 9, it is seen that one sample failed before 55.4 kc, two samples failed at about 55.4 kc, while 18 others had a life greater than 55.4 kc.

In the second case, 25 samples are proof loaded again to 1290 Mpa, and the surviving samples are fatigue tested at 1179 Mpa (0.81 S_m). Here, 4 failed the proof load, 18 failed during fatigue, and 3 samples are suspended at 10^6 cycles. Figure 10 shows the *S-N* scan for these data. The inferred

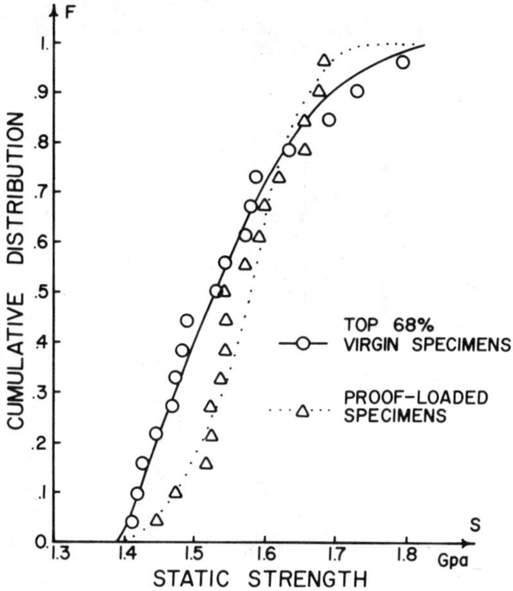

FIG. 6—*Cumulative distribution of post-proof static strength; proof load = 0.95* S_m *(1.393 GPa)*

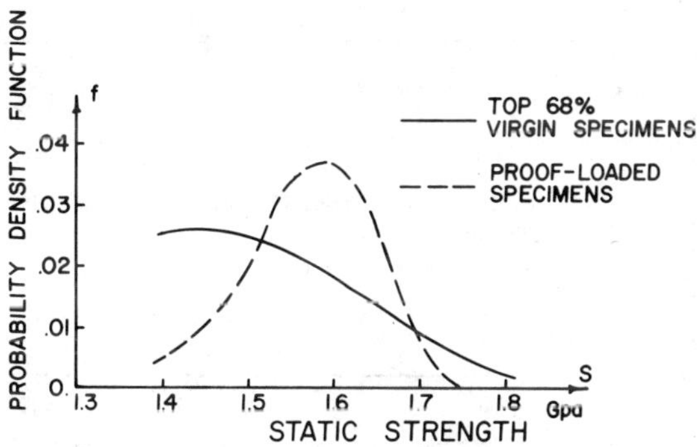

FIG. 7—*Weibull density function of post-proof static strength; proof load = 0.95* S_m *(1.393 GPa).*

minimum life in this case would be 17.8 cycles. Experimentally, life of all the 21 surviving specimens exceeded this minimum life.

Similarly, a total of 25 samples are tested in the third case. The proof load here is 1393 Mpa (0.95 S_m), and the maximum fatigue load is 1179 Mpa (0.81 S_m). In this case, 9 samples failed the proof load, 14 failed during fatigue, and 2 suspended at 10^6 cycles. However, two out of 16 surviving samples failed before the inferred minimum life of 914 cycles, Fig. 11.

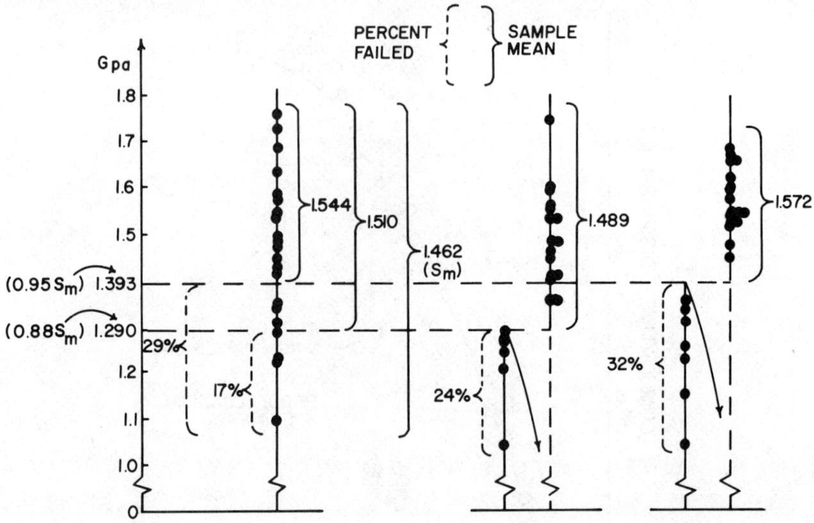

FIG. 8—*Static strength after proof testing.*

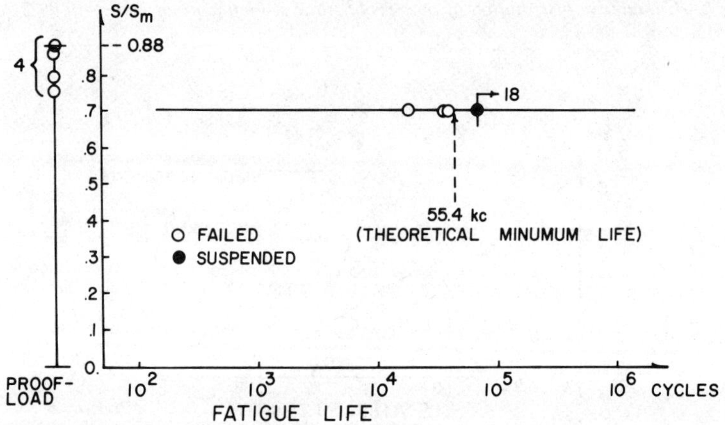

FIG. 9—S-N *scan for fatigue of post-proof specimens; fatigue stress* = *0.71* S_m *(1.034 GPa); proof load* = *0.88* S_m *(1.29 GPa).*

Discussions

As it has been stated earlier, most current fatigue analyses for composite materials are based either implicitly or explicitly upon the "equal-rank" assumption. It is important to verify this assumption experimentally with sufficient test data and test cases. Through the proof-test procedure as described in this study and in view of the test data (limited to unidirectional

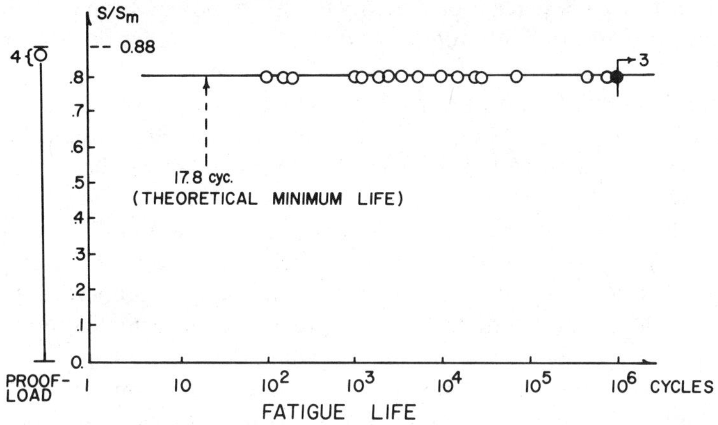

FIG. 10—S-N *scan for fatigue of post-proof specimens; fatigue stress = 0.81* S_m *(1.179 GPa); proof load = 0.88* S_m *(1.29 GPa).*

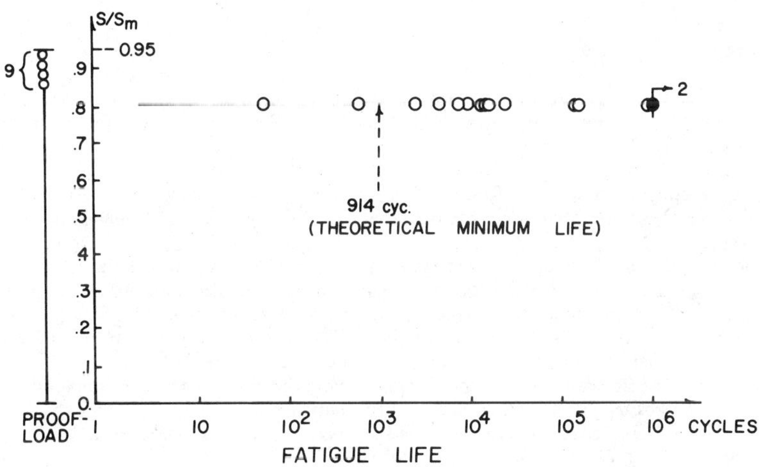

FIG. 11—S-N *scan for fatigue of post-proof specimens; fatigue stress = 0.81* S_m *(1.179 GPa); proof load = 0.95* S_m *(1.393 GPa).*

graphite-epoxy laminates), the "equal-rank" assumption appears to be both reasonable and practical.

Within this context, the test results also indicate that proof load changes only slightly the essential features of the static strength. Specifically, the post-proof specimens generally have a smaller scatter in their strength distribution as compared to the strength distribution of the corresponding top percentile of the virgin specimens. However, all specimens that survived the proof load show a post-proof strength larger than the proof load. This result indicates a 100 percent probability that the proof load can guarantee a minimum strength. In a different viewpoint, this also implies that proof load does not alter the ranking in the strength distribution.

The probability of guaranteeing a minimum life after proof load is less assuring. The results show, however, that the chance is 90 percent or better. Note that two conditions must be met if a minimum life can be guaranteed after proof loading; first, the proof load must not cause damage so as to degrade the fatigue property, and second, the "equal-rank" assumption must be valid. In view of the overall results, it may be stated that proof loading does alter slightly the strength and fatigue properties of the specimens, depending on the level of the proof load. The small alteration in property is due presumably to the proof-loading induced damage.

Since the specimens used here are all unidirectional composites, it is not clear whether the concept of proof test is equally applicable in composite laminates of different fiber orientations and stacking sequences. Such a question clearly needs further study, and its implications could be of considerable practical importance.

References

[*1*] Halpin, J. C., Jerina, K. L., and Johnson, T. A. in *Analysis of Test Methods for High* Modulus Fibers and Composites, ASTM STP 521, American Society for Testing and Materials, 1973, p. 5.

[*2*] Hahn, H. T. and Kim, R. Y., *Journal of Composite Materials,* Vol. 9, 1975, p. 297.

[*3*] Yang, J. N. and Liu, M. D., "Residual Strength Degradation Model and Theory of Periodic Proof Tests for Graphite/Epoxy Laminates," AFML-TR-76-225, Wright-Patterson Air Force Base, Ohio, 1976.

[*4*] Chou, P. C. and Croman, R. in *Composite Materials: Testing and Design* (*Fifth Conference*) *ASTM STP 674*, American Society for Testing and Materials, 1979, p. 431.

[*5*] Chou, P. C. and Croman, R., *Journal of Composite Materials,* Vol. 12, 1978, p. 177.

[*6*] Awerbuch, J. and Hahn, H. T. in *Fatigue of Filamentary Composite Materials, ASTM STP 636,* American Society for Testing and Materials, 1977, p. 248.

[*7*] Alper, J., "An Investigation on the Strength-Life Behavior of a Unidirectional Graphite-Epoxy Composite," Ph.D. thesis, Drexel University, 1980.

[*8*] Bury, K. V., *Statistical Models in Applied Science,* Wiley, New York, 1975.

[*9*] Cohen, A. C., *Technometrics*, Vol. 7, 1965. p. 579.

J. M. Whitney[1]

Fatigue Characterization of Composite Materials

REFERENCE: Whitney, J. M., **"Fatigue Characterization of Composite Materials,"** *Fatigue of Fibrous Composite Materials, ASTM STP 723,* American Society for Testing and Materials, 1981, pp. 133–151.

ABSTRACT: A procedure is outlined that allows the generation of an *S-N* curve with some statistical value without resorting to an extremely large data base. The procedure is based on a power law representation of the *S-N* curve and a two-parameter Weibull distribution for time-to-failure at a specific stress level. A data pooling scheme is also discussed that allows the determination of a fatigue shape parameter that is independent of stress level.

KEY WORDS: fatigue (materials), composite materials, Weibull distribution, data pooling, *S-N* curves

The classical *S-N* curve has been the primary method of characterizing the fatigue behavior of fiber reinforced composites. This method usually consists of determining the number of cycles to failure for a number of stress ranges associated with a particular load history (often constant amplitude tension-tension loading). The resulting *S-N* curve yields an estimate of the mean time-to-failure as a function of stress range. Such a procedure, however, fails to account for the large variation in the time-to-failure at a given maximum stress level. Fatigue data with statistical significance require a large number of replicates at a given maximum stress level in order to measure the distribution of time-to-failure.

In this paper, a procedure that allows the generation of an *S-N* curve with some statistical value without resorting to an extremely large data base is explored in detail. The "wearout" or "strength degradation" model approach [*1–3*][2] provides one means for accomplishing this. Such an approach, however, involves the assumption of a direct relationship between static strength distribution, residual strength distribution after time under a speci-

[1] Materials research engineer, Mechanics and Surface Interactions Branch, Nonmetallic Materials Division, Air Force Materials Laboratory, Wright-Patterson Air Force Base, Ohio 45433.

[2] The italic numbers in brackets refer to the list of references appended to this paper.

fied load history, and distribution of time-to-failure at a maximum stress level. These types of models have to be carefully defined in terms of the load history to be applied to the material. For example, if the load history is tension/compression, then the concept of strength degradation must consider both residual tension and residual compression, along with possible competing failure modes. The alternative approach to *S-N* curve characterization does not require any assumptions relating fatigue life to residual strength. As shown in the Appendix, the procedure does measure parameters that are completely compatible with the "wearout" model approach without any residual strength measurements.

S-N Curve Characterization

An alternative to the "wearout" model approach for fatigue characterization of composite materials has been proposed by Hahn and Kim [*4*]. Their approach involves two basic assumptions: (1) a classical power law representation of the *S-N* curve [*5*], and (2) a two-parameter Weibull distribution of time-to-failure. In mathematical form these assumptions become

$$CNS^b = 1 \tag{1}$$

where S is the stress range, N is the number of cycles to failure, and b and C are material constants. In addition

$$R(N) = \exp\left[-\left(\frac{N}{N_0}\right)^{\alpha_f}\right] \tag{2}$$

where $R(N)$ denotes the reliability of N (probability of survival), N_0 is the characteristic time-to-failure (location parameter), and α_f is the fatigue shape parameter. As illustrated in the Appendix, Eqs 1 and 2 can also be derived from the "wearout" model approach with α_f being related to the shape parameter for static strength [*3*]. Equation 1 can be written in the form

$$N = C^{-1}S^{-b} \tag{3}$$

substituting Eq 3 into Eq 2 and solving for S yields

$$S = K\{[-\ln R(N)]^{-1/\alpha_f b}\}N_0{}^{-1/b}, \qquad \alpha_f \neq f(S) \tag{4}$$

where

$$K = C^{-1/b} \tag{5}$$

When $N = N_0$, $-\ln R(N_0) = 1$, and Eq 4 reduces to

$$S(N_0) = KN_0^{-1/b} \tag{6}$$

A plot of log S versus log N_0 produces a straight line with slope $-1/b$ and a y intercept of log K. Thus, a measure of the distribution of time-to-failure at various stress ranges in conjunction with Eq 2 allows α_f to be determined along with a set of values of N_0, each corresponding to a value of S. Equation 4 can then be utilized to produce an S-N_0 curve for any desired reliability, $R(N)$.

From a practical standpoint, however, one is more interested in obtaining an S-N curve for any desired level of reliability rather than a S-N_0 curve. Writing Eq 6 in the form

$$N_0 = C^{-1}S^{-b} \tag{7}$$

and substituting into Eq 2 yields

$$R(N) = \exp\left[-\left(\frac{N}{C^{-1}S^{-b}}\right)^{\alpha_f}\right] \tag{8}$$

Solving for S leads to the following S-N relationship for any desired level of reliability

$$S = K\{[-\ln R(N)]^{1/\alpha_f b}\}N^{-1/b} \tag{9}$$

Data Reduction Procedure

The data reduction procedure consists of: (a) fitting the time-to-failure data at each stress range to a two-parameter Weibull distribution; (b) use of a data pooling scheme to determine the fatigue shape parameter, α_f; and (c) fit log S versus log N_0 data to a straight line for the determination of b and K.

Let m be the number of stress ranges tested and n_i the number of specimens tested at the i-th stress range, S_i, which leads to the data set

$$N_i(N_{i1}, N_{i2}, \ldots, N_{in_i}), \qquad i = 1, 2, \ldots, m \tag{10}$$

Each stress range is fit to the two-parameter Weibull distribution

$$R(N_i) = \exp\left[-\left(\frac{N_i}{N_{0i}}\right)^{\alpha_{fi}}\right] \tag{11}$$

A number of procedures can be utilized for determining α_{fi} and N_{0i}. One of the methods preferred by statisticians is the maximum likelihood estimator (MLE) that is of the form [*6*]

$$\frac{\sum_{j=1}^{n_i} N_{ij}^{\hat{\alpha}_{fi}} \ln N_{ij}}{\sum_{j=1}^{n_i} N_{ij}^{\hat{\alpha}_{fi}}} - \frac{1}{n_i} \sum_{j=1}^{n_i} N_{ij} - \frac{1}{\hat{\alpha}_{fi}} = 0 \qquad (12)$$

$$\hat{N}_{0i} = \left(\frac{1}{n_i} \sum_{j=1}^{n_i} N_{ij}^{\hat{\alpha}_{fi}} \right)^{1/\hat{\alpha}_{fi}} \qquad (13)$$

where $\hat{\alpha}_{fi}$ and $\hat{N}_{0i}$ denote estimated values of α_{fi} and N_{0i}, respectively. Equation 12 has only one real positive root. As a result, an iterative scheme can be utilized until a value of $\hat{\alpha}_{fi}$ is obtained to any desired number of decimal places. The resulting value of $\hat{\alpha}_{fi}$ obtained from the iterative scheme can then be used in conjunction with Eq 13 to obtain $\hat{N}_{0i}$.

Since fatigue shape parameters are estimated based on a particular sample size, it is anticipated that each value of S_i would produce a different value of $\hat{\alpha}_{fi}$, even though α_f may be independent of stress range. A two-sample test [*7*] is available, however, that allows for testing the equality of shape parameters in two-parameter Weibull distributions with unknown scale parameters. The approach is based on MLE and the results depend on sample size and confidence level desired. Let $\hat{\alpha}_{max}$ and $\hat{\alpha}_{min}$ be the maximum and minimum values obtained for $\hat{\alpha}_{fi}$. For the information tabulated in Ref *7*, it is required that $\hat{\alpha}_{max}$ and $\hat{\alpha}_{min}$ be associated with equal sample sizes, *n*. If $\hat{\alpha}_{max}$ and $\hat{\alpha}_{min}$ are from the same distribution, then it is expected that [*7*]

$$\frac{\hat{\alpha}_{f\,max}}{\hat{\alpha}_{f\,min}} < B(\gamma, n), B > 1 \qquad (14)$$

for a given confidence level, γ, and sample size, *n*. Values of *B* are shown in Table 1 for various sample sizes corresponding to a confidence level of 0.98. This data is taken from Ref *7*. The large values of *B* associated with small sample sizes suggest that significant variations in $\hat{\alpha}_{fi}$ are likely to be encountered with small data sets taken from the same population.

Let us now assume that α_f is independent of stress range. Then a data pooling technique must be utilized in order to determine a single value of α_f for all S_i. Various approaches for obtaining a pooled value of α_f can be found in the literature. The approach used in the present work has been investigated by Lemon [*8*]. This procedure utilizes the normalized data set

$$X(X_{i1}, X_{i2}, \ldots, X_{in_i}), \qquad i = 1, 2, \ldots, m \qquad (15)$$

where

$$X_{ij} = \frac{N_{ij}}{\hat{N}_{0i}} \tag{16}$$

Thus, each set of data at a given stress range is normalized by the estimated characteristic time-to-failure and the results fit to the pooled two-parameter Weibull distribution

$$R(X) = \exp\left[-\left(\frac{X}{X_0}\right)^{\alpha_f}\right] \tag{17}$$

This procedure has the advantage of obtaining a large data base for determining α_f by using a few replicates for a number of values of S_i. In general, for equal accuracy, fewer specimens are needed to determine the location parameter than shape parameter.

For the pooled Weibull distribution, Eq 17, the MLE relationships take the form

$$\frac{\sum_{i=1}^{m}\sum_{j=1}^{n_i} X_{ij}^{\bar{\alpha}_f} \ln X_{ij}}{\sum_{i=1}^{m}\sum_{j=1}^{n_i} X_{ij}^{\bar{\alpha}_f}} - \frac{1}{M}\sum_{i=1}^{m}\sum_{j=1}^{n_i} X_{ij} - \frac{1}{\bar{\alpha}_f} = 0 \tag{18}$$

$$\bar{X}_0 = \left(\frac{1}{M}\sum_{i=1}^{m}\sum_{j=1}^{n_i} X_{ij}^{\bar{\alpha}_f}\right)^{1/\bar{\alpha}_f} \tag{19}$$

where $\bar{\alpha}_f$ and $\bar{X}_0$ are estimated values of α_f and X_0, respectively, and

$$M = \sum_{i=1}^{m} n_i \tag{20}$$

TABLE 1—*Equality of Weibull shape parameters, $\gamma = 0.98$ [7].*

n_i	β
5	3.550
10	2.213
15	1.870
20	1.703
100	1.266

For a perfect fit to the data pooling scheme, the location parameter, X_0, should be unity. The values of $\hat{N}_{0i}$ can be adjusted, however, to produce an exact value of unity for X_0. In particular

$$\bar{N}_{0i} = \bar{X}_0 \hat{N}_{0i} \tag{21}$$

where $\bar{N}_{0i}$ denotes estimated values of N_0 associated with the adjusted two-parameter Weibull distribution

$$R(X) = \exp(-X^{\alpha_f}) \tag{22}$$

The slope of the *S-N* curve, $1/b$, and the *y*-intercept, *K*, can be determined by fitting $\log S_i$ versus $\log \bar{N}_{0i}$ to a straight line. With *K*, *b*, and α_f now determined, Eq 9 can be used to produce an *S-N* curve of any desired reliability.

It should be noted that MLE is asymptotically unbiased, that is, it is a biased estimator for small sample sizes [*6*]. Unbiasing factors are tabulated in Ref *9*. These factors are less than unity as MLE always tends to overestimate the shape parameter. Confidence intervals for both the shape parameter and location parameter as a function of sample size have also been established [*9*]. If conservative estimates are desired for $R(N)$, then a lower bound value of α_f can be utilized. This value of α_f can be used in conjunction with Eq 19 to determine $\bar{X}_0$.

Censoring Procedures

In the case of high cycle fatigue, the time-to-failure may become unacceptably long. This difficulty can be overcome by raising the stress range so that fatigue failures are produced within a reasonable number of cycles. For filament dominated laminates, which tend to have a very flat *S-N* curve, stress levels may have to be raised to an unacceptable level to produce a reasonable time-to-failure for all specimens tested. In particular, it is undesirable to raise the fatigue stress level to such a degree that it significantly overlaps the static strength distribution. In such cases, the probability of a first cycle failure is significant.

If censoring techniques are applied to data reduction procedures, fatigue failures are not required of all specimens. For the data reduction scheme outlined in the present work, Type I censoring seems to be the most desirable in terms of yielding the most information. In the case of Type I censoring, the fatigue test is terminated at a predetermined time (for example, 10^6 cycles) even though all specimens have not failed. The MLE equations for Type I censoring are of the form [*6*]

$$\frac{\sum_{j=1}^{r_i} N_{ij}^{\hat{\alpha}_{fi}} \ln N_{ij} + (n_i - r_i) R_i^{\hat{\alpha}_{fi}} \ln R_i}{\sum_{j=1}^{r_i} N_{ij}^{\hat{\alpha}_{fi}} + (n_i - r_i) R_i^{\hat{\alpha}_{fi}}} - \frac{1}{r_i} \sum_{j=1}^{r_i} \ln N_{ij} - \frac{1}{\hat{\alpha}_{fi}} = 0 \tag{23}$$

$$N_{0i} = \left\{ \frac{1}{r_i} \left[\sum_{j=1}^{n_i} N_{ij}^{\hat{\alpha}_{fi}} + (n_i - r_i) R_i^{\hat{\alpha}_{fi}} \right] \right\}^{1/\hat{\alpha}_{fi}} \tag{24}$$

where n_i now denotes the total number of specimens tested at S_i, r_i is the number of fatigue failures at S_i, and R_i is the number of cycles at which the test is terminated.

The data pooling procedure is now analogous to "progressive censoring" in which a number of samples are removed at predetermined time intervals throughout the duration of the test. The MLE associated with the data pooling procedure in conjunction with censored samples becomes

$$\frac{\sum_{i=1}^{m} \sum_{j=1}^{r_i} X_{ij}^{\bar{\alpha}_{fi}} \ln X_{ij} + \sum_{i=1}^{m} (n_i - r_i) Y_i^{\bar{\alpha}_{fi}} \ln y_i}{\sum_{i=1}^{m} \sum_{j=1}^{r_i} X_{ij}^{\bar{\alpha}_{fi}} + \sum_{i=1}^{m} (n_i - r_i) Y_i^{\bar{\alpha}_{fi}}} - \frac{1}{N} \sum_{i=1}^{m} \sum_{j=1}^{r_i} X_{ij} - \frac{1}{\bar{\alpha}_f} = 0 \tag{25}$$

$$\bar{X}_0 = \left\{ \frac{1}{N} \left[\sum_{i=1}^{m} \sum_{j=1}^{r_i} X_{ij}^{\bar{\alpha}_{fi}} + \sum_{i=1}^{m} (n_i - r_i) Y_i^{\bar{\alpha}_{fi}} \right] \right\}^{1/\bar{\alpha}_{fi}} \tag{26}$$

where

$$Y_i = \frac{R_i}{N_{0i}} \tag{27}$$

and N is the total number of fatigue failures, that is

$$N = \sum_{i=1}^{m} r_i \tag{28}$$

Example Data

Consider the tension-tension (T-T) fatigue data presented by Ryder and Walker [*10*] on quasi-isotropic T300/934[3] graphite-epoxy laminates with the stacking geometry $[45/90/-45/90 - 45/0/45/0]_s$ (Laminate 2). Twenty

[3] Union Carbide's T300 graphite fibers in Fiberite's 934 epoxy resin system.

replicates at three stress levels were used to characterize the scatter in time-to-failure. The effect of sample size on the Weibull parameters can be estimated by using a table of random numbers to select sample sizes of 5, 10, 15, and 20 for the three stress levels and calculating the resulting Weibull parameters. The results are shown in Table 2 along with the pooled parameters $\bar{X}_0$ and $\bar{\alpha}_f$. The Weibull parameters are determined by MLE. Note that the values of $\hat{N}_{0i}$ are not radically effected by sample size, while $\hat{\alpha}_{fi}$ is very sensitive to n. The same trend is noted for the pooled parameters $\bar{X}_0$ and $\bar{\alpha}_f$. It should also be noted that the data for $n = 20$ satisfies the criterion of Eq 14 for data pooling.

Comparison between the normalized data and the Weibull distribution obtained from the pooling procedure is shown in Fig. 1. Three additional stress levels with three replicates each are added to the data pooling process, yielding a total of 69 pooled data points in Fig. 1. Comparisons between data and Weibull distributions obtained from the pooling procedure are shown in Figs. 2 through 4 for the three stress levels with 20 replicates. In Figs. 1 through 4, data points are converted to probabilities of survival from the median rank (MR) defined as

$$MR = \frac{j - 0.3}{n + 0.4} \tag{29}$$

where j is the survival order number (data listed in decreasing order of time-to-failure) and n is the total number of samples tested. Equation 29 is an estimate of the reliability function, R. When probability of failure is desired, Eq 29 is applied with the data listed in increasing time-to-failure.

While the pooled data in Fig. 1 compares favorably with the estimated two-parameter Weibull distribution, similar data correlation for the individual stress levels in Figs. 2 through 4 are much less favorable. This is to be anticipated as small data samples in conjunction with large scatter creates difficulty in fitting the data to any reasonable distribution function. These results are also a good illustration of the desirability of data pooling for obtaining a larger sample size for estimating shape parameters. The characteristic *S-N* curve resulting from the data reduction scheme is shown in Fig. 5 along with a 95 percent survivability curve calculated from Eq 9. The numbers in parentheses correspond to the sample size associated with a specific stress level. Scatter bands on time-to-failure are also shown. The solid dots correspond to fatigue failures outside the 95 percent survivability line. The pooled Weibull parameters along with the *S-N* curve parameters are listed in Table 3.

The *S-N* characterization procedure has been also applied to other data available in the literature and the results are shown in Figs. 6 through 8. As in Fig. 5, sample size at each stress level, scatter bands, and fatigue failures outside of the 95 percent survivability line are shown. In Fig. 6,

TABLE 2—*Effect of sample size on Weibull parameters.*

	Stress Level, MPa (ksi)							
	290(42)		345(50)		400(58)			
n	$\hat{\alpha}_{f1}$	$\hat{N}_{01}$	$\hat{\alpha}_{f2}$	$\hat{N}_{02}$	$\hat{\alpha}_{f3}$	$\hat{N}_{03}$	$\bar{\alpha}_f$	$\bar{X}_0$
5	3.05	1 865 519	44.7	70 199	0.804	3116	1.53	1090
10	1.85	1 485 006	1.14	91 949	1.05	3803	1.23	1007
15	1.47	2 106 780	1.23	87 407	0.858	3206	1.13	1011
20	1.51	1 964 890	1.07	100 815	0.984	4042	1.14	1007

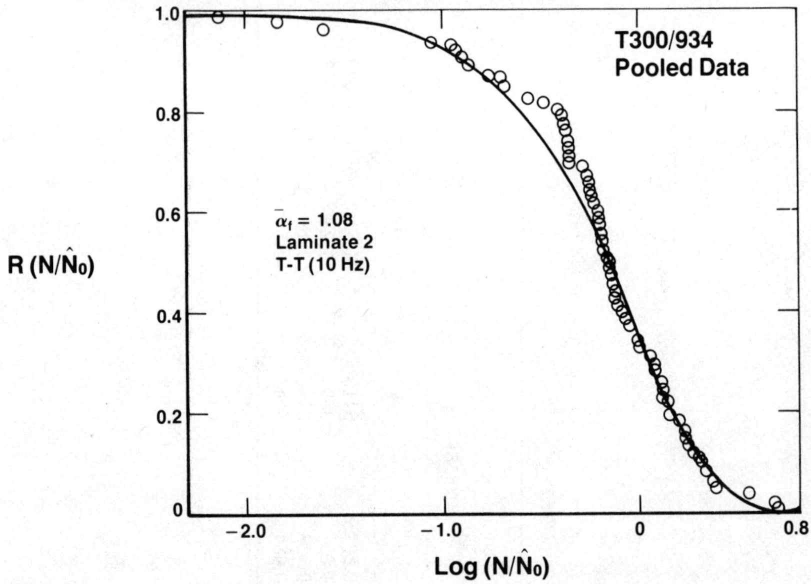

FIG. 1—*Comparison between normalized data and pooled two-parameter Weibull distribution.*

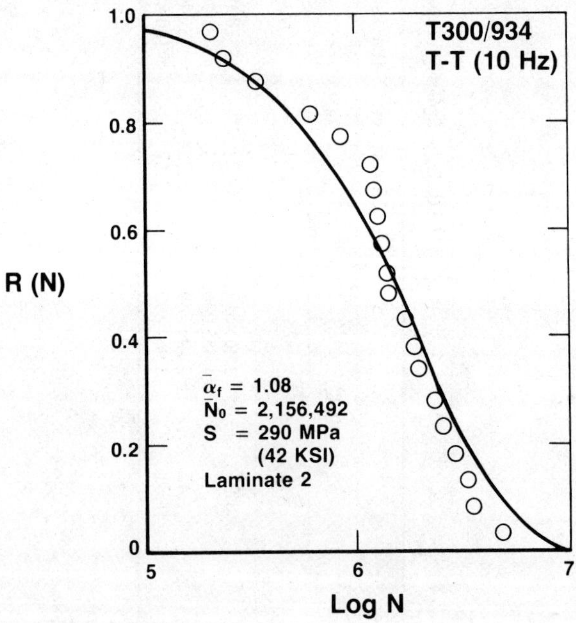

FIG. 2—*Comparison between data and pooled two-parameter Weibull distribution,* $S/\bar{S}_0 = 0.60$.

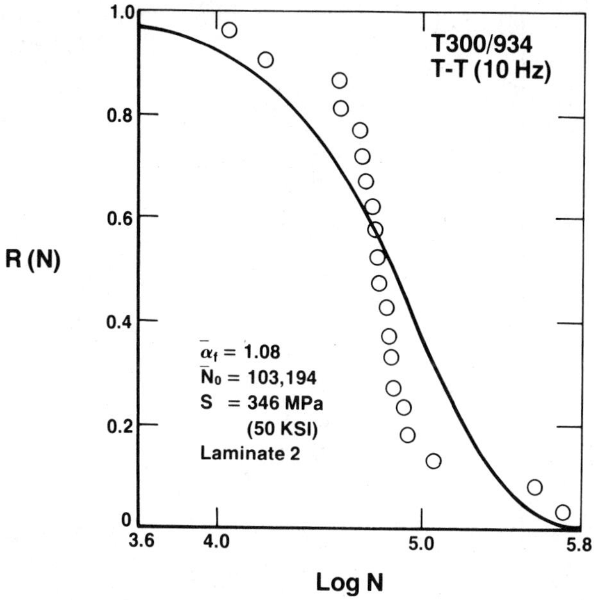

FIG. 3—*Comparison between data and pooled two-parameter Weibull distribution,* $S/\bar{S}_0 = 0.71$.

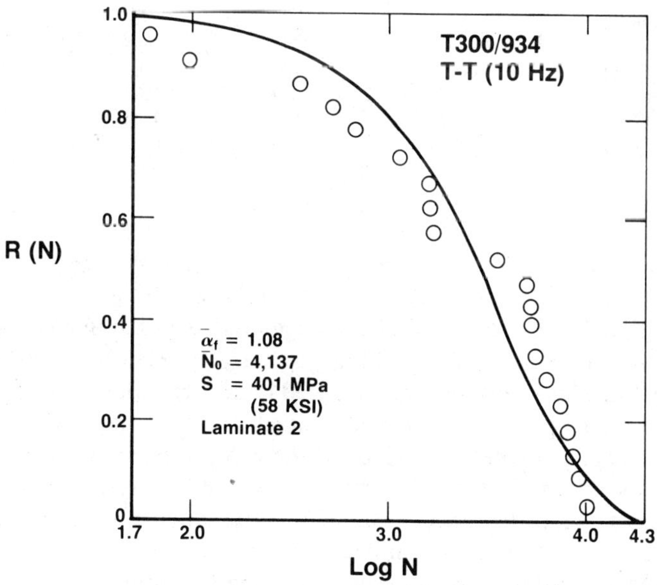

FIG. 4—*Comparison between data and pooled two-parameter Weibull distribution,* $S/\bar{S}_0 = 0.82$.

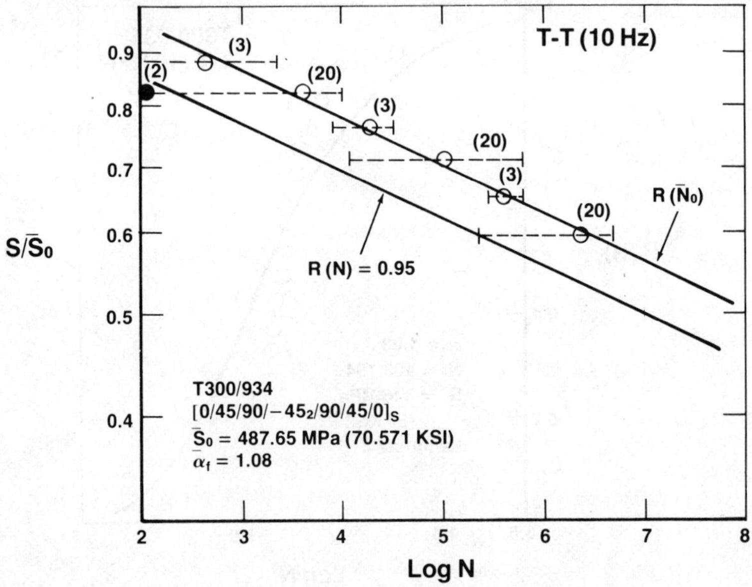

FIG. 5—*Tension-tension* S-N *curves for Laminate 2.*

TABLE 3—S-N *curve and pooled Weibull parameters.*

Laminate	b	K/S_0	$\bar{\alpha}_f$	$\bar{X}_0$
1 (T-T)	66.34	1.066	0.310	0.9490
2 (T-T)	21.70	1.188	1.08	1.014
2 (T-C)	9.705	1.791	1.45	0.9863
3 (T-T)	65.84	1.094	3.82	1.003

tension-compression (T-C) constant amplitude fatigue data from Laminate 2 is illustrated [*10*]. Data in Fig. 7 were also from the work of Ryder and Walker [*11*]. The same graphite-epoxy material as in Laminate 2 is utilized for this T-T data with the stacking geometry $[0/+45/0_2/-45/0_2/+45/0_2/-45/0]_s$. This composite is designated Laminate 1 and presents a very interesting example for the recommended fatigue characterization procedure. Because of the flat nature of the *S-N* curve, fatigue failures were difficult to induce. For the split numbers in parentheses, the first number designates the total number of samples tested at that stress level and the second number refers to the number of fatigue failures recorded. Thus, this data had to be reduced utilizing the censoring procedure outlined. All tests were terminated after 10^6 cycles. It is interesting to note the extremely low value of $\bar{\alpha}_f$ associated with this laminate. This behavior has been previously

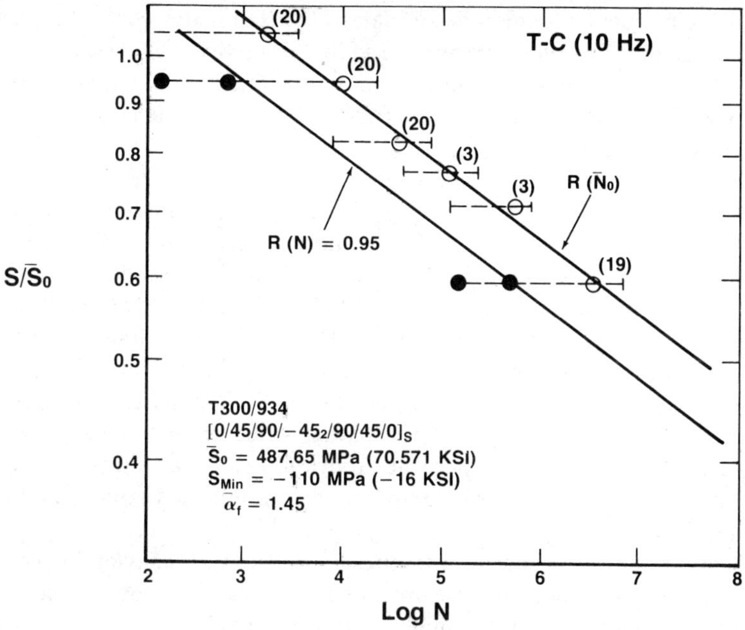

FIG. 6—*Tension-compression* S-N *curves for Laminate 2.*

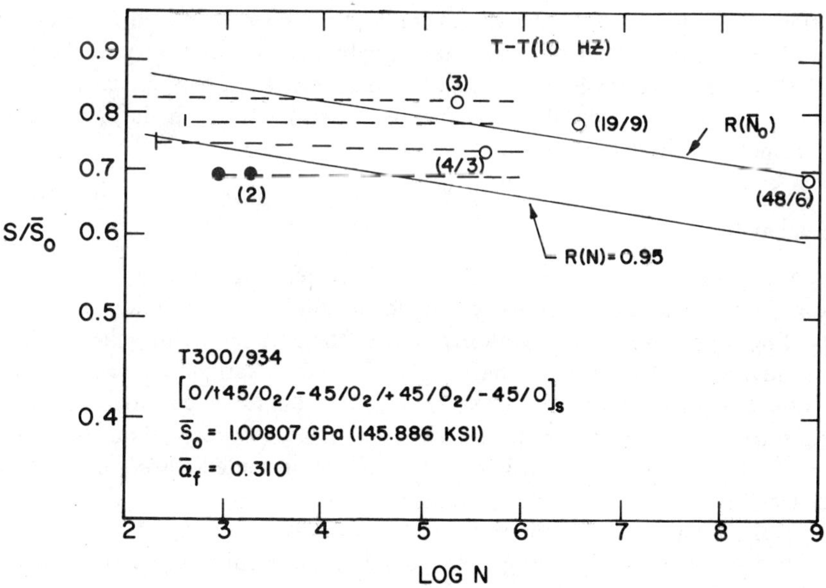

FIG. 7—*Tension-tension* S-N *curves for Laminate 1.*

noted for filament dominated laminates [*12*] (laminates containing a large percentage of 0-deg plies relative to the load direction). The straight line fit of the *S-N* curve is not as good for this laminate as for Laminate 2 in Figs. 5 and 6. The low number of fatigue failures along with the large data scatter are likely reasons for the poorer fit.

The data in Fig. 8 were taken from the work of Yang [*13*]. This constant amplitude T-T data are for a $[\pm 45]_{2s}$T300/5208[4] graphite-epoxy laminate. This orientation was chosen because it induces significant shear stress relative to the fiber direction in each ply. Since there are no 0-deg plies relative to the load direction, this laminate is referred to as matrix dominated. This composite is designated Laminate 3. The laminate numbers have been assigned in descending order of filament dominance. It is interesting to note the relatively high value of $\bar{\alpha}_f$ associated with Laminate 3 compared to Laminates 1 and 2. Because of the higher value of the shape parameter, the amount of shift of the 95 percent survivability line down from the characteristic life line is much less compared to Laminates 1 and 2 in Figs. 5 through 7.

To demonstrate the basic difference in *S-N* curve behavior for constant amplitude T-T loading, the 95 percent survivability lines for the three laminates under consideration are shown in Fig. 8. Graphite fibers are essentially fatigue insensitive. Thus, filament-dominated laminates tend to have a flatter *S-N* curve.

Pooled Weibull parameters along with the *S-N* curve parameters are shown in Table 3 for all laminates and loading conditions under consideration. Note that none of the values for $\bar{\alpha}_f$ have been corrected for the bias of MLE. For these pooled sample sizes, the bias correction is very small.

All stress ranges in Figs. 5 through 9 are normalized by the characteristic static strength, $\bar{S}_0$, obtained by fitting the static strength data to a two-parameter Weibull distribution.

Conclusions

A procedure has been outlined for characterizing the *S-N* behavior of composite laminates with some degree of statistical reliability. The data pooling procedure used in conjunction with the data reduction scheme offers the advantage of a large data base for determining a fatigue shape parameter without requiring large sample sizes at each stress range considered. Use of censoring techniques provides data reduction procedures without requiring all specimens to produce fatigue failures. Data presented show the proposed characterization scheme to be promising.

If the proposed procedure is to be utilized, the fatigue experiments should be planned accordingly. In particular, it would be desirable to use

[4] Union Carbide's T300 graphite fibers in Narmco's 5208 epoxy resin system.

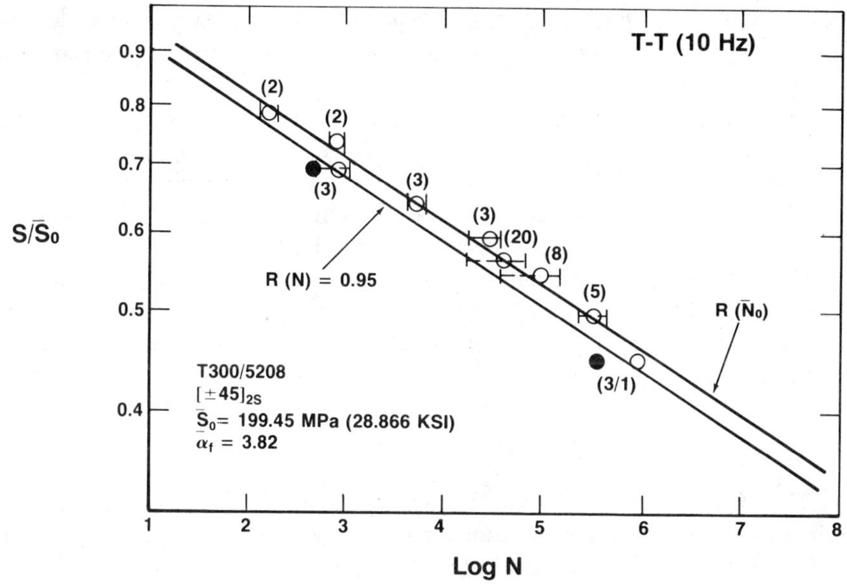

FIG. 8—*Tension-tension* S-N *curves for Laminate 3.*

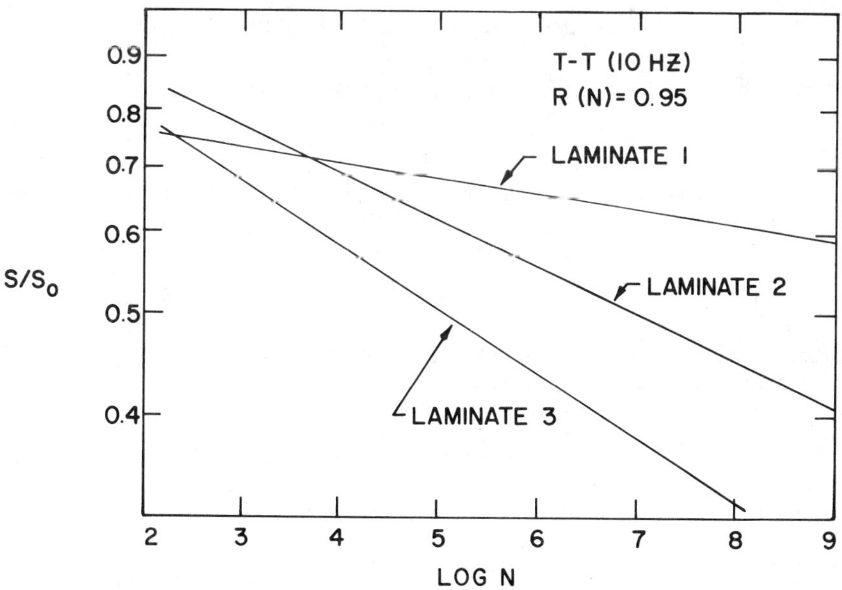

FIG. 9—*Comparison of 95 percent survivability* S-N *curves for tension-tension loading.*

the same sample size for all stress ranges tested. If 60 samples are desired for determining a pooled shape parameter, ten replicates at six different stress ranges is preferable to 20 replicates at three different stress ranges. In the former case, six data points are available for determining the characteristic *S-N* curve, rather than three as would be provided by the latter case.

In the data reduction procedure, use of MLE is strongly recommended for determining Weibull parameters. Confidence intervals and other statistical tools based on MLE have been well established. Furthermore, the data pooling technique for censored data is based on MLE in conjunction with the concept of "progressive censoring."

It should be also noted that the procedure outlined can be easily revised to include *S-N* relationships other than the power law described by Eq 1.

Acknowledgments

The author wishes to acknowledge Professor W. Park of Wright State University for helpful suggestions concerning statistical analysis and G. Sendeckyj of the Air Force Flight Dynamics Laboratory for useful discussions concerning "strength degradation" and "wearout" models.

APPENDIX

For simplicity, consider the case of constant amplitude tension-tension loading. If one assumes a crack growth model as the basis for a "strength degradation" model, the governing equation for residual strength takes the form

$$\frac{dF}{dn} = -\frac{AS^b}{c}F^{(1-c)} b, c > 0 \tag{30}$$

where F is the strength after n cycles and A, b, and c are constants. Thus, there are three parameters that characterize residual strength degradation in this model. For the assumption that the flaw growth is driven by a square root singularity at the crack tip [*1*]

$$b = c + 2 \tag{31}$$

and the number of parameters is reduced to two.

Separating variables in Eq 30 yields

$$\int_{F_0}^{F} y^{(1-c)}dy = -AS^b \int_0^n dx \tag{32}$$

Performing the integration leads to the residual strength equation

$$F_0^c = F^c(n) + AS^b n \tag{33}$$

where F_0 is the static strength. Let us assume that the static strength is described by the two-parameter Weibull distribution

$$R(F_0) = \exp\left[-\left(\frac{F_0}{S_0}\right)^{\alpha_0}\right] \tag{34}$$

where S_0 is the location parameter of the static strength distribution and α_0 is the static strength shape parameter. Substituting Eq 33 into Eq 34 yields

$$R(F) = \exp\left[-\left(\frac{F^c + AS^b n}{S_0{}^c}\right)^{\alpha_0/c}\right] \tag{35}$$

Denoting the number of cycles to fatigue failure by N_f, it is assumed that failure actually occurs when the stress reaches its maximum value during the fatigue cycle. In terms of residual strength, this simply states that the residual strength is reduced to S one cycle prior to failure, as failure occurs on the next cycle when loaded to S. Stated mathematically

$$F(n) = S, \qquad n = N_f - 1 = N \tag{36}$$

and Eq 33 becomes

$$F_0{}^c = S^c + AS^b N \tag{37}$$

The case $N = 0$ corresponds to a one-cycle failure and Eq 37 becomes

$$S = F_0 \tag{38}$$

that simply implies that a first cycle failure is a special case where the maximum stress in the fatigue cycle corresponds to the static strength.

Using Eq 37 in conjunction with Eq 34 leads to the following reliability function for time-to-failure

$$R(N) = \exp\left[-\left(\frac{N + A^{-1}S^{c-b}}{A^{-1}S^{-b}S_0{}^c}\right)^{\alpha_f}\right] \tag{39}$$

where α_f is the fatigue shape parameter

$$\alpha_f = \alpha_0/c \tag{40}$$

Equation 39 can be written in the form of a three-parameter Weibull distribution

$$R(N) = \exp\left[-\left(\frac{N - L_0}{N_0 - L_0}\right)^{\alpha_f}\right] \tag{41}$$

where

$$N_0 = S_0{}^c A^{-1} S^{-b}\left[1 - \left(\frac{S}{S_0}\right)^c\right] \tag{42}$$

$$L_0 = -A^{-1}S^{c-b} \tag{43}$$

The three-parameter Weibull distribution is simply a shifted two-parameter Weibull distribution. It should be noted that L_0 is negative as A and S are positive numbers.

In the present context, this parameter does not, however, have the connotation of a minimum life. It is, in fact, a statement of the probability of a first cycle failure. In particular, if $N = 0$ Eq 41 reduces to the static distribution, Eq 34.

Writing Eq 42 in the form

$$CS^b N_0 = \left[1 - \left(\frac{S}{S_0}\right)^c\right] \tag{44}$$

with $C = AS_0^c$, yields the basic form of the *S-N* curve that is different than Eq 1. It should be noted, however, that the parameter, *c*, is simply a ratio of the static strength parameter to the fatigue shape parameter. Experience has shown this number to be typically greater than 10 for composites. In addition, the stress levels, *S*, of concern are considerably less than the static strength location parameter, S_0. Thus

$$\left(\frac{S}{S_0}\right)^c \ll 1 \tag{45}$$

and Eq 44 becomes

$$CS^b N_0 = 1 \tag{46}$$

which is of the same form as Eq 7. Thus, the "wearout" model approach yields an *S-N* curve, for stress levels of interest, that is of the same form as assumed for the proposed data reduction procedure.

It should be noted that the "strength degradation" model proposed by Yang and Liu [*3*] is not based on a flaw growth law. Instead, they assume

$$AS^b = f(S, \omega r) \tag{47}$$

where ω is the frequency and *r* the stress ratio. They are forced, however, to assume the shape of the *S-N* curve in order to determine *f*. The relationship assumed was of the same form as Eq 46. Thus, from a mathematical standpoint, their model is the same as the crack growth or "wearout" model.

Consider Eq 39 in the form

$$R(N) = \exp\left\{-\left[\frac{N}{N_0} + \left(\frac{S}{S_0}\right)^c\right]^{\alpha_f}\right\} \tag{48}$$

where

$$N_0 = A^{-1} S_0^c S^{-b} \tag{49}$$

For high cycle fatigue, the second term in the exponential is negligible and $R(N)$ can be approximated by the two-parameter Weibull distribution

$$R(N) = \exp\left[-\left(\frac{N}{N_0}\right)^{\alpha_f}\right] \tag{50}$$

Thus, the data reduction procedure proposed measures parameters that are consistent with "wearout" models or "strength degradation" models. Such models must be modified from the details presented here if more complex loadings other than T-T are to be considered. Competing failure modes that may interact are a problem in the case of T-C, for example.

References

[1] Halpin, J. C., Jerina, K. L., and Johnson, T. A. in *Analysis of Test Methods for High Modulus Fibers and Composites, ASTM STP 521,* American Society for Testing and Materials, 1973, pp. 5-64.

[2] Hahn, H. J. and Kim, R. Y., *Journal of Composite Materials,* Vol. 9, No. 3, July 1975, pp. 297-311.

[3] Yang, J. N. and Liu, M. D., *Journal of Composite Materials,* Vol. 11, No. 3, April 1977, pp. 176-203.

[4] Hahn, H. J. and Kim, R. Y., *Journal of Composite Materials,* Vol. 10, No. 2, April 1976, pp. 156-180.

[5] Osgood, C. C., *Fatigue Design,* Wiley-Interscience, New York, 1970.

[6] Mann, N. R., Schafer, R. E., and Singpurwalla, N. D., *Methods for Statistical Analysis of Reliability and Life Data,* Wiley, New York, 1974.

[7] Thoman, D. R. and Bain, L. J., *Technometrics,* Vol. 11, No. 4, Nov. 1969, pp. 805-815.

[8] Wolff, R. V. and Lemon, G. H., *Reliability Prediction for Composite Joints—Bonded and Bolted,* AFML-TR-74-197, March 1976.

[9] Thoman, D. R., Bain, L. J., and Antle, C. E., *Technometrics,* Vol. 11, No. 3, Aug. 1969, pp. 445-460.

[10] Ryder, J. T. and Walker, E. K. in *Fatigue of Filamentary Composite Materials, ASTM STP 636,* American Society for Testing and Materials, 1977, pp. 3-26.

[11] Ryder, J. T. and Walker, E. K., "Ascertainment of the Effect of Compressive Loading on the Fatigue Lifetime of Graphite Epoxy Laminates for Structural Applications," Air Force Technical Report, AFML-TR-76-241, Dec. 1976.

[12] Awerbuch, Jonathan and Hahn, H. T. in *Fatigue of Filamentary Composite Materials, ASTM STP 636,* K. L. Reifanider and K. N. Lauraitis, Eds., American Society for Testing and Materials, 1977, pp. 248-266.

[13] Yang, J. N. and Jones, D. L., *Journal of Composite Materials,* Vol. 12, No. 4, Oct. 1978, pp. 371-389.

Assa Rotem[1] *and H. G. Nelson*[2]

Fatigue Behavior of Graphite-Epoxy Laminates at Elevated Temperatures

REFERENCE: Rotem, Assa and Nelson, H. G., **"Fatigue Behavior of Graphite-Epoxy Laminates at Elevated Temperatures,"** *Fatigue of Fibrous Composite Materials, ASTM STP 723,* American Society for Testing and Materials, 1981, pp. 152–173.

ABSTRACT: The fatigue behavior of multidirectional graphite-epoxy laminates is examined and analyzed in terms of the single lamina behavior. The nonlinear visco-elastic moduli of the epoxy matrix are changed with temperature and thus the characteristics of the laminae are altered. As a result, the static and fatigue strengths of a laminate are changed due to the redistribution of the stress field in the laminate and the weakening effect of temperature on the matrix. The fatigue behavior of a single lamina is characterized by its static strength and its "fatigue function" that expresses the degradation in the strength of the lamina due to cyclic loading. The effect of temperature is introduced through the use of "shifting factors" for both the static strength and the "fatigue function." The laminate strength is predicted by considering the cyclic stress field in each lamina, the interlaminar stresses, and the experimentally determined shifting factors. The first failure of a lamina in a laminate is examined first in terms of stress redistribution and second in terms of total failure and final laminate fracture. The analytically determined results are compared with the actual fatigue behavior of many T300/5208 graphite-epoxy composite laminates. The experimental results are shown to be in good agreement with the theoretical predictions. The temperature "shifting factors" enable one to predict long-term behavior at some temperature from short-time testing at elevated temperatures.

KEY WORDS: composite materials, graphite-epoxy, static strength, fatigue (materials), temperature dependence, shifting functions, failure criterion, life prediction

Nomenclature

a_i temperature shifting factor
b constant
E Young's modulus
f fatigue function

[1] Senior Fellow, National Research Council, on sabbatical leave from the Technion-Israel Institute of Technology, Haifa, Israel.
[2] Chief, Materials Science and Applications Office, Ames Research Center, National Aeronautics and Space Administration, Moffett Field, Calif. 94035.

G shear modulus
k_{ij} expressions
$\overline{M}$ moment per unit length of laminate
$\bar{N}$ force per unit length of laminate
N number of cycles to failure
n cycles per unit time
Q_{ij} elastic constants of lamina
R stress ratio
T temperature
ν Poisson's ratio
θ angle
μ, η expressions
σ normal stress
τ shear stress

Subscripts

A fiber direction
d interlaminar shear
T transverse to fiber direction
o reference level
x x direction
y y direction
z interlaminar normal tension
m matrix
i free index
τ shear

Left subscripts

l laminate
p lamina

Superscripts

c cyclic load
s static failure
u fatigue failure

The fatigue failure of materials is a cornerstone of today's advanced design. Every text book on design considers strength in terms of a fatigue criterion. Fatigue theories have been established based on a concept of a single crack initiating and propagating to final failure. This phenomenon has led in recent years to an improved formulation and use of fracture mechanics to predict fatigue behavior. The minimum fatigue life of a component can be calculated by considering the crack propagation due to cyclic loading up to a critical crack length at which catastrophic failure occurs. These concepts are broadly used today to describe fatigue failure of isotropic materials.

Unfortunately, the single-crack-to-failure concept is useless when applied to fibrous laminated composite materials since a single crack seldom leads to total failure of the composite component. Instead, total failure is usually the result of an accumulation of cracks. Here, strength versus number of cycles to failure (*S-N* curve) is a more suitable description of fatigue behavior. The fatigue strength criteria for failure of composite materials are much more complicated than those of isotropic materials because they must involve additional variables, such as fiber orientation and methods of construction and fabrication.

In order to simplify analysis, fatigue failure criteria can be formulated that are extensions of static failure criteria to fatigue conditions [*1-3*][3]. A static failure criterion is, in general, a combination of the stress components in the lamina and some parameters that must be determined experimentally. The stress components in each lamina are determined by the loading state imposed on the laminate and can be found by using one of the lamination theories for composites [*4*]. A fatigue failure criterion must also contain the cyclic stress components and experimentally determined parameters known as fatigue functions. These functions, which express the degradation in the strength of a lamina as influenced by the number of cycles, are also affected by external factors, such as temperature, frequency of cycling, pattern of loading, and environment. Since each of these factors has some influence on the fatigue functions, it may be possible to find some correlation between these factors. Such a correlation could then serve to extend the range of applicability of the fatigue functions. Specifically, the possibility of extending time (number of cycles) by variations in temperature is of particular interest. The present paper is concerned with this correlation—the influence of temperature on the fatigue function of a fatigue failure criterion.

A Fatigue Failure Criterion

Consider a lamina of a composite material consisting of unidirectional fibers in a continuous matrix. Failure of the lamina under tension-tension fatigue loading can occur by either of two modes: (1) fiber failure or (2) matrix failure. Thus, lamina failure can be expressed either in terms of *fiber* failure [*3*]

$$ {}_{p}\sigma_A{}^{c} \geq \sigma_A{}^{u} = \sigma_A{}^{s}(T) f_A(R, N, n, T) \qquad (1) $$

[3] The italic numbers in brackets refer to the list of references appended to this paper.

or *matrix* failure [*3*]

$$\left(\frac{{}_p\sigma_A{}^c}{\sigma_m{}^u}\,\frac{E_m}{E_A}\right)^2 + \left(\frac{{}_p\sigma_T{}^c}{\sigma_T{}^u}\right)^2 + \left(\frac{{}_p\tau^c}{\tau^u}\right)^2 \geq 1 \tag{2}$$

The cycling loads on the laminate, $\bar{N}^c$ and $\overline{M}^c$, can be decomposed by using lamination theory to principal stresses in each lamina. Failure of the lamina will occur at the lowest values of $\bar{N}^c$ and $\overline{M}^c$ as predicted from Eq 1 or 2.

The first term of Eq 2 represents the tension strength in the matrix along the fiber direction, the second term represents the stress transverse to the fibers, and the third term is the shear stress. For high-performance composite materials, the tension stress in the matrix (in the fiber direction) will be very small, and thus the first term of Eq 2 can be omitted. Therefore, the failure criterion for matrix failure becomes

$$\left(\frac{{}_p\sigma_T{}^c}{\sigma_T{}^u}\right)^2 + \left(\frac{{}_p\tau^c}{\tau^u}\right)^2 \geq 1 \tag{3}$$

The critical transverse and shear fatigue strength parameters can be expressed by the relationships [*3*]

$$\left.\begin{aligned} \sigma_T{}^u &= \sigma_T{}^s f_T(R, N, n, T) \\ \tau^u &= \tau^s f_\tau(R, N, n, T) \end{aligned}\right\} \tag{4}$$

Equations 1 and 4 are the static strength parameters multiplied by the fatigue functions to extend the static failure criterion to the fatigue domain. These fatigue functions are also determined experimentally and are influenced by the temperature, speed of cycling loads, and mean stress level.

Interlaminar failure can also occur in a laminate and is in fact a matrix failure. The criterion for such a failure would be similar to Eq 3 but the acting stresses would be interlaminar tension and interlaminar shear [*3*]

$$\left(\frac{{}_{p'}\sigma_z{}^c}{\sigma_z{}^u}\right)^2 + \left(\frac{{}_{p'}\tau_d{}^c}{\tau_d{}^u}\right)^2 \geq 1 \tag{5}$$

Again, the critical fatigue strength parameters can be expressed by the relationships [*3*]

$$\left.\begin{aligned} \sigma_z{}^u &= \sigma_z{}^s f_z(R, N, n, T) \\ \tau_d{}^u &= \tau_d{}^s f_d(R, N, n, T) \end{aligned}\right\} \tag{6}$$

where f_z and f_d are the interlaminar fatigue functions.

The three failure criteria Eqs 1, 3, and 5 relate to the three possible modes of failure in a laminate: fiber failure, matrix failure, and interlaminar failure. Failure can, of course, occur in each mode separately or in all three modes simultaneously. Optimum design would be the latter situation. The conditions for failure of any laminate can be examined by resolving the applied cyclic external loads on the laminate into the principal stresses acting in the laminae and the interlaminar stresses between them. These stresses are then compared with the critical stresses for each failure mode by a substitution in the relevant failure criterion.

The critical fatigue failure stresses are found by using Eqs 1, 4, and 6, which are the result of the influence of the fatigue functions on the static strength. As we have seen, the fatigue functions are dependent on a number of variables. Assuming all variables are kept constant, with the exception of the number of applied cycles, N, the five fatigue functions (f_i with $i = A, T, \tau, z, d$) can be determined experimentally from the S-N curves (applied cyclic stress versus number of cycles to failure). In this case, the fatigue functions are determined by the ratio of the fatigue strength to the static strength

$$f_i(N) = \frac{\sigma_i^u}{\sigma_i^s} \qquad i = A, T, \tau, z, d \tag{7}$$

If, in addition to the number of cycles, the temperature is changed, there will be, in general, some influence on the static strength and the fatigue functions. A slightly different set of fatigue functions can be determined experimentally from the S-N curves at that temperature. The temperature dependence of the fatigue functions can then be expressed as

$$f_i(N, T) = \frac{\sigma_i^u(T)}{\sigma_i^s(T)} \tag{8}$$

If the change in the fatigue function with temperature is continuous, it can be expressed by a reference fatigue function at some temperature (T_o) multiplied by a scaling function. Thus, Eq 8 could be written

$$f_i[N, T_o, a_i(T)] = \frac{\sigma_i^u(T)}{\sigma_i^s(T)} \tag{9}$$

For each fatigue function there is the respective scaling function or shifting factor, which shifts the reference curve to the desired temperature. This general form of the fatigue function reduces the number of required functions to five (Eq 7) for the three modes (Eqs 1, 3, and 5) of failure. The exact form of the shifting functions is still unknown and can only be experimentally determined.

Most polymeric materials exhibit a change in their mechanical behavior due to a temperature change [5]. The dependence of the stress-strain relationship on temperature is very well established for thermorheologically simple materials [5], where a general shifting factor, WLF, exists. However, the temperature dependence of the failure strength is not as well understood. Some work on unfilled polymers [6] has shown that the failure strength-temperature relationship is somewhat similar to the WLF relationship. Therefore, for a composite laminate having a polymeric material as the matrix, it can be assumed that the temperature may affect the fatigue functions and the static strengths. The influence of the temperature on the static strength can be represented by another scaling or shifting function. For each of the principal static strengths of the laminate, we have

$$1 - a_i^s = \frac{\sigma_i^s(T)}{\sigma_i^s(T_o)} \tag{10}$$

Using Eq 9, the influence of temperature on the fatigue function can be written

$$f_i[N, T_o, a_i(T), a_i^s] = \frac{\sigma_i^u(T)}{\sigma_i^s(T_o)} \tag{11}$$

Rearranging Eq 11 would give an expression for shifting the S-N curve

$$\sigma_i^u(T) = \sigma_i^s(T_o) f_i[N, T_o, a_i(T), a_i^s] \tag{12}$$

The graphic interpretation of Eq 12 showing a shifting of the S-N curve is illustrated in Fig. 1, where Eq 12 has the form

$$\sigma_i^u(T) = \sigma_i^s(T_o)\left[a_i(T)\left(\frac{1}{a_i(T)} - b_i \log N\right) - a_i^s\right]$$

The vertical shift, due to the change in the static strength with temperature, is given by a_i^s and the rotation of the S-N curve, which is due to the change in fatigue function with temperature, is given by a_i. Once the functions a_i^s and a_i are determined experimentally for a given set of laminae, one can construct a fatigue failure surface—a three-dimensional surface formed by stress amplitude, number of cycles to failure, and temperature—for any of the five principal stresses in the laminae. Additionally, the shifting functions can be used to predict the long-term fatigue behavior of a composite at low temperatures from the short-term behavior at elevated temperatures.

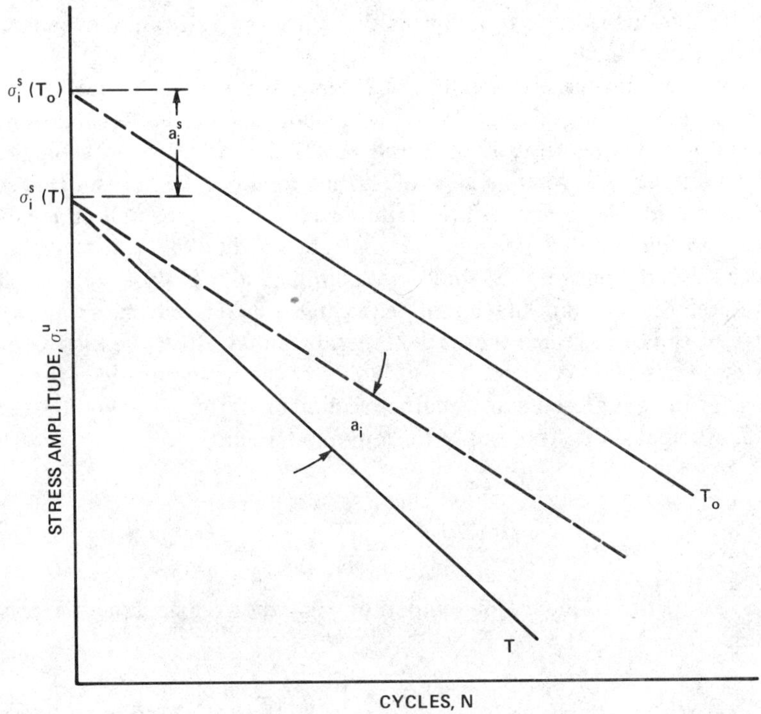

FIG. 1—*Shifting of an* S-N *curve to reference temperature.*

Materials and Procedures

An extensive experimental program was developed to analyze the fatigue behavior of a laminated composite material and to verify and modify the proposed theoretical considerations. The material selected for this study was a graphite-epoxy system of Union Carbide T300 fibers in a Narmco 5208 matrix. This material was chosen because of its extensive use in NASA's Aircraft Energy Efficiency (ACEE) Program [*7*]. The material was fabricated from prepreg tapes by Lockheed Missiles and Space Co., Sunnyvale, Calif. Unidirectional laminate panels, angle-ply laminate panels, and symmetrically balanced laminate panels were fabricated from the prepreg tapes. Each panel consisted of eight layers with an overall average fiber volume fraction of 65 percent. Test specimens having dimensions shown in Fig. 2 were cut from the panels after the end tabs (some made of fiberglass and some of aluminum) were glued to them. Three types of specimens were used: (1) unidirectional, designated $[0]_8$; (2) angle-ply, designated $[\pm 15]_{2s}$, $[\pm 30]_{2s}$, $[\pm 45]_{2s}$, $[\pm 60]_{2s}$, and $[\pm 75]_{2s}$; and (3) symmetrically balanced, designated $[0, \pm 15, 0]_s$, $[0, \pm 30, 0]_s$, $[0, \pm 45, 0]_s$, $[0, \pm 60, 0]_s$,

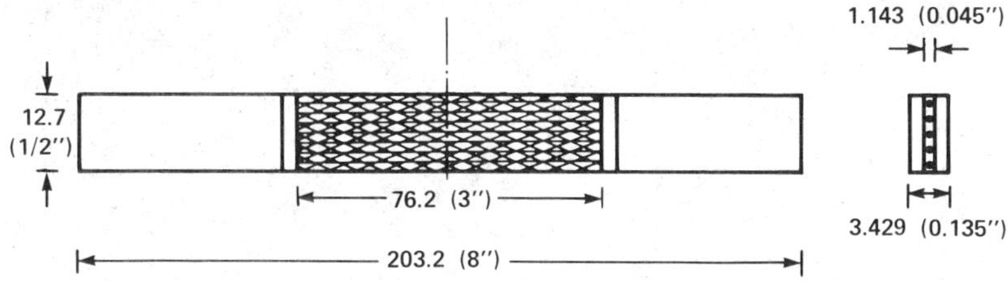

FIG. 2—*Test specimen with end tabs.*

$[0, \pm 75, 0]_s$, $[0, \pm 90, 0]_s$, and $[0, \pm 30, 90]_s$. The unidirectional and angle-ply specimens were used to define the stress-strain relationships of the material as well as the principal strength parameters, the fatigue functions, and the shifting factors. The symmetrically balanced specimens were used to observe the behavior of a typical composite laminate under fatigue conditions and to correlate the behavior with predictions based on the analysis presented here.

All specimens were tested under tension-tension fatigue using an electrohydraulic, servocontrolled test system at R (ratio of minimum applied load to maximum applied load) equal to 0.1, and at frequencies between 10 and 30 cycles/second. Tests were conducted under conditions of load control while monitoring both load and strain during each test. S-N curves were developed for each specimen set at 25°C (77°F), 74°C (165°F), and 114°C (237°F). Specially designed grips facilitated specimen alignment and load transfer. All specimens were equilibrated at temperature for 1 h prior to testing. Strain was measured by means of an extensometer and in some cases by strain gages.

Results and Discussion

Stress-Strain Behavior

The stress-strain behavior was influenced by the temperature, speed of cycling, and level of stress, as would have been expected due to the viscoelastic nature of the matrix material. The static shear stress-shear strain relationships at two temperatures are shown in Fig. 3. These curves are the result of tests conducted on ±45-deg angle-ply specimens using strain-gage rosettes. The tests were performed at a strain rate of 1.5×10^{-4} in./in./s. As can be seen, the modulus clearly is stress dependent. The secant shear modulus as a function of the stress level is shown in Fig. 4. Since the loading levels that cause fatigue failure are quite high, the modulus dependence has to be considered.

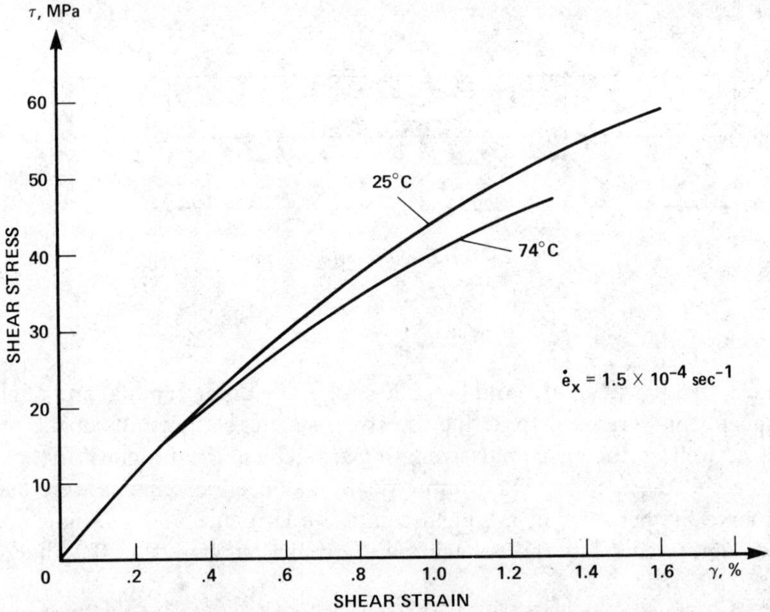

FIG. 3—*Axial shear stress-strain curve.*

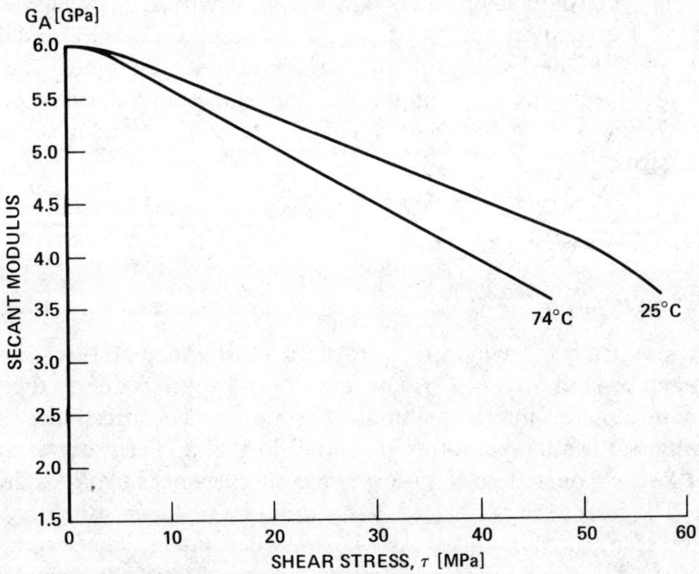

FIG. 4—*Secant shear modulus dependence on stress level.*

The principal stresses in each lamina can be computed easily by

$$\left.\begin{aligned} {}_p\sigma_A{}^c &= {}_pk_{xx}\,{}_p\sigma_x{}^c \\ {}_p\sigma_T{}^c &= {}_pk_{yy}\,{}_p\sigma_x{}^c \\ {}_p\tau^c &= {}_pk_{xy}\,{}_p\tau_x{}^c \end{aligned}\right\} \qquad (13)$$

where ${}_p\sigma_x{}^c$ is the axial stress in the lamina. The axial stress (${}_p\sigma_x{}^c$) is related to the average stress on the laminate (${}_l\sigma_x{}^c$) by

$${}_p\sigma_x{}^c = \frac{{}_l\sigma_x{}^c}{{}_lE_{xx}}\left({}_pQ_{xx} - {}_l\nu_{xy\,p}Q_{xy}\right) \qquad (14)$$

The coefficients ${}_pk_{ij}$ of Eq 13 are functions of the elastic moduli and of the angle, θ, where θ is the angle between fibers in the lamina and the direction of the applied load, x. The ${}_pk_{ij}$ are given by

$$\begin{aligned} {}_pk_{xx} = \frac{1}{2}\Bigg\{&\left[1 + \sec 2\theta - \frac{(\mu + \sec 2\theta)\tan^2 2\theta}{\eta + \tan^2 2\theta}\right] \\ &+ \left[1 - \csc 2\theta - \frac{(\mu - \csc 2\theta)\tan^2 2\theta}{\eta + \tan^2 2\theta}\right] \\ &\times \frac{{}_p\nu_{xy} - {}_l\nu_{xy}}{{}_pQ_{xx}/{}_pQ_{yy} - {}_p\nu_{xy}\,{}_l\nu_{xy}}\Bigg\} \end{aligned} \qquad (15a)$$

$$\begin{aligned} {}_pk_{yy} = \frac{1}{2}\Bigg\{&\left[1 - \sec 2\theta + \frac{(\mu + \sec 2\theta)\tan^2 2\theta}{\eta + \tan^2 2\theta}\right] \\ &+ \left[1 + \csc 2\theta + \frac{(\mu - \csc 2\theta)\tan^2 2\theta}{\eta + \tan^2 2\theta}\right] \\ &\times \frac{{}_p\nu_{xy} - {}_l\nu_{xy}}{{}_pQ_{xx}/{}_pQ_{yy} - {}_p\nu_{xy}\,{}_l\nu_{xy}}\Bigg\} \end{aligned} \qquad (15b)$$

$$\begin{aligned} {}_pk_{xy} = -\frac{1}{2}\Bigg\{&\frac{(\mu + \sec 2\theta)\tan 2\theta}{\eta + \tan^2 2\theta} \\ &- \left[\frac{(\mu - \csc 2\theta)\tan 2\theta}{\eta + \tan^2 2\theta}\right] \times \frac{{}_p\nu_{xy} - {}_l\nu_{xy}}{{}_pQ_{xx}/{}_pQ_{yy} - {}_p\nu_{xy}\,{}_l\nu_{xy}}\Bigg\} \end{aligned} \qquad (15c)$$

$$\mu = \frac{1 - E_A/E_T}{1 + 2\nu_A + E_A/E_T} \tag{16a}$$

$$\eta = \frac{E_A/G_A}{1 + 2\nu_A + E_A/E_T} \tag{16b}$$

The measured moduli, at different temperatures and stress levels, are given in Table 1. The k_{ij} coefficients for the angle ply, calculated using Eqs 15 and 16, are given in Table 2. For this particular material, the results predict that the ±45-deg angle ply has only in-plane shear, and so it is suitable to measure the respective fatigue function.

The dependence of the moduli (in fact, only the shear modulus) on the stress level was verified by measuring the complex modulus, of a ±30-deg angle-ply laminate. This measuring excluded the time influence on the measured modulus. The results, shown in Fig. 5, are in close agreement with the data in Fig. 4, from which it is concluded that the viscoelastic shear behavior is nonlinear.

Static Strength Behavior

After determining the principal stresses, the static strength can be calculated by using the failure criteria Eqs 1, 3, and 5 and the strength parameters given in Table 3. The static strength parameters were calculated from test results of angle-ply laminates shown in Fig. 6. All laminates failed by accumulation of in-plane cracks, as was observed during the tests, except the ±15-deg laminate; it failed by interlaminar cracking.

The failure of the symmetric balanced laminate is more complicated because failure of some laminae is not a total failure. This is shown in Fig. 7 for $[0, \pm 60, 0]_s$ laminate and for $[0, \pm 30, 0]_s$ laminate. In both cases, the laminae with inclined fibers crack first, but the laminate remains intact. Failure occurs when the 0-deg lamina can no longer support the applied load. The strength of the symmetrically balanced laminates $[0, \pm\theta, 0]_s$, as a

TABLE 1—*Elastic moduli.*

Modulus, GPa	Low Stress Level (LSL)		High Stress Level (HSL)	
	$T = 25°C$ (298°K)	$T = 74°C$ (347°K)	$T = 25°C$ (298°K)	$T = 75°C$ (347°K)
E_A	136	136	136	136
E_T	8.729	7.85	8.729	7.85
G_A (secant)	6.0	6.0	3.5	3.22
ν_A, nondimensional	0.36	0.355	0.32	0.32

TABLE 2—*Constituent relation of angle-ply laminate.*

$\pm\theta$-deg	15-deg		30-deg		45-deg		60-deg		75-deg	
	LSL[a]	HSL[b]	LSL[a]	HSL[b]	LSL[a]	HSL[b]	LSL[a]	HSL[b]	LSL[a]	HSL[b]
k_{yy}	−0.04571	−0.05429	−0.8794	−0.1626	0.0786	0.07664	0.51067	0.679375	0.8748	0.9453
k_{xy}	−0.05479	−0.03994	−0.2379	−0.1948	−0.50	−0.50	−0.57119	−0.47379	−0.3508	−0.2287

[a] Low stress level.
[b] High stress level.

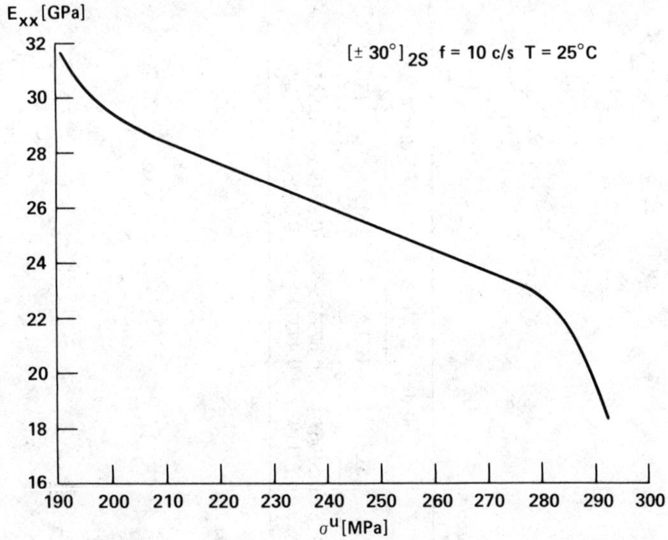

FIG. 5—*Complex modulus dependence on stress level.*

TABLE 3—*Static strength parameters.*

	Static Strength, MPa		
T, °C	25	74	114
σ_A	1510	1510	1510
τ	74	65	56
σ_T	44	40	36

function of θ is shown in Fig. 8. Although the experimental results are somewhat scattered, they fall within a narrow scatter band. For $\theta > 45$-deg, the overall strength is controlled by the 0-deg lamina; at lower θ, these laminae contribute to the strength of the laminate, and dominate it.

Fatigue Behavior

The fatigue strength dependence on the number of cycles is shown in Fig. 9 for a unidirectional laminate. The fatigue function was determined from these results and is given in Table 4. The results from the angle-ply laminates at different temperatures are given in Figs. 10 and 11. The fatigue functions, determined by using Eq 8, are given in Table 4. Using these results and Eq 11, the shifting factors, $a_i(T)$ and a_i^s, can be determined. Also notice that fatigue strength degradation does not start with the first

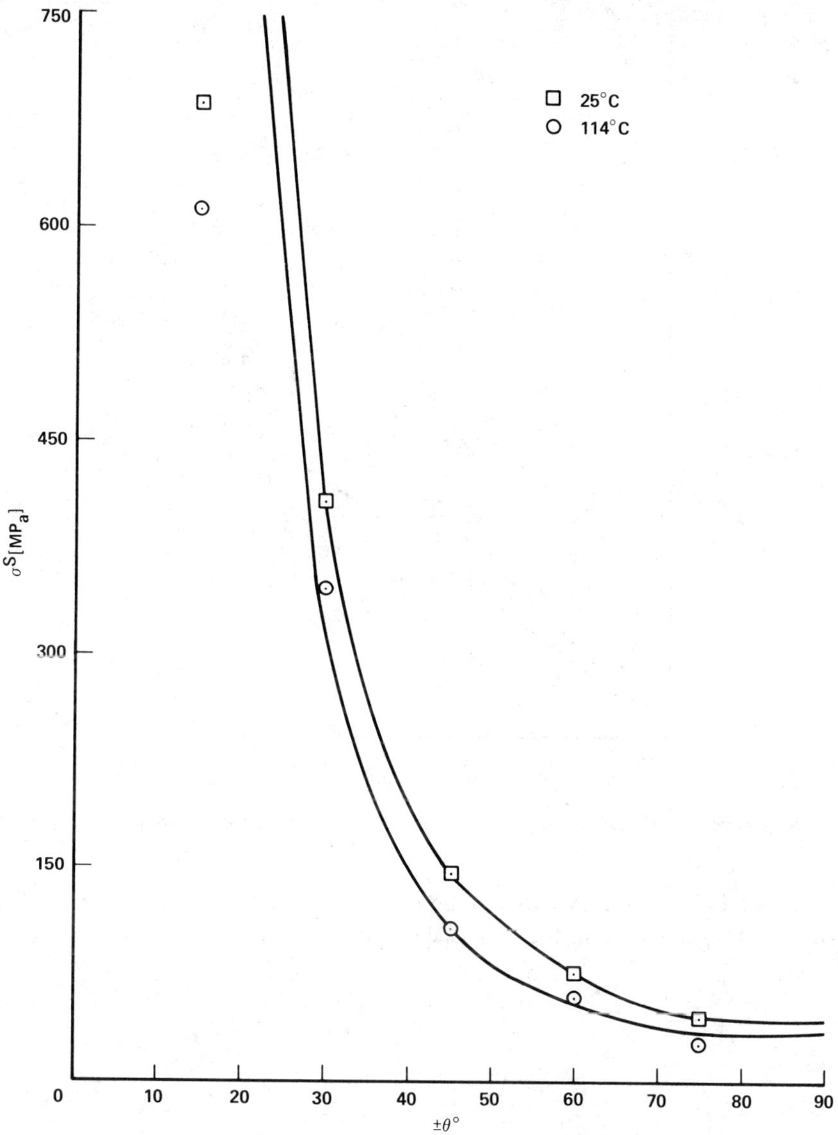

FIG. 6—*Static strength of some angle-ply laminates.*

few cycles, but only after a few hundred cycles. In fact, there is sometimes a strengthening effect with the first few cycles. In order to simplify the functions, the static strength appearing in the fatigue failure criteria will not be used. Instead, one can introduce an artificial static strength defined as the

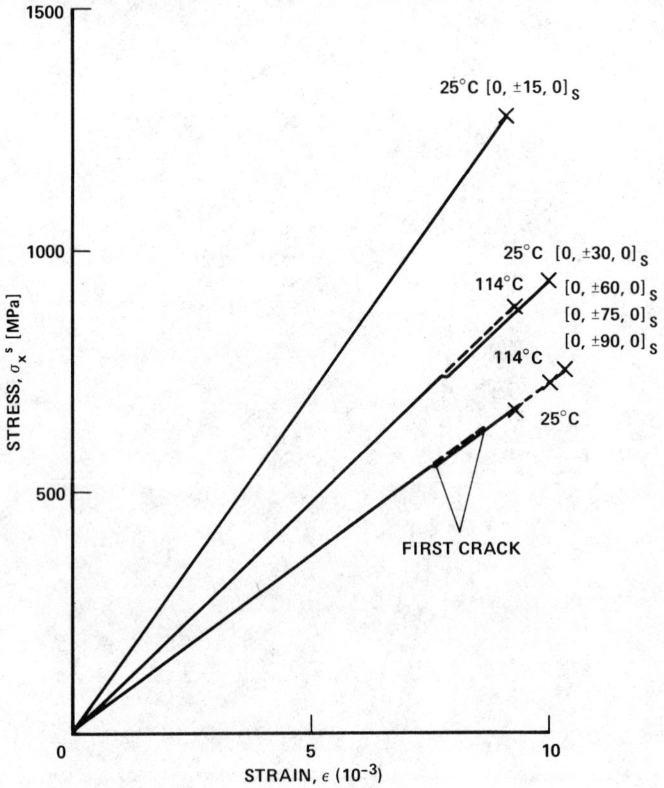

FIG. 7—*Stress-strain curves of various symmetrically balanced laminates.*

extrapolation of the *S-N* curve to the point of one cycle ($N = 1$). If $\sigma_i^{s\prime}$ is the artificial static strength, then Eq 12 could be rewritten as

$$\sigma_i^u(T) = \sigma_i^{s\prime}(T_o)f_i \tag{12}$$

and

$$f_i = a_i(T)\left(\frac{1}{a_i(T)} - b_i \log N\right) - a_i^{s\prime} \tag{17}$$

From the data in Figs. 10 and 11, we can calculate the modified static strengths, fatigue functions, and shifting factors given in Table 5. It is surprising to note that $a_i(T) = 1$, which means that the slope of the fatigue functions is not a function of the temperature, provided the modified static strengths are used. The shifting of these strengths was found to conform with

$$1 - a_i^{s'} = \left(\frac{T_o[K]}{T[\mathrm{K}]}\right)^{1/2} \tag{18}$$

The fatigue function of the ±15-deg laminate is different. This laminate fails by interlaminar cracking and is found to be (from the results shown in Fig. 10)

$$f_i = 1 - 0.100 \log N \quad \text{at} \quad T = 25°\mathrm{C} \tag{19}$$

Since this laminate fails by delamination, this function is for an interlaminar fatigue failure. The predicted failure by in-plane cracking gave a much higher value (1850 MPa) as seen in Fig. 6.

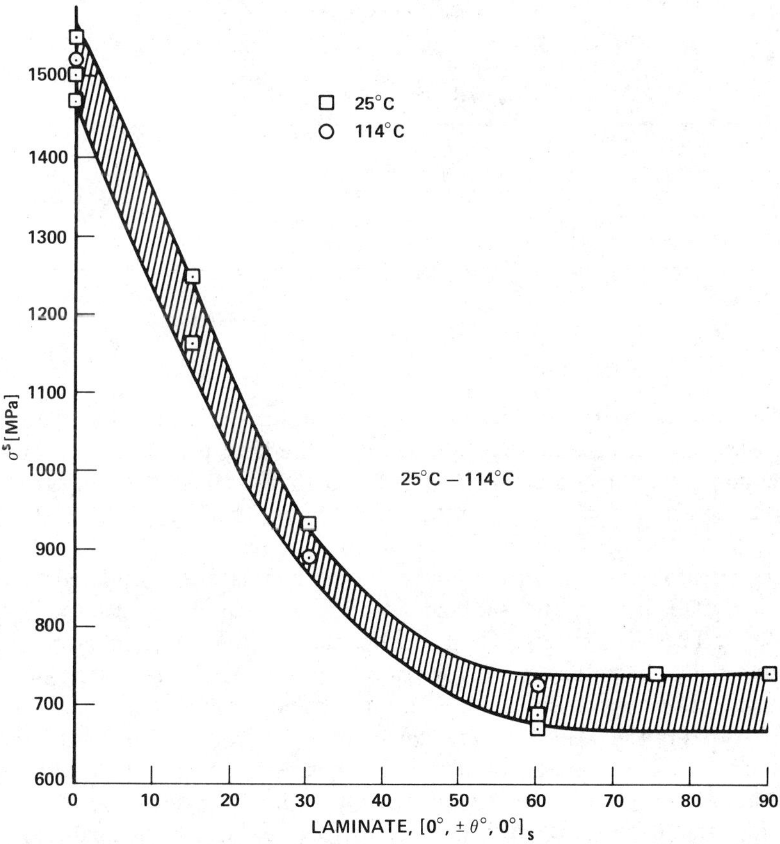

FIG. 8—*Static strength of various symmetrically balanced laminates.*

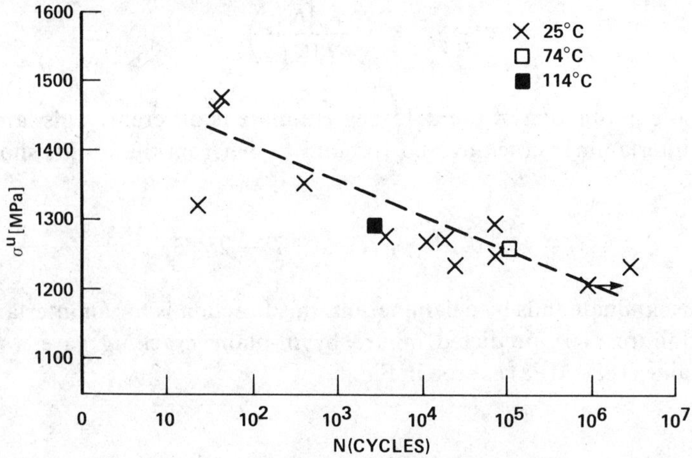

FIG. 9—S-N *curve of unidirectional lamina.*

TABLE 4—*Fatigue functions* f_i.

T, °C	25	74	114
f_A	$1 - 0.033 \log N$	$1 - 0.033 \log N$	$1 - 0.033 \log N$
f_τ	$1.1 - 0.081 \log N$	$1.15 - 0.086 \log N$	$1.3 - 0.092 \log N$
f_T	$1.12 - 0.0897 \log N$	$1.13 - 0.090 \log N$	$1.14 - 0.0904 \log N$

Fatigue Failure Predictions

The fatigue failure predictions of a structural material, such as the symmetrically balanced laminates, are based on the failure criteria, the fatigue functions, and the temperature shifting factors, as just determined. Examples considered here are the $[0, \pm 15, 0]_s$, $[0, \pm 30, 0]_s$, $[0, \pm 60, 0]_s$, $[0, \pm 75, 0]_s$, $[0, \pm 90, 0]_s$, and $[0, \pm 30, 90]_s$ laminates.

The stress condition in each lamina of the laminate is considered by the use of Eqs 13 through 16, and then substituted in the failure criteria Eqs 1, 3, and 5. The $[0, \pm 60, 0]_s$, $[0, \pm 75, 0]_s$, and $[0, \pm 90, 0]_s$ laminates failed on the inclined laminae at very low applied stress, but the 0-deg laminae continued to carry the load. Thus, the predicted fatigue life of these laminates is that of the 0-deg lamina. This is shown in Fig. 12. Predicted fatigue (dashed line) is in good agreement with the experimental results. Also note that temperature has no effect on the fatigue life as was predicted. Next, consider the $[0, \pm 15, 0]_s$ laminate. The stress field in the inclined laminae is different from a simple angle-ply as can be seen from Eq 14, but the difference is minute. Failure is predicted to occur by delamination followed by

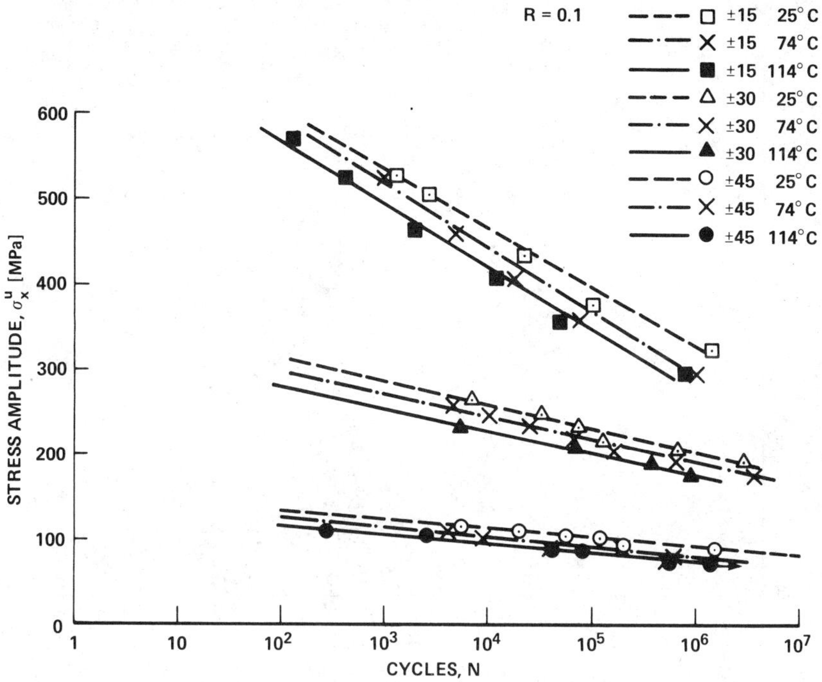

FIG. 10—S-N *curve of some angle-ply laminates.*

catastrophic failure of the 0-deg lamina, which cannot take the extra load. The experimental results and the predictions are shown in Fig. 13 for two temperatures. Also shown in this figure are the results for the $[0, \pm 30, 0]_s$ laminate. In this case, failure was predicted to occur mainly by in-plane shear. After the first cracks occur, the constraint of the 0-deg lamina is relieved and the ±30-deg laminae now act as a single angle ply. Failure by cracking of these laminae at a higher number of cycles is followed by catastrophic failure of the 0-deg laminae. But, in both cases, the viscoelastic nature of the matrix material must be considered. Since the stress amplitude on the whole laminate is kept constant, the angle-ply laminae was deformed by creep and subsequently the modulus was decreased. This is shown for illustration in Fig. 14, where the modulus change as a function of cycles of ±30-deg angle ply is plotted. As a result of creep, the stress amplitude on the angle ply is decreased and the stress amplitude on the 0-deg laminae is somewhat increased. As a consequence, the fatigue function of the laminate is less stiff than that of the angle ply. The temperature dependence of these combined fatigue functions obey the same law; they are shown in Fig. 13.

The results for the $[0, \pm 30, 90]_s$ laminate are shown in Fig. 15. The predicted failure mechanism of this laminate was the failure of 90-deg laminae

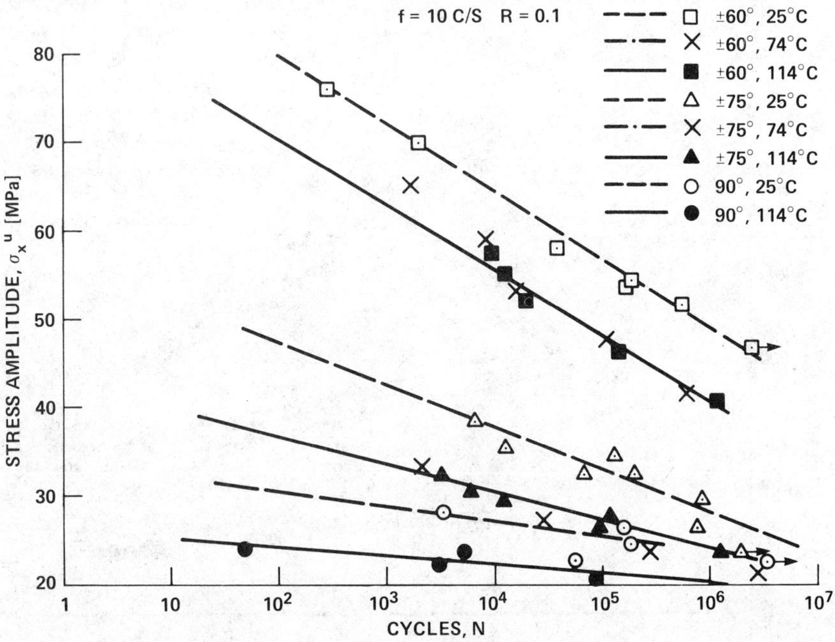

FIG. 11—S-N *curves of some angle-ply laminates.*

TABLE 5—*Modified static strength and fatigue functions.*

		f_i		
T, °C	25	25	74	114
σ_A	1510 MPa	$1 - 0.033 \log N$	$1 - 0.033 \log N$	$1 - 0.033 \log N$
τ	85 MPa	$1 - 0.0735 \log N$	$1 - 0.0735 \log N$	$1 - 0.0735 \log N$
σ_T	48 MPa	$1 - 0.080 \log N$	$1 - 0.080 \log N$	$1 - 0.080 \log N$

first. Then, at higher stress levels, the ±30-deg laminae fail due to in-plane shear, which releases the constraint of the 0-deg laminae. With this change in stress field, the ±30-deg laminae are subjected to higher stress resulting in failure by in-plane shear. Subsequently, the 0-deg laminae cannot carry the extra load, and they fail catastrophically. This mechanism of failure implies that the fatigue failure is governed by the in-plane shear cracking of the ±30-deg laminae. Since the 90-deg laminae failed first, there was an extra load on the ±30-deg laminae, which caused a faster failure, or a stiffer fatigue function. This is shown in Fig. 15. As before, the temperature effect obeys the same law and the shifting factor was found to agree with the experimental results.

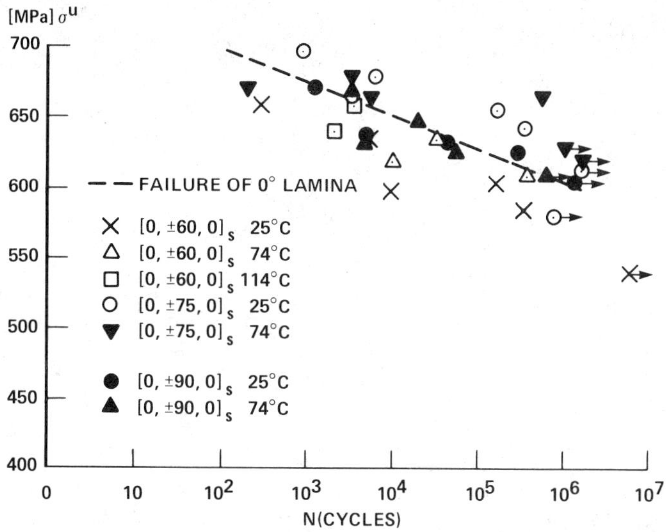

FIG. 12—S-N *curves of some symmetrically balanced laminates that fail by fiber fracture.*

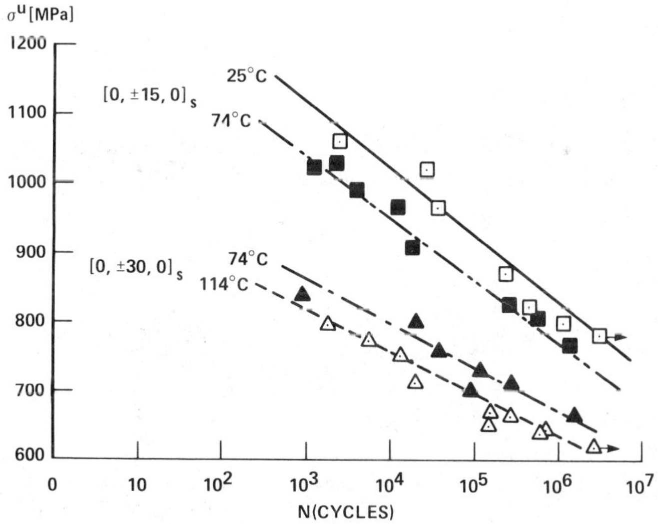

FIG. 13—S-N *curves of some symmetrically balanced laminates, shifted with temperature.*

Conclusions

The fatigue behavior of a graphite-epoxy laminated composite material, at elevated temperatures, was analyzed theoretically and experimentally. By decomposing the stress field in the laminate to five principal stresses,

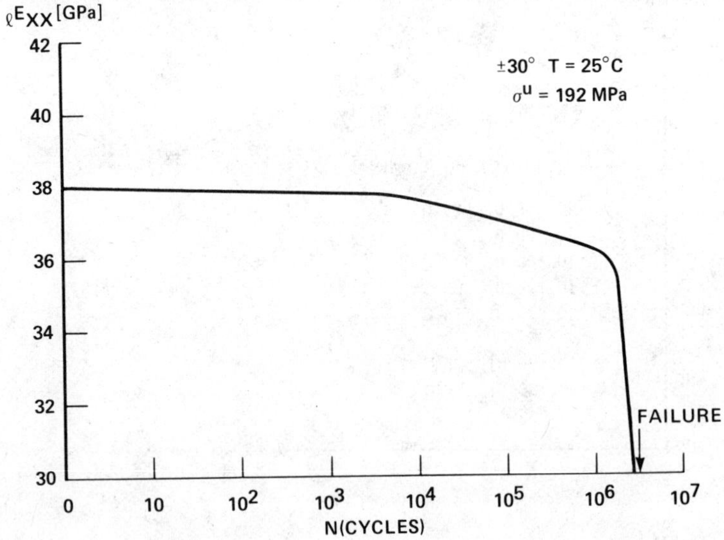

FIG. 14—*Complex modulus changes with cycling.*

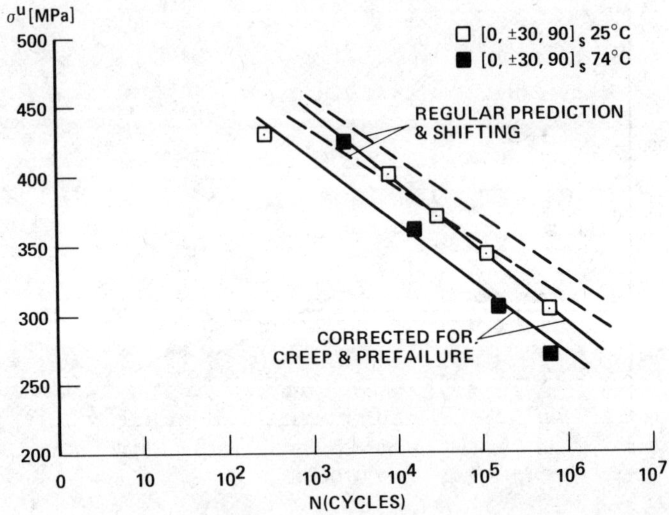

FIG. 15—S-N *curves of symmetrically balanced laminate, shifted with temperature.*

failure criteria were formulated that can be associated with fatigue functions and temperature shifting factors. The fatigue functions describe the degradation of the principal strengths with the number of cycles, and the temperature shifting factors describe the changes in the fatigue strength criteria with temperature. Thus, one can characterize the overall fatigue behavior with a single curve. Using lamination theory, expressions were formulated for finding the principal stresses in a symmetrically balanced laminae under uniaxial cyclic loading. By introducing an imaginary static strength, which is shifted due to temperature by a very simple law, the fatigue functions were found to be independent of temperature. Even though the calculation of the stress amplitudes in each lamina of the laminate is very complicated, due to the creep behavior of some laminae, the shifting of the fatigue behavior with temperature remains simple. These results may permit the prediction of long-time, low-temperature behavior from data obtained in short-time, high-temperature testing.

References

[*1*] Hashin, Z. and Rotem, A., *Journal of Composite Materials,* Vol. 7, Oct. 1973, pp. 448-464.
[*2*] Rotem, A. and Hashin, Z., *Journal,* American Institute of Aeronautics and Astronautics, Vol. 14, No. 7, July 1976, pp. 868-872.
[*3*] Rotem, A., *Journal,* American Institute of Aeronautics and Astronautics, Vol. 17, No. 3, March 1979, p. 271.
[*4*] Ashton, J. E. and Whitney, J. M., *Theory of Laminated Plates,* Technomic Publications, U.S.A., 1970.
[*5*] Ferry, J. C., *Viscoelastic Properties of Polymers,* Wiley, New York, 1961.
[*6*] Landel, R. F. in *Fracture Processes in Polymeric Solids,* B. Rosen, Ed., Interscience, New York, 1964, pp. 361-485.
[*7*] Aircraft Field Conservation Technology, Task Force Report, NASA Office of Aeronautics and Space Technology, 10 Sept. 1975.

M. S. Rosenfeld[1] *and L. W. Gause*[1]

Compression Fatigue Behavior of Graphite/Epoxy in the Presence of Stress Raisers*

REFERENCE: Rosenfeld, M. S. and Gause, L. W., **"Compression Fatigue Behavior of Graphite/Epoxy in the Presence of Stress Raisers,"** *Fatigue of Fibrous Composite Materials, ASTM STP 723,* American Society for Testing and Materials, 1981, pp. 174–196.

ABSTRACT: Graphite/epoxy composites are being applied to aircraft structures because of their demonstrated capability to reduce weight and increase life. Although composites have better fatigue properties than metals, their behavior differs significantly. Unlike metals, composites exhibit excellent tensile fatigue behavior—the constant amplitude fatigue strength at 10^7 cycles being close to static ultimate. Compression and reversed loading fatigue behavior in the presence of stress raisers, however, has not been well characterized. This program was therefore undertaken to determine the characteristics of composites under these conditions. Two stress raisers were considered; the first was an open hole and the second was damage resulting from low velocity, hard object impact.

For the open hole specimen, constant and variable amplitude tests were performed to determine the significance of compression fatigue and to investigate the failure mechanism. These tests indicated that the fatigue life under compression and reversed loading is less than for tension-tension loading and will be an important design consideration in future composite applications. The failure mechanism appeared to be local progressive failure of the matrix near the stress raiser resulting in delamination, and final failure by fiber buckling. The variable-amplitude loading results also demonstrated the unconservativeness of Miner's rule for making analytical predictions.

In order to determine the nature and significance of impact damage to generic composite structural elements, low velocity impact and residual properties tests were conducted on solid laminate specimens of 42 and 48 plies thick and on honeycomb sandwich specimens with 12-ply composite face sheets. Results of post impact properties tests indicate subvisual damage can degrade compression static and fatigue strength, although subvisual damage will not propagate under moderate (0.003) cyclic strain.

KEY WORDS: Composite materials, graphite/epoxy laminates, compression fatigue, impact damage, fatigue (materials)

*The opinions and assertions expressed in this paper are the private ones of the authors and are not to be construed as official or reflecting the views of the Department of the Navy or the Naval services at large.

[1]Aero research engineer and aerospace engineer, respectively, Aero Structures Division, Aircraft and Crew Systems Technology Directorate, Naval Air Development Center, Warminster, Pa. 18974.

The application of graphite/epoxy laminates to aircraft structures has demonstrated the capability to save weight and improve aircraft performance. The fatigue characteristics of these laminates have been determined for tension-tension loading, and they far exceed the comparable behavior of metals. However their effect on structural life under compressive and reversed loading is not well characterized. Consequently, it was decided to investigate the compression fatigue characteristics of graphite/epoxy laminates using a specimen with a centrally located open hole to determine the mechanism of failure and to obtain some preliminary data on the fatigue life characteristics in a laboratory environment. This program was purely exploratory to determine if a more extensive and elaborate program would be required.

In the initial phase of this program [*1*],[2] a limited number of tests were performed for $R = 0$, $-\infty$, and -1 loading to determine the significance of the compression loading and to investigate the failure mechanism. These tests indicated a significant life reduction for both $R = -\infty$ and $R = -1$ loading with respect to tension-tension loading with the life reduction for $R = -1$ being greater. The mechanism of failure appeared to be local progressive fatigue failure of the matrix near a stress riser thus causing fiber splitting, progressive delamination and resulting fiber buckling that then precipitates laminate failure.

A limited number of tests were also performed to assess the effect of spectrum loading on the laminate fatigue characteristics. A typical fighter spectrum was applied and represented the loading applied to the compression skin. The results indicated that the design limit load gross stress for a quasi-isotropic laminate should not exceed -193.1 MN/m^2 (-28.0 ksi) to attain the desired factored life of 12 000 flight hours for a fighter airplane. Furthermore, the results also indicated that the Miner cumulative damage rule is extremely unconservative for graphite/epoxy laminates. The results from the open hole specimen tests raised the question of the effect of fastener constraint on the laminate fatigue characteristics. This question and the effect of loading rate were then investigated and are reported herein.

During the time that this program was underway, a program to determine the effects of low velocity impact damage on graphite/epoxy solid laminates and composite-faced honeycomb sandwich skin panels was also undertaken. One of the major problems in assessing the effects of the impact damage was to establish a logical and consistent method for measuring the effect of the damage. Since the mechanism of compression fatigue failure is based upon delamination and since the damage imposed by low velocity impact is primarily delamination or debonding or both, it was felt that cyclic loading in compression would be a good way to assess the effects of impact damage. Two methods of applying the cyclic compressive loading were used and the results are discussed herein.

[2]The italic numbers in brackets refer to the list of references appended to this paper.

Procedure

Compression Fatigue Tests

Test Specimens—The specimens used in the spectrum loading phase of the program were surplus from the original program [*1*]. These specimens were fabricated from Narmco 5209 prepreg using MODMOR Type II high-strength graphite fibers. The material was laid up in panels using a $[(0/\pm45)_S]_3$ balanced layup with the 0-deg fibers on the outside. The panels were cured in an autoclave at 394 K (250°F) and 0.55 MN/m^2 (80 psi) and were not postcured. The panels were then cut into 25.4-mm (1-in.) wide specimens that were reinforced with glass fiber reinforced plastic (GFRP), as shown in Fig. 1, to provide lateral restraint against buckling. Similar specimens were also fabricated from a graphite/polysulfone laminate to obtain comparative data.

The specimens for the idle fastener tests were fabricated from Hercules AS/3501-6 prepreg. However, these specimens were made with the three outer plies interchanged so that the outer plies were $\pm$45-deg rather than 0-deg to delay delamination. The panels from which these specimens were made were cured in an autoclave at 439 K (330°F) and 0.55 MN/m^2 (80 psi). All these specimens had the GFRP restraint. One half of these specimens were made with 6.35-mm (1/4-in.) diameter hexagon head fasteners installed

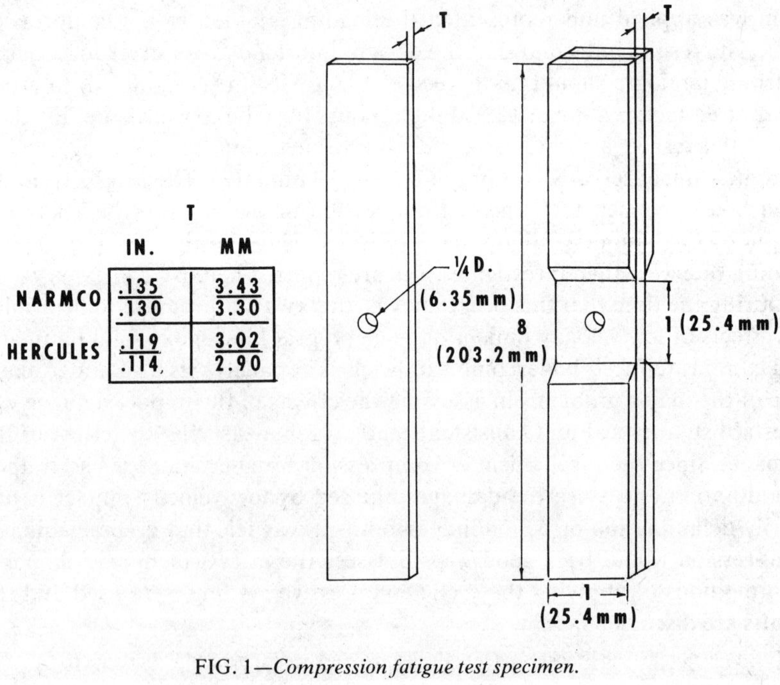

FIG. 1—*Compression fatigue test specimen.*

in the holes with washers under the heads and nuts. The remainder of the specimens were made with 6.35-mm (1/4-in.) diameter 100-deg c's'k head fasteners with a washer under the nut. All fasteners were a snug fit in the hole and were preloaded to a torque of 6.78 Nm (60 in · lb) prior to test.

Equipment—All tests were performed in an MTS closed-loop servo-hydraulic test machine equipped with "Alignomatic" self-aligning hydraulic grips. The test loads were monitored throughout by an MTS amplitude measurement unit and a Hewlett Packard oscilloscope.

Test Procedure—Prior to each test, the test machine grips were aligned, using a rigid dummy specimen. Each test specimen was then installed in the grips, with a compressive load of approximately 22 N (50 lb) applied while locking the grip jaws.

Spectrum fatigue tests were performed using the MIL-A-8866 fighter spectrum with the 1 g level flight load as the base and 7.33 g as the limit load factor. The loading was primarily in compression and represented that which would be applied to the compression skin during the specified maneuver loads. The spectrum loads were applied in a low-high sequence for an equivalent block size of 50 flight hours; 20 blocks represented the 1000-h spectrum specified in MIL-A-8866. Tests were performed at an average cyclic frequency of 1 Hz. Tests were performed to failure or runout at various values of the limit load gross area stress.

The effect of fastener constraint was investigated under constant-amplitude fatigue for completely reversed loading, $R = -1$. Tests were performed at a frequency of 5 Hz.

Impact Program

Test Specimens—The graphite/epoxy material used in this study and the material called for in current Navy flight structure designs is Hercules AS/3501-6. Two types of specimens were made; composite-faced honeycomb sandwich specimens representative of typical full depth honeycomb secondary structures, and solid laminate plates of two layup constructions representative of primary structures.

The sandwich specimens investigated were rectangular sandwich beams 76.2 mm (3 in.) wide by 355.6 mm (14 in.) long as shown in Fig. 2. Face sheets were 8-ply laminates of $[0/\pm45/0]_s$ layup. Large panels were fabricated by hand layup and autoclave cured at 585 kPa (85 psi) and 450 K (350°F) for 120 min and post cured at 478 K (400°F) for 2 h in an air circulation oven. These laminated panels were then bonded to honeycomb core to form two large honeycomb panels from which the individual test specimens were machined. The adhesive used was FM-123-5 film adhesive. The core materials were 12.7 mm (0.50 in.) thick and consisted of HRP-3/16-5.5 in the central 76.2 by 76.2-mm (3 by 3-in.) test section and AL-1/8-5052-.003-12 elsewhere. These specimens, manufactured by General Dynamics, are of the

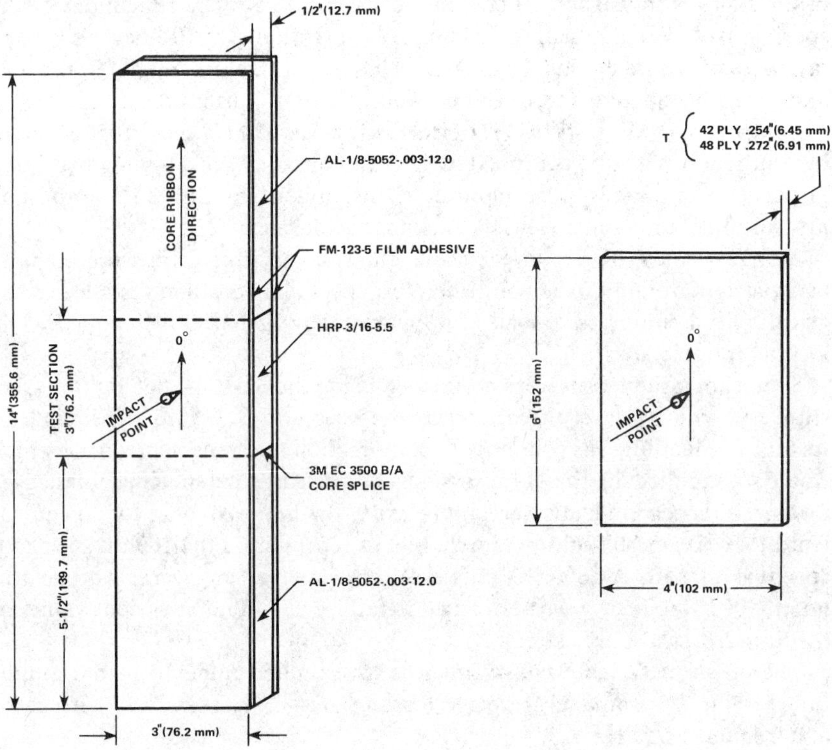

FIG. 2—*Impact test specimens.*

same construction as the sandwich specimens studied in a concurrent program [*2*].

Solid laminate plate specimens, also shown in Fig. 2, were manufactured using the same cure cycle as with the sandwich beams. Two layup constructions were studied:

Laminate A: 48 ply $[\pm 45/0_2/\pm 45/0_2/\pm 45/0/90]_{2S}$

Laminate B: 42 ply $[+45/90/-45/+22.5/-67.5/-22.5/+67.5/\pm 45/+67.5/+22.5/-67.5/-22.5/\pm 67.5/\pm 22.5/0_2/\pm 22.5]_S$

The B laminates were made from prepreg material of 0.15 mm (6 mil) ply thickness while A laminates were fabricated from the normal 0.13 mm (5.2 mil) ply thickness material. Individual test specimens, 102 mm (4 in.) wide by 152 mm (6 in.) long, were machined from a large panel of each layup sequence.

Impact Damage Procedure—Drop weight tests were performed employing variable weight steel indenters of 6.35-mm (1/4 in.) and 25.4-mm (1-in.) tip radii to simulate typical tools. The indenter could be dropped from various heights and was aligned so that it struck the specimen once at its center and was caught on rebound.

The sandwich specimens were simply supported as beams with a 304.8-mm (12-in.) span. The indenter was dropped from a height of 1.89 m (6.21 ft) to given an impact velocity of 6.1 m/s (20 ft/s). Both radii indenters were used and the indenter mass was varied to induce different damage levels, from subvisual to complete penetration of the top facesheet.

Each plate specimen was supported along its edges as a simply supported plate. Only the 6.35-mm (1/4-in.) indenter was used. An impact damage threshold was determined for each layup sequence by incrementally increasing the drop height of a 454 g (1 lbm) indenter until damage could be detected either visually or by ultrasonic inspection. The damage threshold was additionally determined for Laminate B using a 908 g (2 lbm) indenter. Several impacts were made above the threshold level to examine impact threat versus damage magnitude relationships.

Damage Evaluation Methods—Specimens were inspected both visually and ultrasonically to determine the onset and extent of damage. Sandwich specimens were inspected by a hand-held ultrasonic pulse-echo technique using the portable AN/GSM-238 Ultrasonic Flaw Detector Set. This equipment is currently available in Navy fleet maintenance organizations; therefore damage detected during this test program should be representative of damage levels detectable in Navy field service.

The ultrasonic inspection technique used with the plate specimens monitored the acoustic amplitude corresponding to a reflection thickness resonance of the back surface echo and used this amplitude to modulate the intensity of a C-scan [*3*]. Using this technique, it was possible to detect delaminations as small as 3 mm (1/8 in.) in diameter occurring at any depth throughout the thickness. Although the C-scan does not record the depth of the damage, the equipment operator can read damage depth from an oscilloscope display of the ultrasonic pulse-time response when this information is needed.

For all tests, ultrasonic damage areas were calculated considering the damage to be elliptical in shape.

$$\text{damage area} = \frac{\pi}{4} AB$$

where

A = maximum length of damage region, and
B = maximum width of damage region.

Post-Impact Compression Tests—To determine the effects of various levels of impact damage on the graphite/epoxy specimens tested, post-impact static and compressive fatigue tests were performed.

Two types of loading fixtures were used with the sandwich specimens; a modified four-point bending fixture and uniaxial loading by an MTS 100 kip servo-hydraulic test machine. The four-point bend test fixture, shown in Fig. 3, incorporated off-center load introduction at the top support to impose a small shear stress through the specimen test section. The specimens were oriented to place the damaged face sheet in compression, with the damaged region at mid-span. The ratio of the average shear stress through the sandwich cross section to the in-plane face sheet stress at the damaged region is 1:588. Test specimens were loaded cyclically by a hydraulic actuator to a stress ratio for the test face sheet of $R = -\infty$ at a frequency of 1.25 Hz. Tests were performed to a maximum compressive stress at the center of the test face of 405 MPa (58.7 ksi); corresponding strain is 0.0059. This reference stress level was selected by taking two-thirds of 90 percent of the *B*-basis tensile strength of the laminate. No tests were performed above this level that represents the design ultimate stress for a fully bonded (no manufactured holes) structure. Due to the off-center loading, the maximum shear stress in the core was 689 kPa (100 psi). To facilitate testing, any specimen that sustained 60 000 cycles without failure was considered to be a run-out. Specimens were removed for visual and ultrasonic examination at 2000, 4000, 10 000, 20 000, 40 000, and 60 000 cycles to observe and measure damage growth.

The four-point-bending test fixture imposes a load state on the sandwich

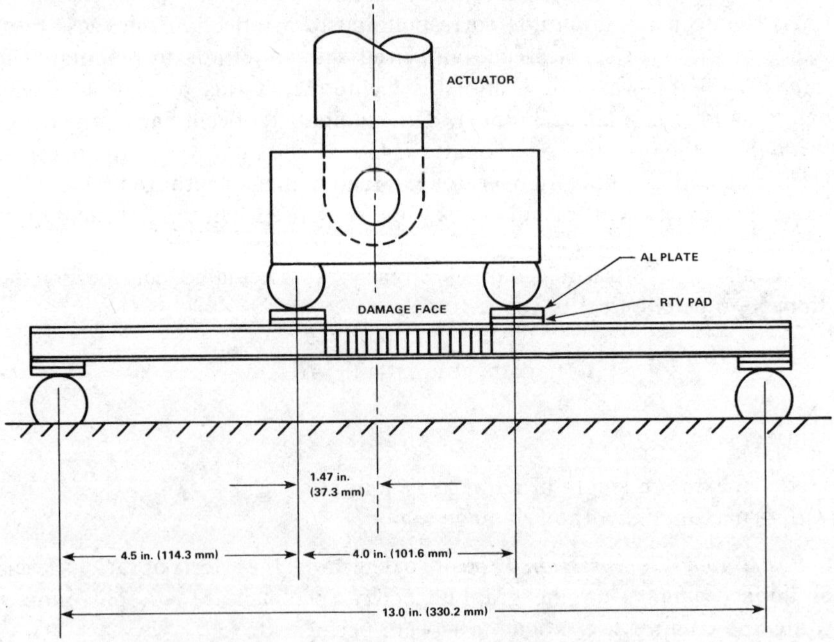

FIG. 3—*Four-point bend compression fatigue test set-up.*

beam that would be encountered in secondary structures such as flaps, spoilers, elevators, etc. In these cases, the sandwich is loaded by bending so that the upper and lower faces are oppositely stressed. Sandwich panels in structural applications such as fuselage and wing skins, however, would be loaded in-plane, so that both faces are loaded in the same sense. Four damaged sandwich beams were therefore tested in uniaxial compression at $R = -\infty$ in a 100 kip MTS fatigue machine to determine if there was a difference in the fatigue life of the damaged face sheet when compared with results of identically damaged panels in four-point-bending fatigue tests.

The solid laminate plate specimens were tested in uniaxial compression only. No tests were performed at strains above 0.0050; current composite designs for thick skin primary structure typically use 0.0040 as the design ultimate strain [*4*]. Constant amplitude $R = -\infty$ fatigue tests were performed in an MTS 100-kip fatigue machine. Specimens were stabilized against buckling by aluminum plates and held in hydraulic self-aligning grips as shown in Fig. 4. This gripping arrangement was adequate for the strain levels considered in this program, but had higher strains been required, another loading and stabilizing system would have been called for as a slight out-of-plane displacement of the grip heads develops at strains above 0.0056.

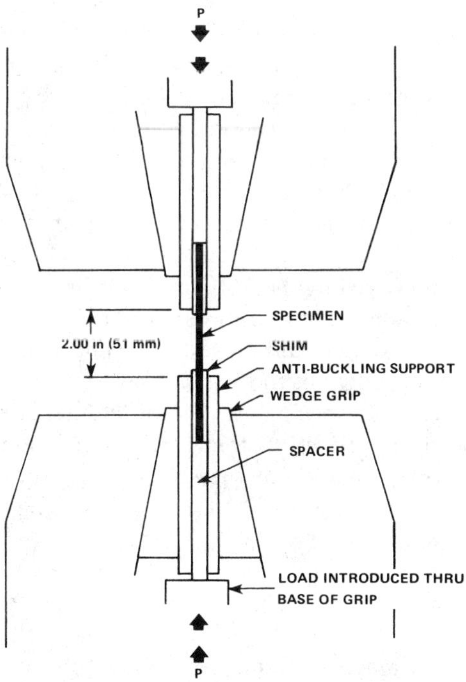

FIG. 4—*Compression fatigue test set-up for impact damaged specimens.*

Results

Compression Fatigue Tests

The results of the spectrum fatigue tests of the Narmco 5209 graphite/epoxy open-hole specimens are shown in Fig. 5. All lines in this figure are least square fits to the data. The results of the previous tests [*1*] performed at an average cyclic frequency of 3 Hz are superimposed for comparison. For a desired fighter life of 12 000 h, the limit load stress is reduced approximately 8 percent when the average cyclic frequency is increased from 1 to 3 Hz. Thus, it appears that the compressive fatigue characteristics of graphite/epoxy laminates might be strain rate dependent. This was also noted by Ryder [*5*] and should be investigated further, since strain rate sensitivity could affect static mechanical properties as well as fatigue characteristics.

Because compression fatigue failures are matrix dominated failures, it was felt that a less brittle matrix might improve the fatigue behavior in compression. Under another program, some test panels were fabricated using the same Hercules AS fibers but with a polysulfone instead of epoxy matrix.

Identical open hole specimens were fabricated from one of the graphite/polysulfone panels and were tested for the same spectrum at an average frequency of 3 Hz. The results are compared with the graphite/epoxy results in Fig. 5. It is evident that the polysulfone matrix did not work as well as the epoxy matrix. Although delamination did not seem to occur as for the epoxy matrix, the polysulfone specimens appeared to blister under load that accounted for the reduced fiber restraint.

The results of the $R = -1$ constant amplitude tests for the idle fastener specimens are shown in Fig. 6 along with the open hole specimen data [*1*] for comparison. The straight lines are fitted to the data by linear regression. Although the lines for the hex-head and countersunk-head bolts are different, it is believed that larger data samples might make the two lines coincident. The constraint offered by the fasteners in the holes appears to increase the stress for a given fatigue life by approximately 10 percent in the region of 10^4 cycles. Figure 7 shows typical failures of the idle fastener specimens. All failures occurred at the edges of the washers rather than at the center of the hole as for the open hole specimens. The difference in failure location is caused by the constraint offered by the clamping action of the bolt and nut that alters the buckling mode shape thereby shifting the failure location and reducing the column effective length resulting in the higher nominal axial stress for a given life.

Impact Program

Impact Damage—Figure 8 presents the results of impact testing for the honeycomb sandwich beams. Initial damage consisted of crippling of the core directly under the impact point with no apparent damage to the com-

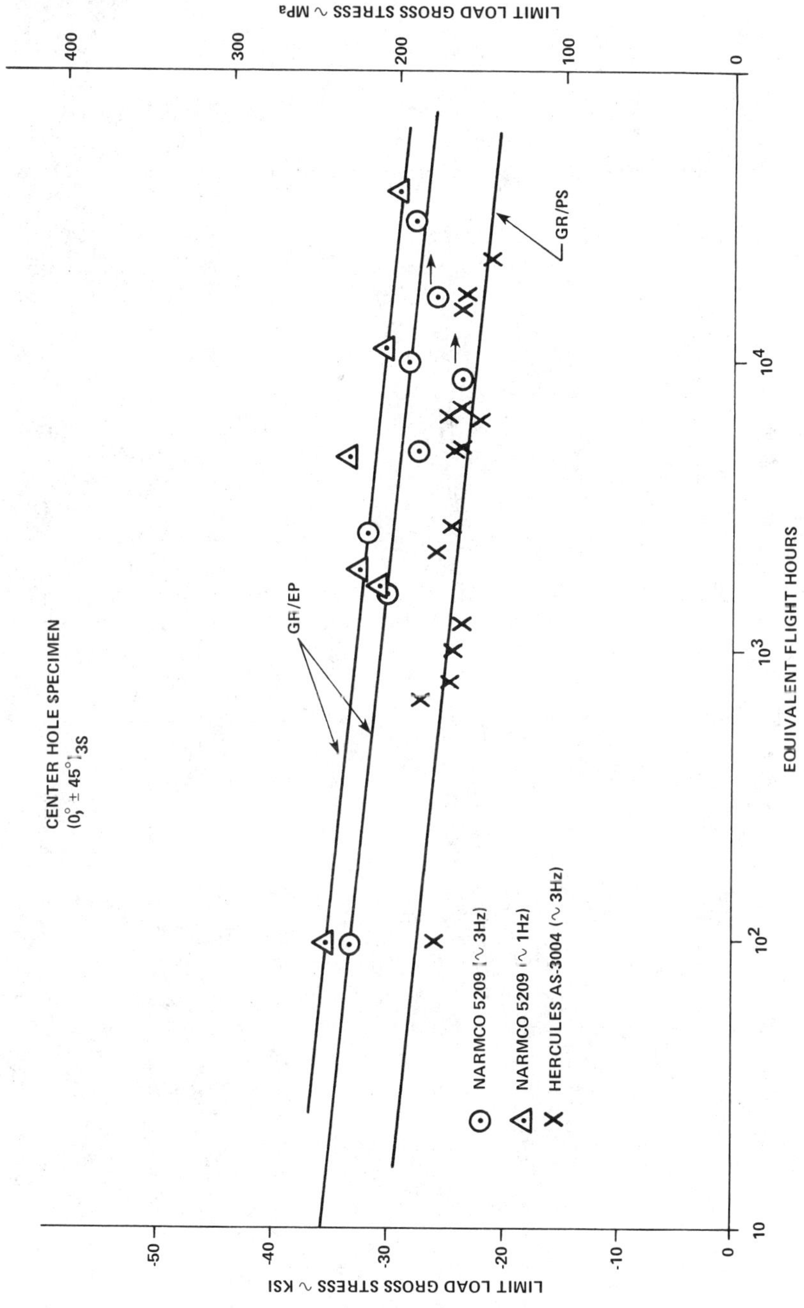

FIG. 5—*Life to failure, fighter spectrum, compression skin.*

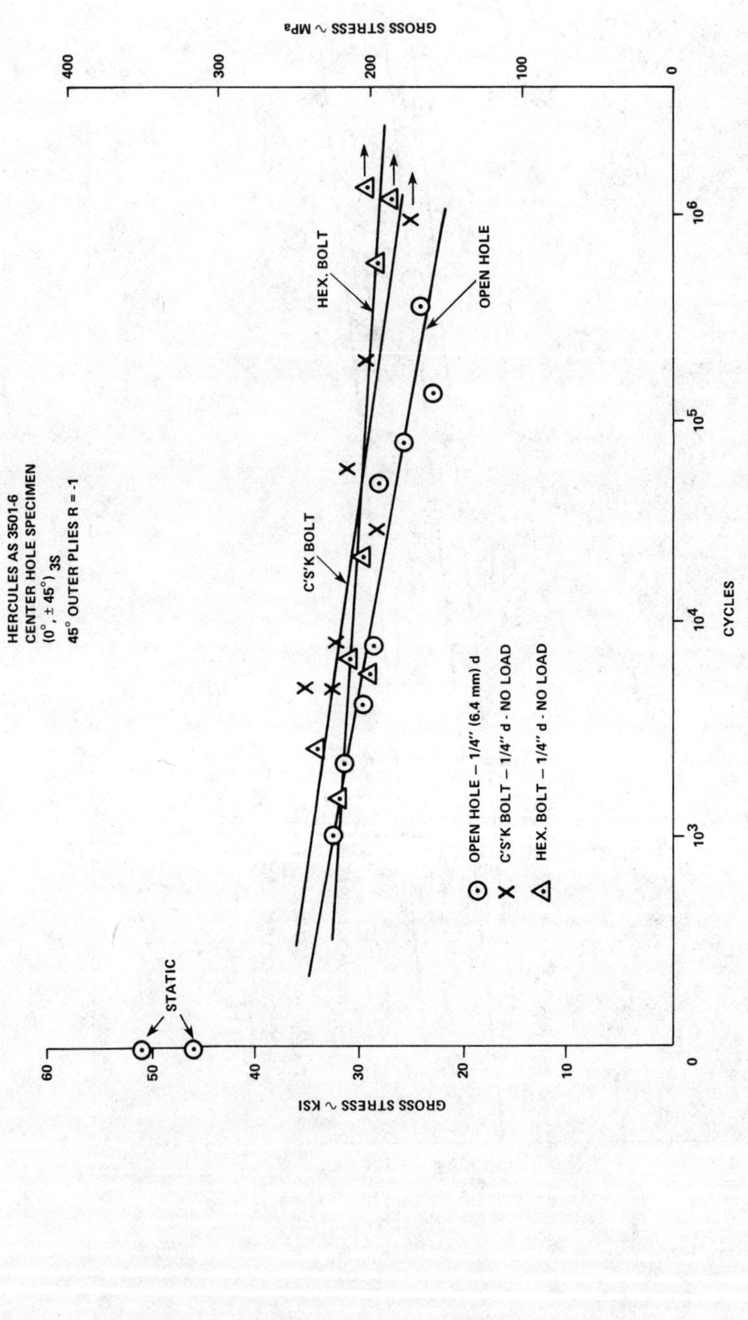

FIG. 6—*Life to failure, center hole specimens* R = −*1.*

FIG. 7—*Idle fastener specimen failures;* (left) *hex head and* (right) *100-deg countersink head.*

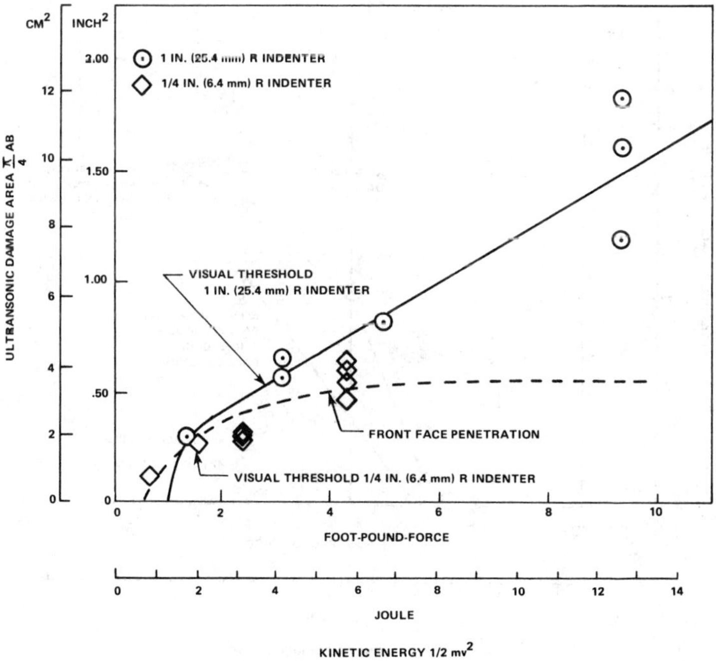

FIG. 8—*Impact energy versus damage area, sandwich beams.*

posite face sheet. Higher impact energies resulted in delamination damage to the face sheet, and when the impact level was sufficient to cause face sheet bending failure, the damage became readily visible. The support provided to the face sheet by the core limits the maximum damage area to less than two times the indenter diameter even with total face sheet penetration.

Results of impact on the solid laminate plates are shown in Fig. 9. The laminate experienced no apparent effect from impacts up to 6.8 J (5 ft·lb) when incipient damage occurs. Ultrasonic inspection showed incipient damage to be slight delamination, 1.59 mm (1/16 in.) in diameter, located approximately 10 percent of the depth below the impact point. By varying the impact parameters of mass and velocity, it was possible to induce different levels of damage from just perceptible to large delaminations accompanied by obvious front and back face visible damage. Increased energy impacts resulted in greater damage. Laminate B specimens sustained more damage than Laminate A specimens for equivalent energy impacts. Experimental scatter in damage areas, particularly the 11.1 J (8.22 ft·lb) impacts on B laminates, was extreme. Figure 10 shows a C-scan of typical impact damage.

A more detailed evaluation of impact damage to the sandwich beams and laminated plates can be found in previous work [*6, 7*].

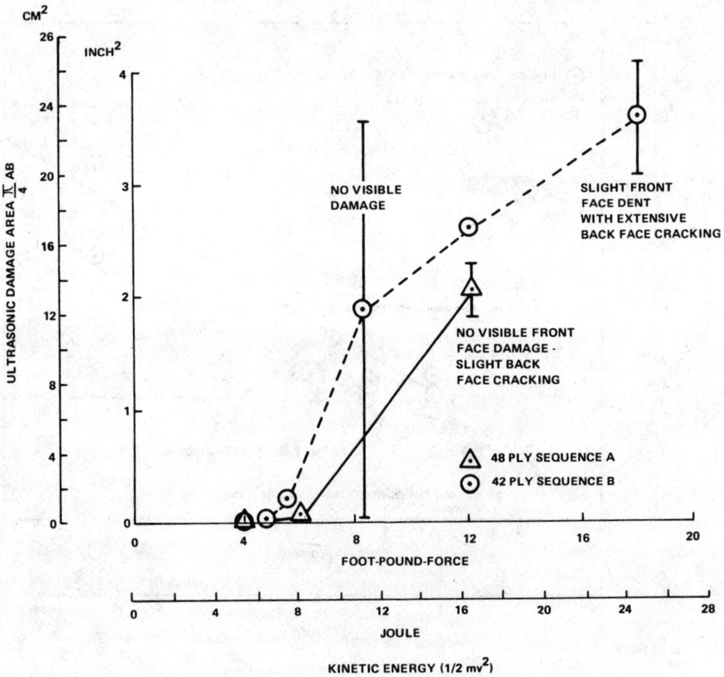

FIG. 9—*Impact energy versus damage area, laminated plate specimens.*

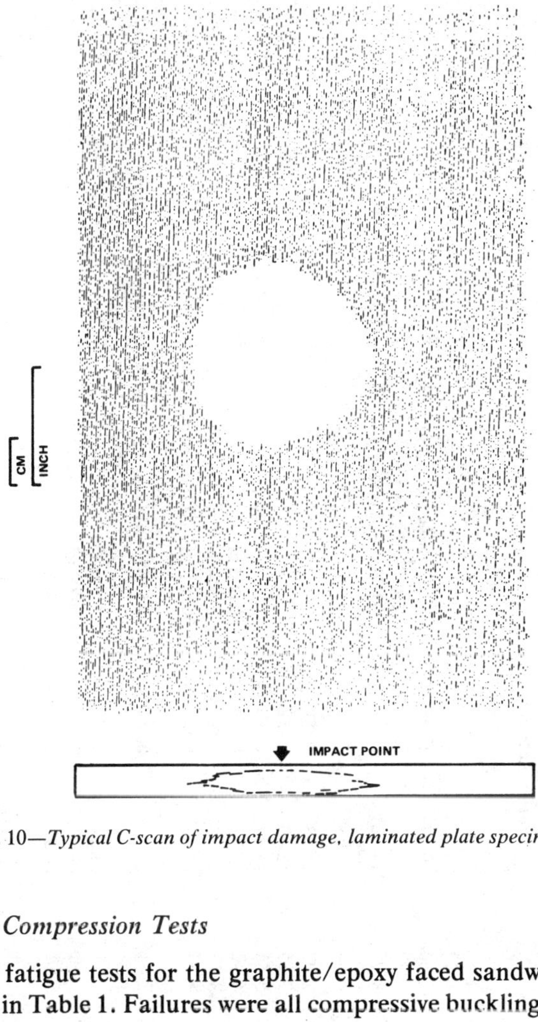

FIG. 10—*Typical C-scan of impact damage, laminated plate specimens.*

Post-Impact Compression Tests

Results of fatigue tests for the graphite/epoxy faced sandwich beams are summarized in Table 1. Failures were all compressive buckling failures of the face sheet through the damage section. Of the 14 specimens tested under four-point bending, only three cases of damage growth were noted. This growth consisted of a slight increase in the ultrasonically detectable damage region or minor delamination of the surface ply or both. In general, the size of damage did not appear to grow with cycling, although there may be a weakening of the matrix around the damage region or a growth of damage within the initial damage region, or both, which contributed to eventual failure.

The results of the axial compression fatigue tests were in sharp contrast to those of the four-point-bend fatigue tests. Dramatic delamination of the surface 0-deg ply outward from the damage was observed in axial testing. Static strength of a damaged sandwich panel loaded in axial compression was also

TABLE 1—*Impact damaged graphite/epoxy faced sandwich beams fatigue test results;* R = $-\infty$, *frequency* = *1.25 Hz.*

Specimen	Indenter Radius, mm	Indenter Radius, (in.)	Impact Energy, 1/2 mV², J	Impact Energy, 1/2 mV², (ft·lb)	Maximum Compression Strain	Cycles to Failure	Comments
				Four Point Bending Fatigue Tests			
Control	...	...	...	...	0.00593	100 000 nf[a]	...
25	25.4	(1)	1.86	(1.37)	0.00594	60 000 nf	...
26	25.4	(1)	6.74	(4.97)	0.00486		static failure
27	25.4	(1)	12.64	(9.32)	0.00427		static failure
28	25.4	(1)	4.22	(3.11)	0.00593	1 113	
29	6.4	(1/4)	0.92	(0.68)	0.00593	60 000	damage growth at 20 000 cycles ΔA = 0.05 in. (1.3 mm) ΔB = 0.05 in. (1.3 mm)
30	6.4	(1/4)	2.07	(1.53)	0.00593	7 411	...
31	6.4	(1/4)	3.25	(2.40)	0.00593	5	...
32	6.4	(1/4)	5.86	(4.32)	0.00439		static failure
33	25.4	(1)	4.22	(3.11)	0.00474	60 000 nf	...
34	6.4	(1/4)	3.25	(2.40)	0.00474	1 894	...
35	6.4	(1/4)	3.25	(2.40)	0.00356	60 000 nf	NDI damage growth ΔA = 0.14 in. (3.6 mm) $\Delta B = 0$
36	25.4	(1)	12.64	(9.32)	0.00356	4 516	damage growth at 2000 cycles surface ply delamination $\Delta A = 0 \Delta B$ = 0.2 in. (5.1 mm)
37	25.4	(1)	12.64	(9.32)	0.00386	66	...
38	6.4	(1/4)	5.86	(4.32)	0.00356	60 000 nf	...
39	6.4	(1/4)	5.86	(4.32)	0.00386	965	...
				Uniaxial Compression Fatigue Test			
Control	...	...	...	...	0.00572	100 000 nf[a]	...
40	6.4	(1/4)	5.86	(4.32)	0.00368	...	static failure
41	6.4	(1/4)	5.86	(4.32)	0.00268	60 000 nf	damage growth after 10 000 cycles $\Delta A = 0 \Delta B$ = 0.08 in. (2.0 mm)

							additional growth after 20 000 cycles $\Delta A = 0.06$ in. (1.5 mm) $\Delta B = 0.15$ in. (3.8 mm)
					0.00400	25	retest at higher strain
							damage growth upon first cycle $\Delta A = 0 \Delta B = 0.55$ in. (14.0 mm)
							delamination of surface ply
42	6.4	(1/4)	5.86	(4.32)	0.00326	102 000 nf	damage growth upon load growth delamination of surface ply
							damage growth after 100 cycles $\Delta A = 0 \Delta B = 0.44$ in. (11.2 mm)
							additional growth after 22 000 cycles $\Delta A = 0 \Delta B = 0.20$ in. (5.1 mm)
							additional growth after 42 000 cycles $\Delta A = 0 \Delta B = 0.31$ in. (7.9 mm)
43	6.4	(1/4)	5.86	(4.32)	0.00350	10 000 nf	damage growth within first 100 cycles delamination of surface ply $\Delta A = 0 \Delta B = 0.34$ in. (8.6 mm)
					0.00400	50	retest at higher strain
							no damage growth prior to failure

[a]nf indicates specimen did not fail.
[b]ΔA = increase in damage length, ΔB = increase in damage width.

less than for a similarly damaged panel in the four-point-bend test. Figure 11 traces the growth of the surface ply delamination for one specimen. It is believed that cycling tends to reduce the effect of the stress concentration as two specimens that had been cycled at strains below that of the static failure point were then able to withstand limited cycling above the static failure strain.

The reason for the difference in damage growth between axial and four-point-bend tests is attributed to the curvature of the sandwich face imposed by the bending moment. The curvature constrains the 0-deg surface ply in the direction normal to the face and prevents this ply from buckling upward away from the face sheet inhibiting delamination. The axial loading imposes no such constraint. The results of this work suggest that the delamination results from the tensile stress normal to the face.

Life to failure results presented in Fig. 12 show that visually damaged specimens can withstand fatigue cycling to a strain of 0.0033 for at least 10^5 cycles. It is interesting to note that the effects on compressive fatigue life of the damage caused by the 0.31 kg (0.70 lb), 6.35-mm (1/4-in.) radius indenter impacting the sandwich beam at 6.1 m/s (20 ft/s) is roughly equivalent to the effects of the 0.680 kg (1.5 lb), 25.4-mm (1-in.) radius indenter impacting the beam at the same speed.

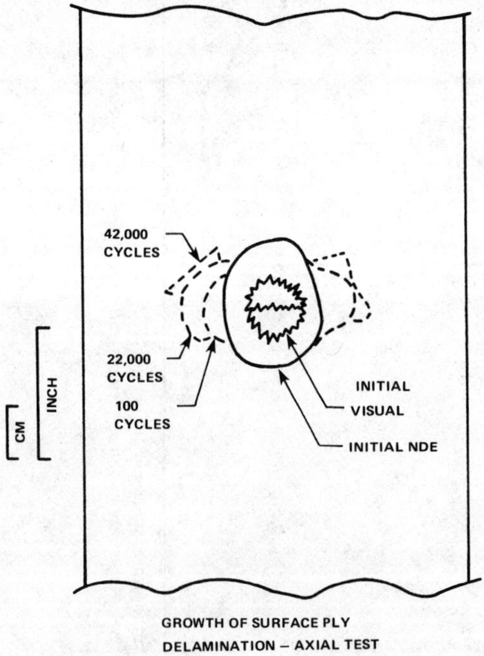

FIG. 11—*Damage growth of surface 0-deg ply delamination for Specimen 42.*

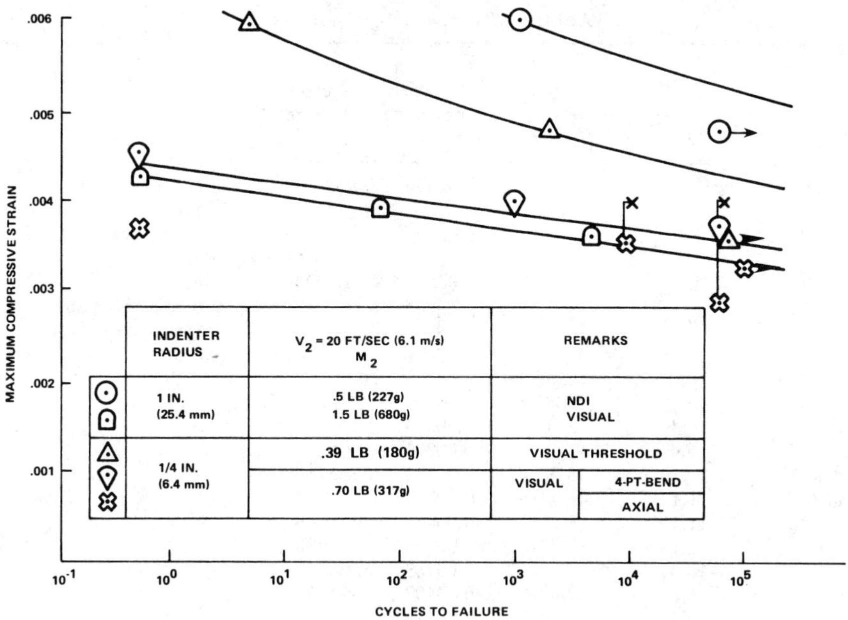

FIG. 12—*Life to failure, impact damaged sandwich beams* R − − ∞.

Results of compression tests of solid laminate plates are presented in Table 2. The effects of two levels of impact were investigated using the 48-ply A laminates; incipient and 12.9-cm^2 (2-in.^{2}) delamination area damage with no front face visible damage. These damage levels correspond to impact energies of 8.22 J (6.06 ft·lb) and 16.35 J (12.06 ft·lb), respectively. Laminate B specimens were fatigue tested with three levels of impact damage; 11.6-cm^2 (1.8-in.^{2}) delamination with no visible damage, 16.1-cm^2 (2.5-in.^{2}) delamination with slight back face visible damage, and 22.6-cm^2 (3.5-in.^{2}) delamination with both front and back face visibly damaged. Corresponding impact energy levels are 11.1 J (8.22 ft·lb), 16.3 J (12.0 ft·lb), and 24.4 J (18.0 ft·lb). The Laminate A specimen with incipient level impact damage was cycled more than 10 000 times at each of the 0.0040 and 0.0050 strain levels with no effect. The results of the compression fatigue cycling for all other specimens are presented in Fig. 13. The failure mode for both Laminates A and B was compressive buckling through the damaged section. Although the damaged areas of Laminate B specimens were larger than for Laminate A specimens for equivalent impact conditions, the Laminate B specimens were more tolerant of the damage when compared on the basis of maximum compressive strain. The Laminate A specimens with 12.9-cm^2 (2-in.^{2}) delamination had a static compressive strength corresponding to a strain of 0.00378 with fatigue run-out (2 000 000 + cycles) at a strain level of

TABLE 2—*Static and fatigue test results.*

Specimen	Impact Energy, 1/2 mV^2 J	(ft·lb)	Maximum Gross Compressive Strain	Cycles to Failure, at 1.5 Hz	Comments
A5	...	...	0.00559	static[a]	no effect; validate test fixture
A2	8.2	(6.06)	0.00400	10 900 nf[b]	no effect; retested at 0.0050 strain
A2	...	...	0.0050	10 000 nf	no effect
A1, A2	16.4	(12.06)	0.0040		specimens accidentally destroyed due to fatigue machine malfunction
A3	16.4	(12.06)	0.00378	static	
A4	16.4	(12.06)	0.00300	1 836	slight damage
A5	16.4	(12.06)	0.00243	2 000 000 nf	growth 1/16 in. (< 2 mm)
A6	16.4	(12.06)	0.00270	61 255	...
B5	11.1	(8.22)	0.00482	static	...
B6	11.1	(8.22)	0.00452	3 069	...
B7	11.1	(8.22)	0.00438	496 509	...
B8	11.1	(8.22)	0.00465	2 184	...
B9	16.3	(12.00)	0.00463	static	...
B10	16.3	(12.00)	0.00407	533	...
B4	24.4	(18.00)	0.00417	static	...
B2	24.4	(18.00)	0.00367	655	...

[a] Static indicates monotonic gradually increasing load cycle.
[b] nf indicates specimen did not fail.

0.00243. The Laminate B static compressive strength for all damage levels tested corresponded to strain loads exceeding 0.0040. Frequent ultrasonic C-scan inspections during testing revealed no significant damage growth due to fatigue cycling in any of the specimens, whether or not the specimens ultimately failed in fatigue.

The mechanism of load transfer around the impact damage region was examined during the static test of Specimen B5, which had been damaged by a 11.14 J (8.22 ft·lb) impact. This specimen was instrumented with a strain gage array and loaded to failure. Results are presented in Fig. 14. The main effect of the damage is to allow transverse (out-of-plane) displacement of the plies in the immediate damaged region, and therefore allow local buckling leading to catastrophic failure of the laminate as indicated by the strain gage readings directly over the impact point. It is believed that the subvisual damage (delamination) does not break the graphite fibers. Failures under compressive load result from matrix damage and the resultant inability of the matrix to stabilize the fibers against buckling.

Discussion

The results of compression testing indicate significant reductions in the strength and fatigue properties of composite laminates due to impact in-

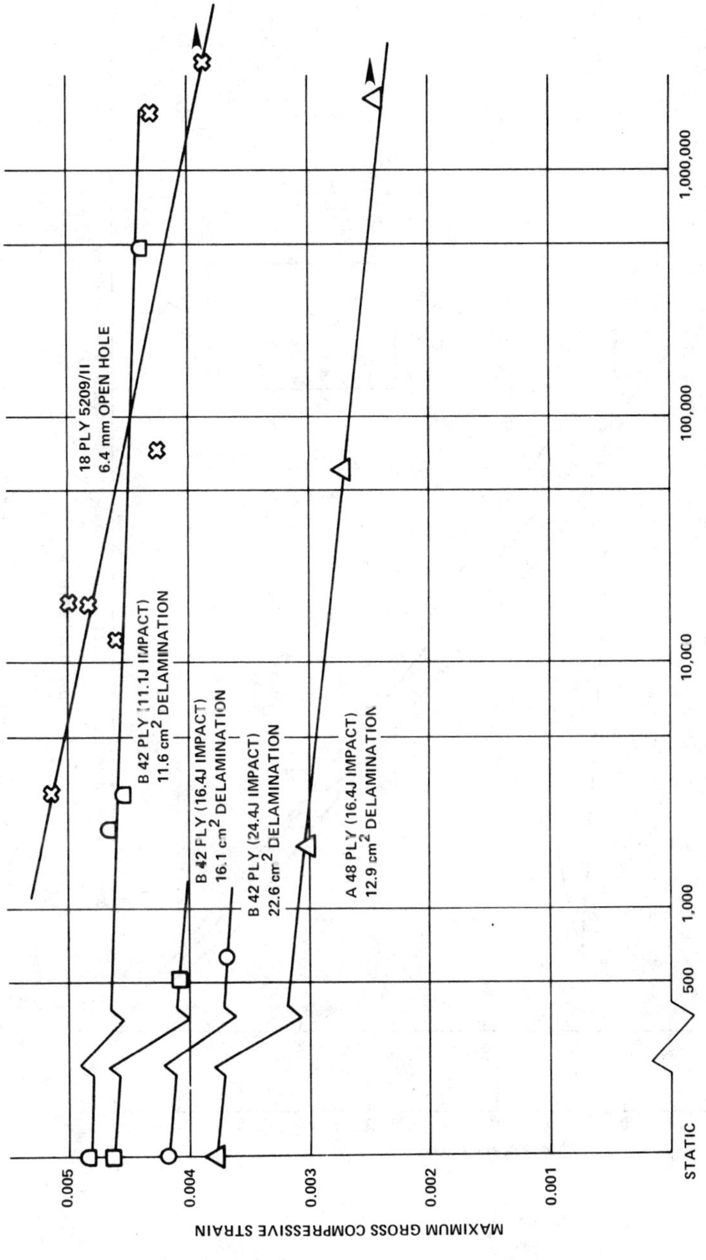

FIG. 13—*Life to failure, laminated plate specimens* R = − ∞.

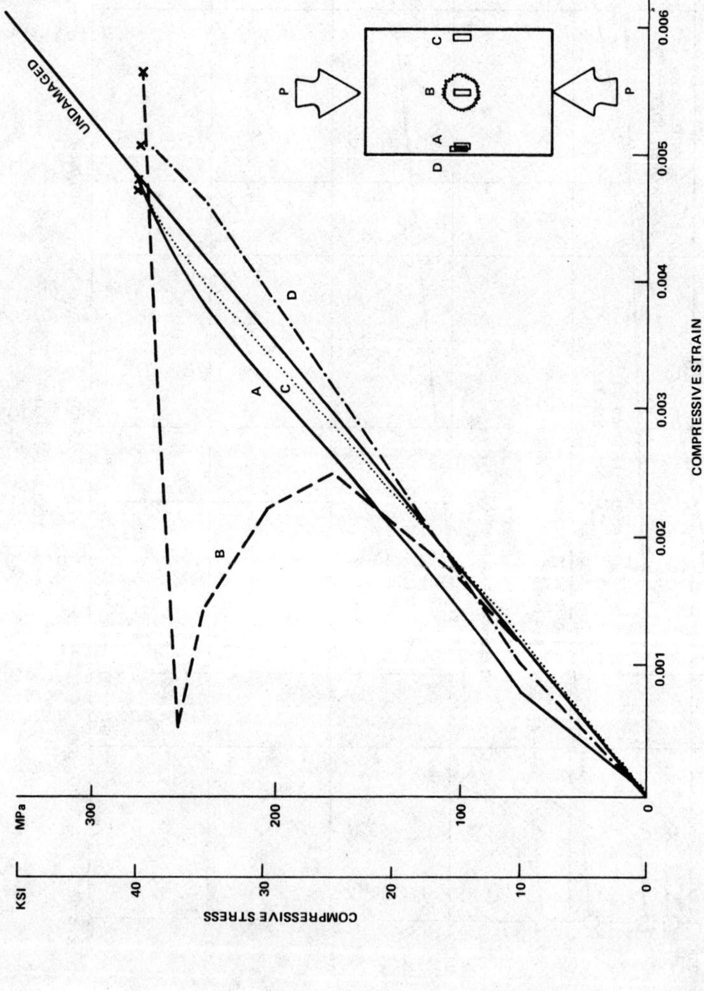

FIG. 14—*Static compressive stress, strain curve, 42-ply laminated plate.*

duced damage. In order to assess the implication of these results, it is necessary to evaluate the data in the context of the structural applications envisioned for the material. Current structural configurations incorporating advanced composite material are generally stressed skin on spar type construction, with mechanical fasteners employed to attach the skin to the substructure. Thus, the design allowables are driven by the stress raiser of the fastener holes as discussed previously. Impact damage that is readily visible is not considered critical as it would be immediately noted and repaired. It is necessary to examine the effects of subvisual impact damage and determine its severity as compared with that of a typical fastener hole. Figure 13, in addition to the fatigue results for the solid laminate impact specimens, presents the compression-compression fatigue data from Ref *1* for 12-ply $(0/\pm45)_{3S}$ Narmco 5209/II test specimens with 6.35 mm (1/4-in.) diameter open holes. Keeping in mind the difference in material and layup sequence, it is evident that the impact damage is at least as severe, and in general is more severe than the open hole. This may necessitate even further reduction of strain allowables to be used in the design of graphite/epoxy structures.

Conclusions

This program was established for the purpose of determining the effects of some additional variables on the compressive fatigue behavior of filamentary graphite laminates. The limited results lead to the following conclusions:

1. The compressive fatigue characteristics of filamentary graphite laminates may be strain rate dependent.
2. The clamping constraint provided by mechanical fasteners increases the stress for a given fatigue life by approximately 10 percent in the region of 10^4 cycles.
3. The use of a polysulfone matrix in lieu of an epoxy matrix reduces the allowable stress for a given life.
4. Graphite/epoxy structures are susceptible to damage under realistic impact threats. Impact can cause damage to the composite matrix that, while not visually detectable, allows subsequent laminate failure due to local buckling under axial compressive loading.
5. Growth of impact induced matrix damage is not necessarily a self-similar increase in the damage zone, but also includes increase in the severity of damage within the original area.
6. The effect of impact damage to graphite/epoxy structures should be evaluated by cyclic axial compression loading.

References

[*1*] Rosenfeld, M. S. and Huang, S. L., *Journal of Aircraft*, Vol. 15, No. 5, May 1978, pp. 264-268.

[2] Adsit, N. R. and Waszczak, J. P., "Investigation of Damage Tolerance of Graphite/Epoxy Structures and Related Design Implications," Technical Report No. NADC-76387-30, Naval Air Development Center, Warminster, Pa. 1976.
[3] Scott, W. R., "Ultrasonic Spectrum Analysis for NDI of Layered Composite Materials," Technical Report No. NADC-75324-30, Naval Air Development Center, Warminster, Pa., 1975.
[4] Weinberger, R. A., Somoroff, A. R., and Riley, B. L., "U.S. Navy Certification of Composite Wings for the F-18 and Advanced Harrier Aircraft," in *A Collection of Technical Papers on Structures and Materials*, AIAA/ASME 18th Structures, Structural Dynamics and Materials Conference, San Diego, Calif., 1977, Vol. A, pp. 396–407.
[5] Ryder, J. T. and Walker, E. K., "Ascertainment of the Effect of Compressive Loading on the Fatigue Lifetime of Graphite Epoxy Laminates for Structural Applications," Technical Report AFML-TR-76-241, Air Force Materials Laboratory, Dayton, Ohio, 1976.
[6] Gause, L. W. and Huang, S. L., "Compression Fatigue of Impact Damaged Graphite Epoxy Sandwich Beams," Technical Report No. NADC-77305-60, Naval Air Development Center, Warminster, Pa., 1978.
[7] Gause, L. W., "Low Speed, Hard Object Impact on Thick Graphite/Epoxy Plates," Technical Report No. NADC-78051-60, Naval Air Development Center, Warminster, Pa., 1978.

E. P. Phillips[1]

Effects of Truncation of a Predominantly Compression Load Spectrum on the Life of a Notched Graphite/Epoxy Laminate

REFERENCE: Phillips, E. P., **"Effects of Truncation of a Predominantly Compression Load Spectrum on the Life of a Notched Graphite/Epoxy Laminate,"** *Fatigue of Fibrous Composite Materials, ASTM STP 723,* American Society for Testing and Materials, 1981, pp. 197–212.

ABSTRACT: The fatigue behavior of a notched, graphite/epoxy (T300/5208) laminate subjected to predominantly compressive loading was explored in a series of constant amplitude and transport wing spectrum tests. Results of these tests indicated that (1) the amount of local (near the notch) buckling allowed in the tests significantly affected fatigue life; (2) spectrum truncation of either the high- or low-load end of the spectrum produced lives greater than those obtained in the baseline, complete-spectrum test, but life was much more sensitive to truncations at the high-load end; and (3) the Palmgren-Miner linear cumulative damage theory always predicted lives much longer than the actual spectrum loading test lives.

KEY WORDS: composite materials, fatigue (materials), compression tests, graphite composites, epoxy laminates, constant life fatigue diagrams, variable amplitude loading tests, spectrum loading tests

In most airframe development programs, the results of structural fatigue tests are used to decide if the structure satisfies the fatigue life design requirement. Years of experience with tests on aluminum coupons and structures have led to satisfactory procedures for defining suitable test load spectra, that is, spectra that yield a representative life estimate in a reasonably short test time. In particular, a general understanding of the effects of truncating load spectra at the high- and low-load levels has evolved. For composite materials, however, a similar base of test experience has not yet been accumulated.

The current work was undertaken to explore (1) the effect of spectrum truncations on the fatigue life of notched, quasi-isotropic graphite/epoxy

[1]Research engineer, NASA Langley Research Center, Hampton, Va. 23665.

coupons and (2) the capability of the Palmgren-Miner linear cumulative damage theory to predict the truncation effects. Since cyclic compressive loading is generally more detrimental to composites than cyclic tension [*1-3*],[2] a load spectrum representative of the upper surface of a transport wing was used in the current work. Constant amplitude tests were conducted to explore the sensitivity of fatigue life to the amount of local (near the notch) buckling allowed in the tests and to provide data for cumulative damage calculations.

Experimental Procedure

Materials and Specimens

Specimens were cut from 16-ply, $[45/0/-45/90]_{2s}$ (quasi-isotropic) sheets made from T300/5208 graphite/epoxy unidirectional tape. The cured laminate had an average ply thickness of 0.14 mm, a fiber volume of 64 percent, and a void content of 0.36 percent. Specimens were stored and tested in an ambient laboratory air environment. Measurements taken on several specimens during the test program indicated a 0.6 to 0.7 percent moisture content by weight.

The test specimen configuration is shown in Fig. 1. The central 6.35-mm hole was made by a diamond-coated, ultrasonically vibrating drill. Generally, this drilling procedure produced very clean hole surfaces, but some in-

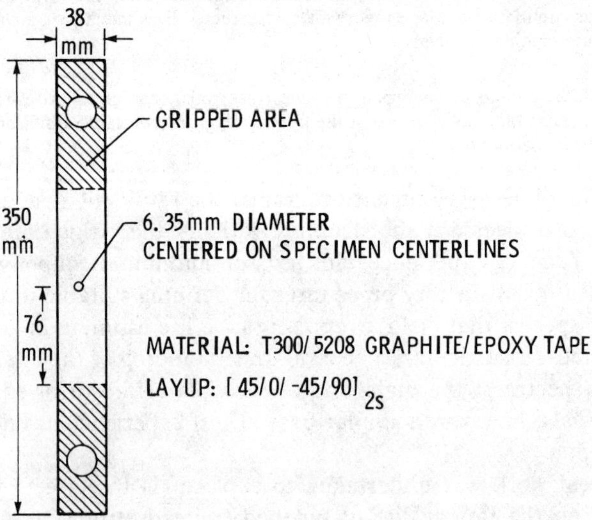

FIG. 1—*Specimen configuration.*

[2]The italic numbers in brackets refer to the list of references appended to this paper.

terlaminar cracking was detected between the first two plies on the drill exit surface of most of the specimens. Nondestructive examinations showed the depth of the delaminations did not exceed 0.3 mm, and subsequent test observations indicated the delaminations did not play a significant role in the failure process.

Test Machines

Specimens were tested to complete rupture in axial-load, closed loop, servohydraulic testing machines having about 45 kN force capacity. Constant amplitude and variable amplitude tests were run at a nearly uniform loading rate so that loading frequency ranged from about 3 to 20 Hz depending upon load amplitude. Load sequences were generated by a small, on-line digital computer.

Load Spectra

The standardized test TWIST (Transport WIng Standard Test) [*4*] was used in the spectrum loading tests. TWIST was developed jointly by the Laboratorium für Betriebsfestigkeit in Germany and the National Lucht-en Ruimtevaartlaboratorium in The Netherlands. A list of the most significant features of this test is given in Table 1 and a tabulation of the load spectrum is given in Table 2. In Table 2, note that some of the alternating (half the cycle range) loads were large enough to cause excursions into tension even though the flight mean load was compressive.

In addition to the baseline TWIST spectrum, several truncated versions were used. The truncated versions were created by deleting the following alternating flight-load levels: (1) the lowest, (2) the two lowest, (3) the four lowest, (4) the two highest, and (5) the four highest and the lowest. In the

TABLE 1—*Description of TWIST (Transport WIng Standard Test).*

Significant Features
Flight-by-flight loading.
Constant flight mean load.
All ground loads represented by a single load event equal to minus one-half the flight mean load.
Each flight load reversal occurred at 1 of 20 discrete levels; 10 levels above the flight mean and 10 levels below the flight mean (see Table 2).
Ten flight types (severities)—each characterized by the number of load levels involved and the number of occurrences of each level.
Sequence of flight types and sequence of loads within each flight determined by random draw without replacement.
Random draw of flight loads restricted so that successive loads must generate a mean-level crossing, but no other restrictions on magnitudes of successive loads; that is, the sequence is generated in the random half-cycle fashion.
Load sequence repeats after 4000 flights; that is, the block length is 4000 flights.

TABLE 2—*Frequency of occurrence of flight types and load cycles within each flight in TWIST.*

Flight Type	Frequency of Occurrence of Each Flight Type in One Block of 4000 Flights	Frequency of Occurrence of Flight Load Cycles[a] at the Ten Alternating Load Levels										Total Number of Cycles per Flight
		1.60[b]	1.50	1.30	1.15	0.995	0.84	0.685	0.53	0.375	0.222	
A	1	1	1	1	4	8	18	64	112	391	900	1500
B	1	...	1	1	2	5	11	39	76	366	899	1400
C	3	...	...	1	1	2	7	22	61	277	879	1250
D	9	...	...	...	1	1	2	14	44	208	680	950
E	24	...	...	...	...	1	1	6	24	165	603	800
F	60	...	...	...	...	...	1	3	19	115	512	650
G	181	...	...	...	...	...	...	1	7	70	412	490
H	420	...	...	...	...	...	...	...	1	16	233	250
I	1090	...	...	...	...	...	...	...	...	1	69	70
J	2211	...	...	...	...	...	...	...	...	...	25	25
Total number of load cycles per block of 4000 flights		1	2	5	18	52	152	800	4170	34 800	358 665	...
Cumulative number of load cycles per block of 4000 flights		1	3	8	26	78	230	1030	5200	40 000	398 665	...

[a]In this table, the frequency of occurrence numbers indicate full cycles.
[b]Ratio of alternating load to the flight mean load.

truncation process, flights were never completely eliminated even though all of the flight loads included in the baseline version of the flight were scheduled to be deleted (see Tables 1 and 2). For such flights, the truncated version of the flight consisted of a ground-air-ground cycle that was bounded by the ground load and the highest flight load in the baseline version of the flight (see Fig. 2). The sequence of flight types remained the same as the baseline TWIST sequence in all tests, but the sequence of loads within flights changed as a result of each spectrum truncation.

Specimen Antibuckling Procedure

Specimens were restrained from column buckling during compressive loading by sandwiching the specimen between two aluminum plates (coated with a trifluoroethylene resin plastic). This antibuckling procedure was chosen because it was the simplest of the methods described in the literature [*1–3,5,6*] and because no method had a clear advantage. In the instances where the "plate-sandwich" antibuckling procedure was used in studies reported in the literature [*1,3,5*], the degree to which local (near the notch) buckling was allowed varied over a wide range. That is, either the plates covered the entire specimen surface or some portion of the plates was cut away to free the test section. Because the size of the windows (cutouts) in the antibuckling plates could reasonably be expected to influence fatigue life, preliminary tests were conducted to explore the magnitude of the effect.

Results and Discussion

Specimen Buckling Restraint Tests

The four antibuckling plate configurations used in the tests and corresponding median test lives for one stress level are shown in Fig. 3. Test results are tabulated in Table 3. As expected, the more specimen surface covered by the antibuckling plates, the longer the fatigue life. The life for tests in which the antibuckling plates had no window was about 30 times that for tests in which the windows were 32-mm wide and either 32- or 64-mm long. For higher compressive loading, the difference in lives was even more pronounced (see Table 3). The results in Fig. 3 illustrate the need to consider the influence of the antibuckling procedures in making comparisons among compressive-loading fatigue data from the literature.

The antibuckling plates having a 32- by 32-mm window were selected for all subsequent tests in the current work. This selection was based on the feeling that local buckling associated with delamination around the hole should not be restrained, but that the specimen should be restrained from general-section buckling so that testing can cover the full material strength range. In the current tests, the delamination zone exceeded the 19-mm window size in

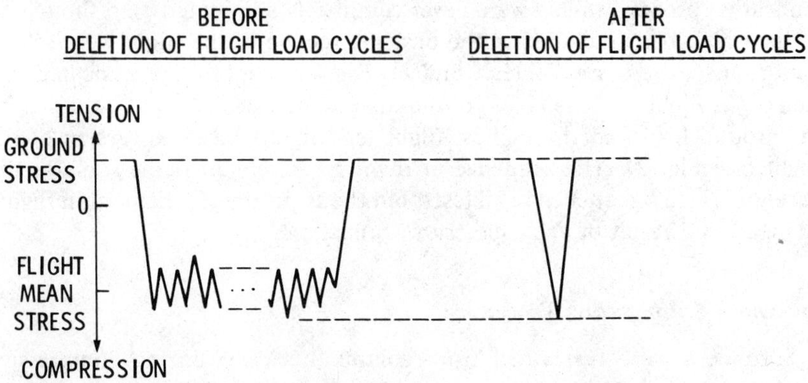

FIG. 2—*Example of truncation procedure for flights in which all flight load cycles were scheduled for deletion.*

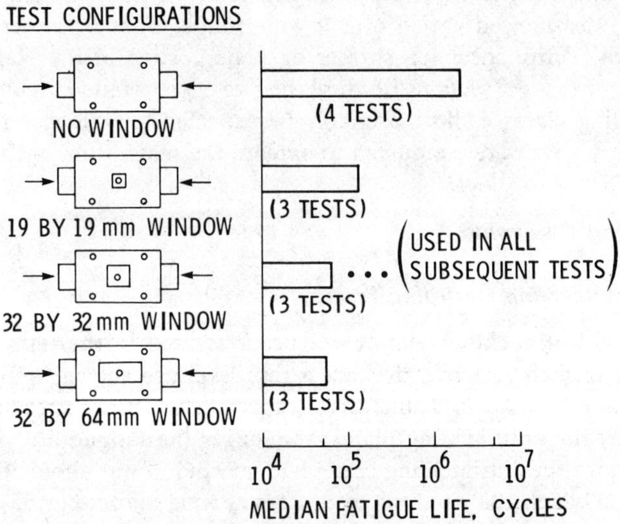

FIG. 3—*Effect of antibuckling plate window size on fatigue life. (Constant amplitude,* R = *−2, minimum gross-section stress = −207 MPa.)*

many tests but rarely exceeded the 32-mm size. The choice of any antibuckling procedure is, of course, rather arbitrary in the sense that none of the procedures will realistically simulate buckling conditions in all structural configurations.

TABLE 3—*Results of tests[a] to explore the effect of local buckling restraint on fatigue life.*

Dimensions of Windows in Antibuckling Plates, mm	Minimum Gross-Section Stress, MPa	Fatigue Life, cycles
No window	−248	158 851 233 299 254 487
No window	−207	194 000 1 112 149 2 861 708 5 881 796
19 by 19	−207	102 517 136 372 349 650
32 by 32	−248	1 939 2 710 8 141
32 by 32	−207	61 914 66 029 132 274
32 by 64	−207	45 065 55 742 112 071

[a]Constant amplitude loading, $R = -2$.

Constant-Amplitude Tests

Constant-amplitude tests were conducted at several R values

$$\left(R = \frac{\text{minimum stress}}{\text{maximum stress}}\right)$$

to provide data for the cumulative damage calculations. Results of the tests are tabulated in Table 4 and shown in Fig. 4 in the form of a constant life diagram. Results of tension and compression static strength tests of this specimen configuration are also plotted in Fig. 4. As expected, the constant life diagram shows that compressive mean stresses had a deleterious effect on fatigue life. All of the constant life lines proceed down and to the left from the $R = -1$ line (that is, in the compressive mean stress half of the diagram). This trend in the constant life lines means that for a constant alternating stress, the higher the compressive mean stress, the lower the life. By

TABLE 4—*Results of constant amplitude fatigue tests.*[a]

R, minimum stress/maximum stress	Minimum Gross-Section Stress, MPa	Fatigue Life, cycles
+3	−289	1 600
		3 334
+3	−248	737 968
		1 131 736
+3	−207	>10 000 000
		>10 000 000
∞	−289	1 298
		1 844
∞	−248	15 924
		78 132
∞	−207	150 098
		527 518
		1 076 384
∞	−165	>10 000 000
		>10 000 000
−5	−289	392
		1 070
−5	−248	33 976
		43 417
−5	−207	395 282
		546 434
−2	−289	318
		350
		370
−2	−248	1 939
		2 710
		8 141
−2	−207	61 914
		66 029
		132 274
−2	−165	3 133 789
		3 917 639
		4 925 000
−1	−238	444
−1	−207	22 263
		25 550
−1	−165	247 012
		317 110
−1	−134	>10 000 000
−0.5	−119	123 618
−0.5	−103	1 433 070
		3 876 274
		3 954 528

[a]Window in antibuckling plate was 32 by 32 mm.

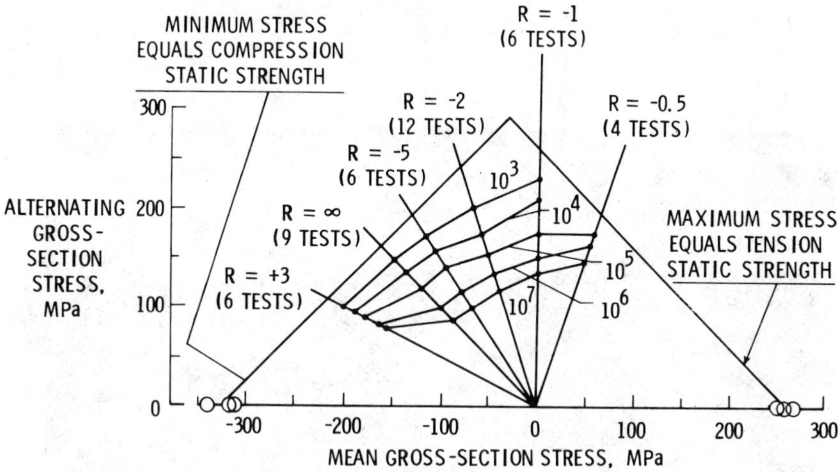

FIG. 4—*Constant-life diagram constructed from the results of the constant amplitude tests.*

contrast, the opposite trend is evident over at least a portion of the tensile mean stress half of the diagram; that is, the higher the tensile mean stress, the higher the life. In general, the trends in the data show that the laminate tested in the current program is more susceptible to compressive fatigue loading than to tensile fatigue loading.

The failure process in all compression-dominated tests was progressive delamination followed by failure in a crippling mode. In most tests, the surface plies split and buckled near the hole at about 10 to 20 percent of the life. The shadow moiré [*7*] photograph in Fig. 5 illustrates this failure mode. The area of out-of-plane displacement is outlined by the closely spaced fringes near the hole. The more widely spaced fringes away from the hole are "initial condition" fringes and do not represent out-of-plane displacement of the specimen. During each test, delaminations initiated and grew at all of the ply interfaces at the hole. The extent of the delaminated zone near the end of a test (at 93 percent of the life) is shown in the dye-enhanced (tetrabromoethane dye) X-ray radiograph [*8*] in Fig. 6. At equal lives, the extent of delamination at rupture (especially of the surface ply) was greater for tests with the greater tensile stress in the loading cycle. This is illustrated in the photograph of failed specimens in Fig. 7.

Spectrum Loading Tests

The flight mean stress (spectrum reference level, see Table 2) in the current tests was −111 MPa (based on the gross section). This stress level was chosen because it produced test lives representative of transport aircraft design goals (about 60 000 flights of the baseline spectrum). A few

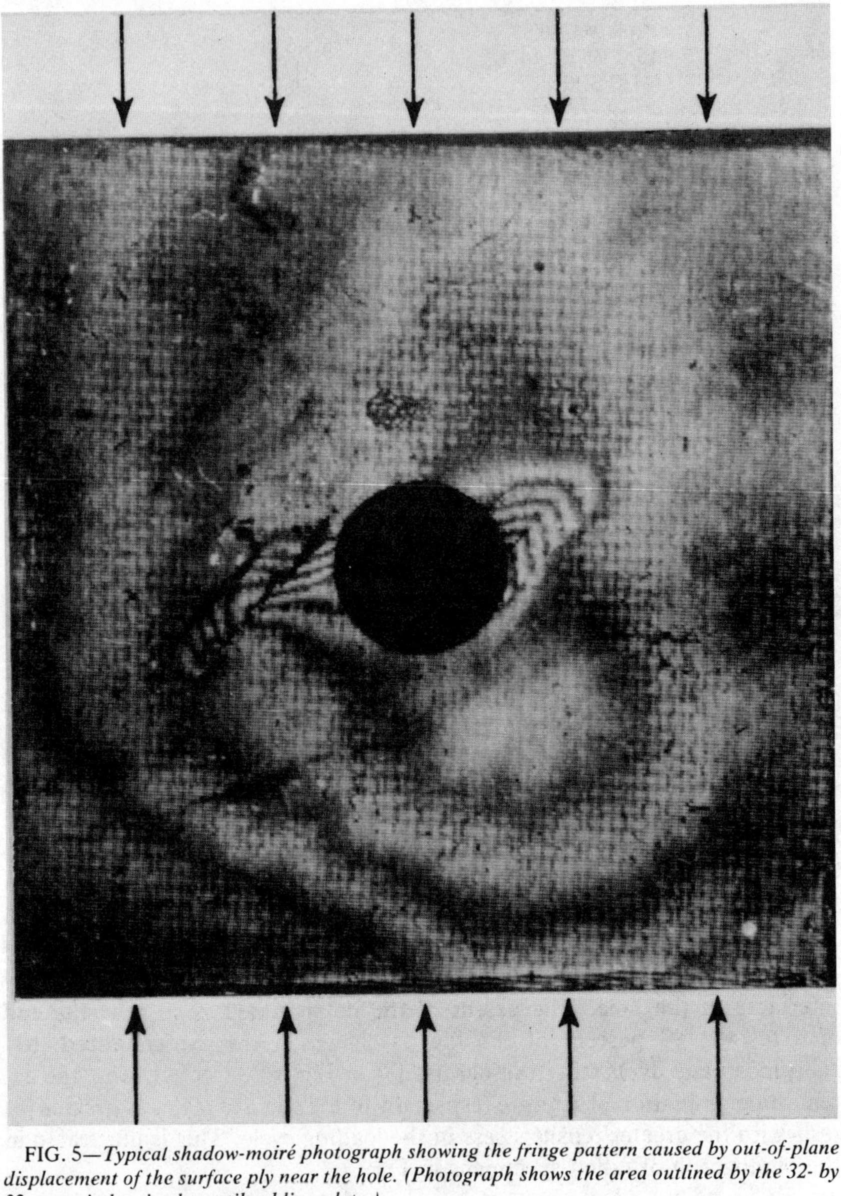

FIG. 5—*Typical shadow-moiré photograph showing the fringe pattern caused by out-of-plane displacement of the surface ply near the hole. (Photograph shows the area outlined by the 32- by 32-mm window in the antibuckling plate.)*

preliminary tests at a flight mean stress of −95 MPa had produced lives of about 10^6 flights. At the −111 MPa flight mean stress, the maximum stress in the baseline spectrum was −289 MPa, or 89 percent of the median compressive static strength of the fatigue specimen as determined from static

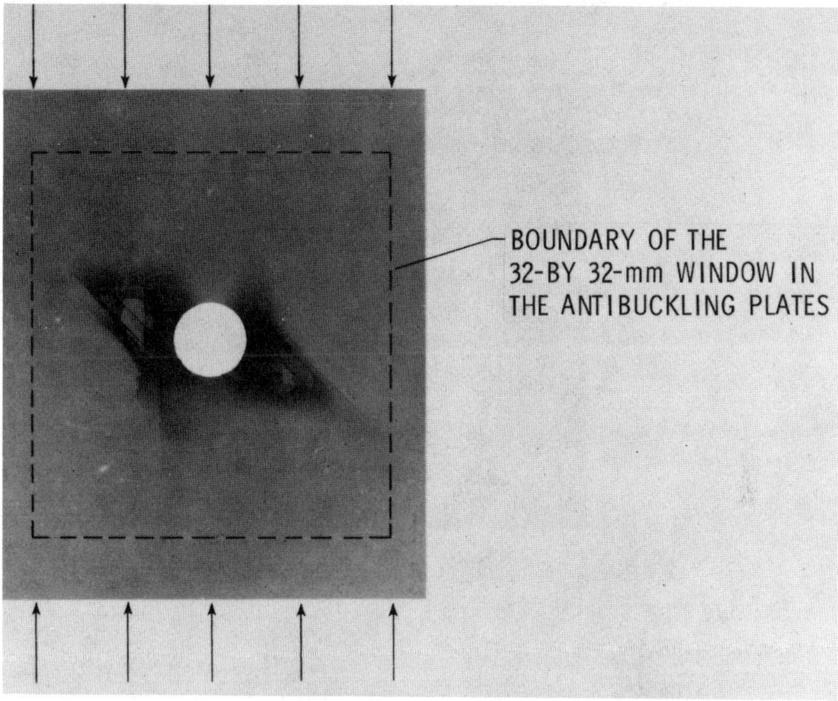

FIG. 6—*Dye-enhanced radiograph showing the extent of the delamination and splitting near the end (at 93 percent of the life) of a test.*

tests using antibuckling plates having a 32- by 32-mm window. Corresponding gross-section strain levels for the flight mean and maximum stresses in the spectrum were 0.002100 and 0.005450, respectively. The appearance of the failures and the extent of delamination at failure in the spectrum loading tests were similar to those for predominantly compression constant amplitude tests at the same maximum compressive load.

The results of the tests to explore truncation effects are shown in Fig. 8 and tabulated in Table 5. A comparison of the median lives (in flights) indicates that progressively larger truncations of the spectrum at either the high- or low-load end produced a trend toward lives greater than those obtained in the baseline spectrum test. However, life was much more sensitive to truncations at the high-load end. Omission of the four highest load levels plus the lowest level produced lives 11 times the baseline spectrum life. With these omissions, the effect on life appears to be largely due to omission of the high loads (which constitute less than 0.007 percent of the total number of load cycles in the spectrum), since omission of the lowest load alone did not produce lives longer than the baseline spectrum. Indeed, omission of the four lowest load levels produced lives only two times the baseline spectrum life

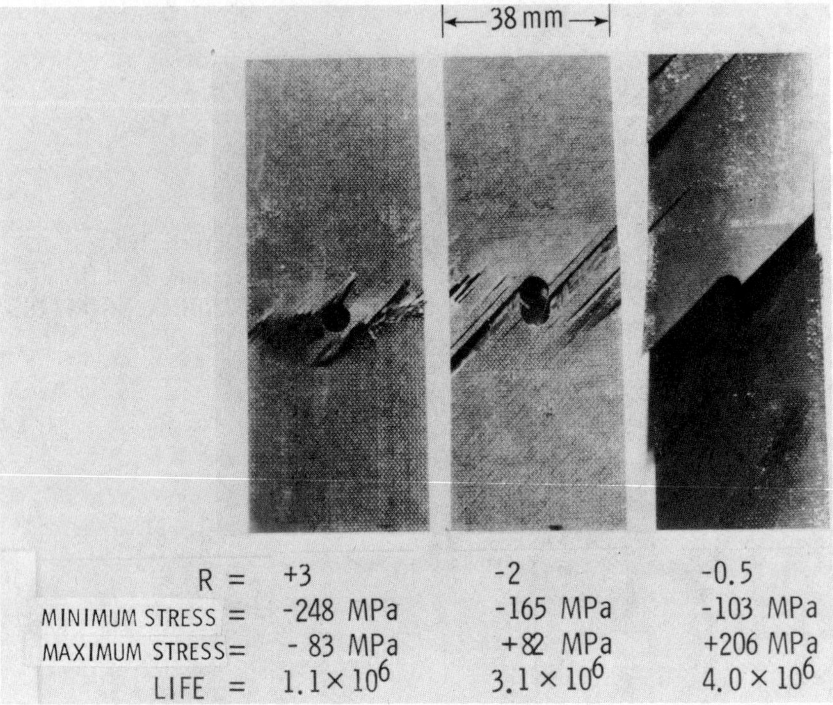

FIG. 7—*Appearance of specimens after test at various* R *values.*

even though the omitted load cycles constitute more than 99 percent of the loads in the baseline spectrum.

In defining the test spectrum for wing structures, normal practice for metallic structures has been to substantially truncate the high-load end of the spectrum because the high loads produce large retardation effects that lead to unrealistically long estimates of fatigue life. By contrast, the data of Fig. 8 suggest that for composite structures the high loads in the spectrum do not produce retardation and are the dominant damage-producing cycles in the spectrum. Therefore, for tests of composite structures, the high-load end of the spectrum should not be truncated to the same extent as has been the practice for metallic structures because that practice would lead to unrealistically long estimates of fatigue life. On the other hand, truncation of the low-load end of the spectrum shows promise for achieving large reductions in test time without significantly changing the test result.

Linear Cumulative Damage Evaluation

To explore the usefulness of the Palmgren-Miner linear cumulative damage theory for composite applications, the median lives from the spec-

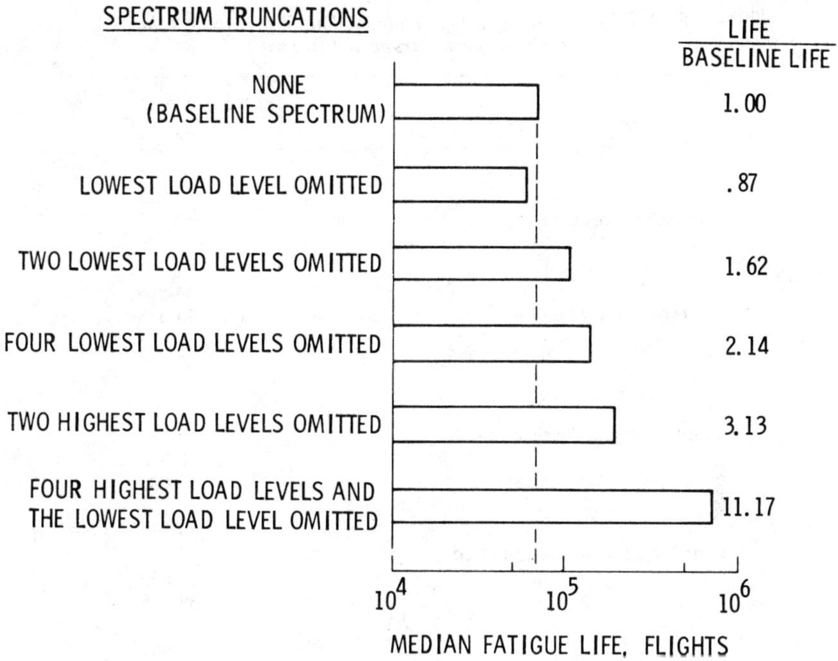

FIG. 8—*Effect of spectrum truncations on fatigue life. (Transport upper-wing-surface spectrum, flight mean gross-section stress = −111 MPa, four tests per spectrum.)*

trum loading tests were compared to the corresponding lives calculated from the theory. The cumulative damage calculations were based on the constant amplitude data represented in Fig. 4 and on the "rainflow" method [*9*] of defining the random load history in terms of constant amplitude cycles.

Predicted and measured lives are plotted in Fig. 9 for comparison. As can be seen, the linear damage theory always predicted longer lives than the actual test lives; that is, predictions were always on the unsafe side. The ratios of predicted to test life ranged from 6.6 to 19.8. Rosenfeld and Huang [*1*] and Schutz and Gerharz [*3*] have reported the same trend for tests on other graphite/epoxy laminates using fighter-wing load spectra.

The results in Fig. 9 also show that the linear damage theory failed to predict the trend toward longer life for progressively larger truncations of the spectrum at the low-load end. For truncations at the high-load end of the spectrum, however, the theory did predict the correct trend in life.

Concluding Remarks

The fatigue behavior of a notched, graphite/epoxy (T300/5208), $[45/0/-45/90]_{2s}$ laminate subjected to predominantly compressive loading

TABLE 5—*Results of the spectrum loading fatigue tests*[a] *using the baseline and truncated spectra.*

Spectrum Truncations	Fatigue Life, flights
None (baseline spectrum)	61 655
	61 655
	65 655
	165 655
Lowest load level omitted	34 855
	53 655
	57 655
	157 655
Two lowest load levels omitted	82 855
	86 855
	118 855
	137 655
Four lowest load levels omitted	34 855
	134 855
	137 655
	138 855
Two highest load levels omitted	49 652
	130 855
	267 982
	311 359
Four hightest and the lowest load levels omitted	508 105
	710 935
	711 276
	1 107 180

[a]Flight mean gross-section stress = −111 MPa, window in anti-buckling plate was 32 by 32 mm.

was explored in a series of constant amplitude and transport wing spectrum tests. These exploratory tests support the following conclusions and observations:

1. The amount of local (near the notch) buckling allowed in the tests significantly affected fatigue life. This result indicates that the influence of antibuckling procedures used in a test must be accounted for to make meaningful comparisons among compressive-loading fatigue data from the literature.

2. Spectrum truncation at either the high- or low-load end of the spectrum produced lives greater than those obtained in the baseline, complete-spectrum test. However, life was much more sensitive to truncations at the high-load end. The results suggest that in defining test spectra for composite wing structures, the high-load end of the spectrum should not be truncated

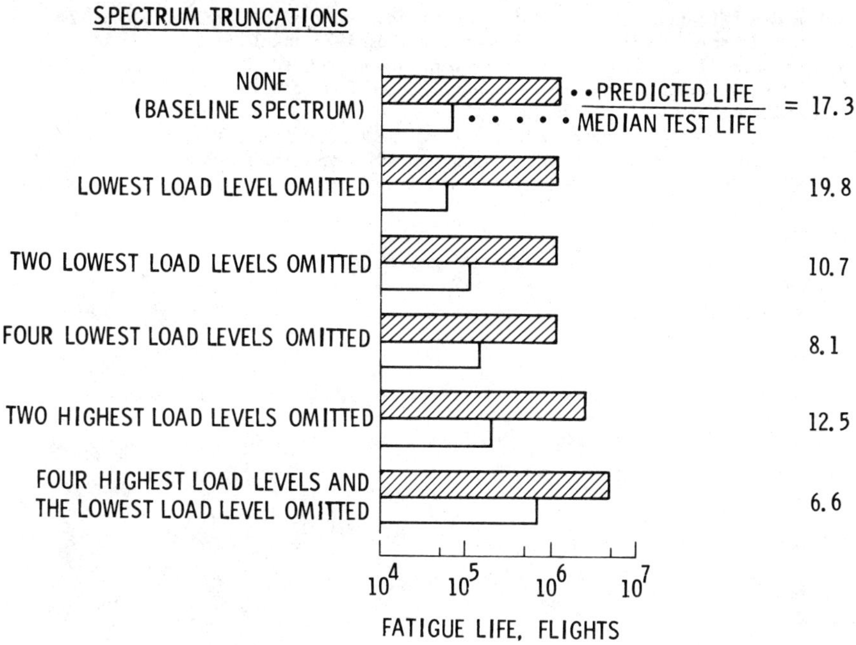

FIG. 9—*Comparisons of lives predicted by the linear cumulative damage theory to actual test lives.*

to the same extent as has been the practice for metallic structures. Also, truncation of the low-load end of the spectrum shows promise for achieving large reductions in test time without significantly changing the test result.

3. The Palmgren-Miner linear cumulative damage theory always predicted lives much longer than the actual spectrum loading test lives; that is, predictions were always on the unsafe side. However, the theory predicted the correct trends for effects of truncating the high-load end of the spectrum.

References

[*1*] Rosenfeld, M. S and Huang, S. L., "Fatigue Characteristics of Graphite/Epoxy Laminates Under Compression Loading," Technical Papers, Volume A, American Institute for Aeronautics and Astronautics, Inc., New York, 1977, pp. 423–427.

[*2*] Ramani, S. V. and William, D. P. in *Fatigue of Filamentary Composite Materials, ASTM STP 636*, American Society for Testing and Materials, 1977, pp. 27–46.

[*3*] Schutz, D. and Gerharz, J. J., *Composites*, Vol. 8, No. 4, Oct. 1977, pp.245–250.

[*4*] de Jonge, J. B., et al, "A Standardized Load Sequence for Flight Simulation Tests on Transport Aircraft Wing Structures," NLR-TR-73029U, National Lucht-en Ruimtevaart-laboratorium, Amsterdam, The Netherlands (also LBF-Bericht-FM-106, Laboratorium fur Betriebsfestigkeit, Darmstadt, German Federal Republic), 1973.

[*5*] Walter, R. W., et al in *Fatigue of Filamentary Composite Materials, ASTM STP 636*, American Society for Testing and Materials, 1977, pp. 228–247.

[6] Ryder, J. T. and Walker, E. K. in *Fatigue of Filamentary Composite Materials, ASTM STP 636,* American Society for Testing and Materials, 1977, pp. 3-26.
[7] Theocaris, P. S., *Journal of Scientific Instruments,* Vol. 41, No. 3, 1964, pp. 133-138.
[8] Chang, F. H., et al in *Composite Reliability, ASTM STP 580,* American Society for Testing and Materials, 1975, pp.176-190.
[9] Endo, T., et al, "Damage Evaluation of Metals for Random or Varying Loading—Three Aspects of Rain Flow Method," Proceedings, 1974 Symposium on Mechanical Behavior of Materials, Volume I, The Society of Materials Science, Kyoto, Japan, 21-24 Aug. 1974, pp. 371-380.

J. N. Yang[1] *and D. L. Jones*[1]

Load Sequence Effects on the Fatigue of Unnotched Composite Materials

REFERENCE: Yang, J. N. and Jones, D. L., **"Load Sequence Effects on the Fatigue of Unnotched Composite Materials,"** *Fatigue of Fibrous Composite Materials, ASTM STP 723,* American Society for Testing and Materials, 1981, pp. 213-232.

ABSTRACT: A comprehensive version of a fatigue and residual strength degradation model previously developed by the authors is proposed. This theoretical model is capable of predicting the statistical distributions of the fatigue life and the residual strength as well as the effect of load sequence, under variable amplitude or spectrum loadings, on these distributions. The model has been verified by the use of existing test data on glass/epoxy laminates. It is shown that the correlation between the model and the test results is good.

KEY WORDS: composite materials, fatigue life, residual strength degradation, statistical fatigue theory, load sequence effect, glass/epoxy, variable amplitude loading, spectrum loading, fatigue (materials)

The degradation of the residual strength and the fatigue behavior of composite laminates under service loads is of vital importance to the design of aircraft structures. As a result, intensive research on this subject has been performed in recent years [1-19].[2] Because of the statistical nature of the fatigue behavior of composite materials, emphasis has been placed on the statistical approach. Since service loadings are usually simplified by the use of spectrum or variable amplitude block loadings, any viable fatigue model should be able to predict the statistical fatigue behavior, as well as load sequence effects, under spectrum loadings. Unfortunately, most of the research efforts have been restricted to constant amplitude fatigue with the exception of Refs *10-13.* In order to predict the load sequence effect on the fatigue life, a deterministic model has been proposed in Refs *11-13,* while in Ref *10* a statistical model has been addressed.

A fatigue and residual strength degradation model proposed previously

[1] Professor and associate professor, respectively, School of Engineering and Applied Science, The George Washington University, Washington, D.C. 20052.

[2] The italic numbers in brackets refer to the list of references appended to this paper.

has been generalized to include a wider class of composite materials by introducing an additional parameter, ν. In particular, the composites having a strong degradation rate of the residual strength, such as glass/epoxy, can be taken into account. However, the advantages of the previous model [*7-10*] have been maintained, including the ability to account for the load sequence effect on the fatigue behavior of composite laminates in a straightforward manner. Likewise, by further making various approximations, the present model degenerates into various models proposed by Halpin and Waddoups [*1,2,15,16*], Hahn and Kim [*4,5*], Yang et al [*7-10*], and Broutman and Sahu [*11-13*].

The load sequence effect is found to be contributed by two different physical phenomena: (1) the difference in fatigue failure stress, referred to as the boundary effect, and (2) the memory effect of materials with respect to previous loadings. A model to account for the memory effect of materials, along with test results, will be presented elsewhere. When the load sequence effect is due to the difference in fatigue stress levels, it is shown herein that: (1) the Miner's damage sum at fatigue failure, D, is a statistical variable; (2) D is always greater than or equal to unity for the high-low load sequence; and (3) D is always smaller than or equal to unity for the low-high load sequence. The deviation of the Miner's damage sum, D, from unity increases as the difference in the stress level increases or as the degradation of the residual strength becomes stronger.

The present model has been verified by published data on E glass/epoxy laminates, and a simplified version of this model has also been verified by the test results of graphite/epoxy angle-ply laminates [*10*]. It is shown that the correlation between the theoretical model and the experimental data is reasonable.

Theoretical Derivation

The static ultimate strength is assumed to follow the two-parameter Weibull distribution with the shape parameter, α, and the scale parameter, β. For simplicity, let $R(0) = R_0$ be the static ultimate strength normalized by the scale parameter, β. Then, the distribution function of $R(0)$ is

$$F_{R(0)}(x) = P[R(0) \leq x] = 1 - \exp(-x^{\alpha}) \tag{1}$$

The residual strength degradation rate at the nth load cycle is assumed to be a power function of the current residual strength, $R(n)$, and the current maximum cyclic stress, σ, that is

$$dR(n)/dn = -f(\sigma, R_0)/\nu R^{\nu-1}(n) \tag{2}$$

in which $R(n)$ is also normalized by β, v is a parameter, and $f(\sigma, R_0)$ is a function of the maximum cyclic stress, σ, and the ultimate strength, $R_0 = R(0)$, to be determined later.

Integration of Eq 2 from n_1 to n_2 cycles yields

$$R^v(n_2) = R^v(n_1) - f(\sigma, R_0)(n_2 - n_1) \tag{3}$$

For $n_1 = 0$ and $n_2 = n$, Eq 3 reduces to

$$R^v(n) = R^v(0) - f(\sigma, R_0)n \tag{4}$$

The fatigue failure is assumed to occur when the residual strength, $R(n)$, is reduced to the maximum stress level, σ (also normalized by β), that is

$$R(n) = \sigma \quad \text{and} \quad n = N \tag{5}$$

in which N is a statistical variable denoting the fatigue life.

Substitution of the fatigue failure condition of Eq 5 into Eq 4 leads to the following expression

$$f(\sigma, R_0) = [R^v(0) - \sigma^v]/N \tag{6}$$

Since the static ultimate strength is assumed to follow the two-parameter Weibull distribution with the lower bound at zero, there always exists a probability that the specimen may fail instantly when the maximum cyclic stress, σ, is greater than the ultimate strength, $R(0)$. Such a failure is called an instantaneous failure. The probability of instantaneous failure, p_0, follows from Eq 1 as

$$p_0 = P[R(0) \leq \sigma] = 1 - \exp(-\sigma^\alpha) \tag{7}$$

It is reasonable to assume that the fatigue life can be represented by a Weibull distribution. Because of the probability of instantaneous failure as well as the fact that the fatigue life, N, is always greater than zero, the fatigue life distribution should follow the three-parameter Weibull distribution [see Refs *8-10*]

$$\begin{aligned} F_N(n) &= 1 - \exp\left[-\left(\frac{n}{\tilde{N}} + \sigma^{\alpha/\alpha_f}\right)^{\alpha_f}\right] \quad &&\text{for} \quad n \geq 0 \\ &= 0 &&\text{for} \quad n < 0 \end{aligned} \tag{8}$$

In Eq 8, α_f is the shape parameter and $\tilde{N}$ is the scale parameter (or characteristic life), which is a function of the cyclic stress range, S.

It can be verified easily from Eqs 7 and 8 that the instantaneous failure probability, p_0, corresponds to $n = 0$ in Eq 8, that is, $F_N(0) = P[N \leq 0] = 1 - \exp(-\sigma^\alpha) = p_0$. Thus in the distribution function of the fatigue life, Eq 8, there is a discontinuity $p_0 = 1 - \exp(-\sigma^\alpha)$ at $n = 0$ that represents precisely the probability of instantaneous failure. Hence, the probability density function of N has a singularity at $n = 0$, that is, $p_0\delta(0)$, where $\delta(0)$ is the Dirac delta function as shown in Fig. 1. In Fig. 1, the shaded area under the density function of $R(0)$ is p_0. Note that σ is normalized by β.

It is assumed that the strength degradation follows a deterministic path, that is, a specimen with a higher ultimate strength results in a longer fatigue life as shown in Fig. 1. In a statistical sense, it implies that the ultimate strength, $R(0)$, and the fatigue life, N, are completely correlated. When both statistical variables are completely correlated, they are functionally related. The functional relationship between the ultimate strength, $R(0)$, and the fatigue life, N, can be derived using the corresponding distribution functions given by Eqs 1 and 8, with the result

$$N = \tilde{N}\{[R(0)]^{\alpha/\alpha_f} - \sigma^{\alpha/\alpha_f}\} \tag{9}$$

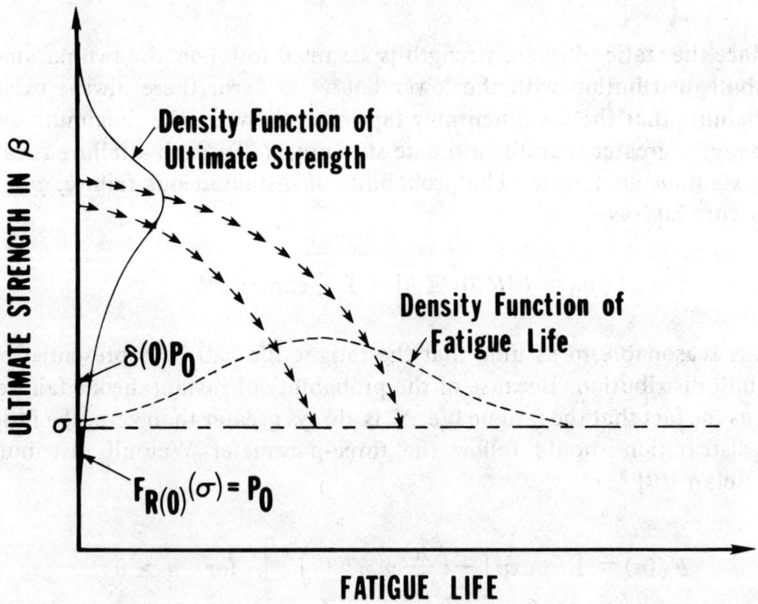

FIG. 1—*Illustration of the probability density functions of the ultimate strength and the fatigue life.*

To verify Eq 9, the distribution function $F_N(n)$ of the fatigue life can be derived using the ultimate strength distribution given by Eq 1 through the transformation of Eq 9. The resulting fatigue life distribution can easily be shown to be identical to that given by Eq 8.

By substitution of Eq 9 into Eq 6, the function of $f(\sigma, R_0)$ is obtained as follows

$$f(\sigma, R_0) = \frac{1}{\tilde{N}} \frac{R^v(0) - \sigma^v}{R^c(0) - \sigma^c} \tag{10}$$

in which $R_0 = R(0)$ and

$$c = \alpha/\alpha_f \tag{11}$$

The characteristic fatigue life, $\tilde{N}$, varies with respect to the fatigue loading, in particular the stress range, S. The characteristic fatigue life, $\tilde{N}$, can be expressed in terms of the classical S-N curve as

$$KS^b\tilde{N} = 1 \quad \text{or} \quad \tilde{N} = 1/KS^b \tag{12}$$

Residual Strength Degradation Model

The residual strength degradation model is thus obtained by substituting Eq 10 into Eq 3 as follows

$$R^v(n_2) = R^v(n_1) - \frac{R^v(0) - \sigma^v}{R^c(0) - \sigma^c}(n_2 - n_1)KS^b \tag{13}$$

in which $c = \alpha/\alpha_f$ (Eq 11) is the ratio of the shape parameter of the ultimate strength to that of the fatigue life, and $\tilde{N} = 1/KS^b$ (Eq 12) is the S-N curve of the characteristic fatigue life.

For $n_1 = 0$ and $n_2 = n$, Eq 13 reduces to the following

$$R^v(n) = R^v(0) - \frac{R^v(0) - \sigma^v}{R^c(0) - \sigma^c}KS^bn \tag{14}$$

It is observed from Eq 14 that the rate of degradation of the residual strength depends on the parameter v. Figure 2 depicts the effect of v on the strength degradation, that is, higher values of v are associated with lower rates of degradation.

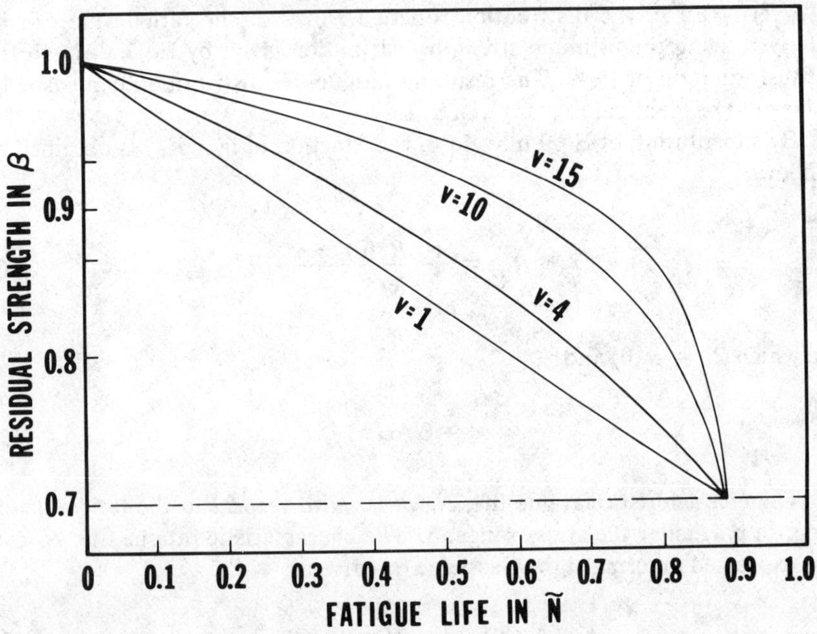

FIG. 2—*Illustration of the effect of* v *on the strength degradation.*

Special Cases

For graphite/epoxy laminates, test results indicate that the strength degrades very slowly [*1,4–10*] such that the value of v is very close to that of c. Hence for $v = c$, Eq 13 becomes

$$R^c(n_2) = R^c(n_1) - KS^b(n_2 - n_1) \tag{15}$$

Equation 15 is exactly identical to the model proposed by Yang et al [*7–10*]. Note that $R(n)$ and $R(0)$ are normalized by β in Eq 15 while they are not in Refs *7–10*.

In addition to the equality of v and c, if b is approximated by $c + 2$, that is, $b = c + 2$, then Eq 14 becomes

$$R^c(n) = R^c(0) - KS^{c+2}n \tag{16}$$

Equation 16 is exactly identical to the model proposed by Halpin and Waddoups [*1,2,15,16*].

Under constant amplitude loading where the applied stress is fixed (not allowed to vary), and $v = c$, Eq 14 becomes

$$R^c(n) = R^c(0) - \Phi n \tag{17}$$

Equation 17 is exactly identical to the model proposed by Hahn and Kim [5] for the stress rupture problem, where Φ is a constant.

Furthermore, Eq 14 degenerates into the deterministic fatigue model proposed by Broutman and Sahu [*11-13*] when the following approximations are made: (1) the statistical variables $R(0)$ and $R(n)$ are approximated by their corresponding median values (50 percent points), $\hat{R}(0)$ and $\hat{R}(n)$, respectively; (2) the term $\hat{R}^c(0) - \sigma^c$ is approximated by unity; (3) $\tilde{N}$ is approximated by the median fatigue life $\hat{N}$; and (4) v is equal to unity, that is, $v = 1.0$. With the approximations made above, Eq 14 becomes a deterministic equation

$$\hat{R}(n) = \hat{R}(0) - \frac{n}{\hat{N}}[\hat{R}(0) - \sigma] \tag{18}$$

that is proposed in Refs [*11-13*].

Distributions of Fatigue Life and Residual Strength

Incorporation of the fatigue failure condition given by Eq 5 into Eq 14 yields

$$N = \tilde{N}[R^c(0) - \sigma^c] \tag{19}$$

in which $\tilde{N} = 1/KS^b$. Equation 19 is identical to Eq 9 as expected. The distribution function of the fatigue life, N, can be derived from that of $R(0)$ given by Eq 1 through the transformation of Eq 19; with the result

$$\begin{aligned} F_N(n) &= 1 - \exp\left[-\left(\frac{n}{\tilde{N}} + \sigma^c\right)^{\alpha_f}\right]; \quad && n \geq 0 \\ &= 0 \quad ; && n < 0 \end{aligned} \tag{20}$$

which is identical to Eq 8 as expected (also identical to those given by Refs *8-10*).

It should be pointed up that unlike the two-parameter Weibull distribution in which the statistical variability (dispersion) is completely defined by the shape parameter, the statistical dispersion of the three-parameter Weibull distribution given by Eq 20 depends on both α_f and σ. Thus, although the shape parameter, α_f, in Eq 20 is assumed to be fixed for all stress levels of fatigue loading, the statistical dispersion of the fatigue life represented by Eq 20 varies with respect to the fatigue stress level, σ. In fact, it can be shown from Eq 20 that the statistical dispersion decreases as the maximum stress level, σ, increases. This is consistent with the test results presented in Refs *11–13*. Likewise, this phenomenon is also well-known for the fatigue data of metallic materials.

The distribution function, $F_{R(n)}(x)$, of the residual strength, $R(n)$, can be obtained from that of $R(0)$ given by Eq 1 through the transformation of Eq 14 in the following manner.

$$F_{R(n)}(x) = P[R(n) \leq x] = P[R(0) \leq y] = F_{R(0)}(y) \tag{21}$$

in which x and y are, respectively, the corresponding values of $R(n)$ and $R(0)$. They are related through Eq 14 by replacing $R(n)$ and $R(0)$, respectively, by x and y

$$x^{\nu} = y^{\nu} - \frac{n}{\tilde{N}} \frac{y^{\nu} - \sigma^{\nu}}{y^{c} - \sigma^{c}} \tag{22}$$

It then follows from Eqs 21 and 1 that

$$F_{R(n)}(x) = F_{R(0)}(y) = 1 - \exp(-y^{\alpha}) \tag{23}$$

Solving for y in terms of $F_{R(n)}(x)$ from Eq 23 results in

$$y = (-\ln[1 - F_{R(n)}(x)])^{1/\alpha} \tag{24}$$

Thus, the distribution function of the residual strength $R(n)$ is obtained implicitly in Eqs 22 and 24, that is

$$y = (-\ln[1 - F_{R(n)}(x)])^{1/\alpha}, \quad x = \left[y^{\nu} - \frac{n}{\tilde{N}} \frac{y^{\nu} - \sigma^{\nu}}{y^{c} - \sigma^{c}}\right]^{1/\nu} \quad \text{for } x > \sigma \tag{25}$$

Equation 25 is convenient for the numerical computation of $F_{R(n)}(x)$. The computational procedures start by assigning a series of values for $F_{R(n)}(x)$, such as 0.05, 0.1, . . . 0.9, 0.95, etc. Then, the corresponding values of x and y can easily be computed from Eq 25. For instance, the median strength x_{50} (50 percent point) is obtained from Eq 25 by setting $F_{R(n)}(x) =$ 0.5 and replacing x by x_{50}.

The distribution function of $R(n)$ evaluated at σ, that is, $F_{R(n)}(\sigma)$, given by Eq 25 is exactly the probability that the specimen will not survive n load cycles. It can easily be shown from Eqs 25 and 20 that $F_{R(n)}(\sigma) = F_N(n)$ as expected.

Load Sequence Effects

The effect of load sequence on the fatigue of composite materials may be attributed to two different physical phenomena; (1) the difference in fatigue failure stress and (2) the material memory.

Assuming that fatigue fracture occurs when the residual strength is reduced to the maximum cyclic stress level, the fatigue process will be terminated (that is, fracture occurs) at different strength levels under high-low and low-high sequences. Such an effect is herein referred to as the boundary effect, and it will be derived later. The boundary effect can easily be observed from the fatigue life data and the degree of the load sequence effect can be measured using Miner's damage sum.

The memory effect of materials resulting from previously experienced loadings may be nonlinear in nature. A typical example in metallic materials is the residual stress existing in the plastic zone of the crack tips resulting from overload. Further investigation is needed for composite materials in this regard.

A Sequence of Two Fatigue Loadings

Let the specimen be subjected to a sequence of two fatigue loadings with maximum stresses, σ_1 (stress range, S_1), and σ_2 (stress range S_2), respectively, for n_1 and n_2 cycles. Then, it follows from Eq 14 that the residual strength, $R(n_1)$, after the application of the first fatigue loading is

$$R^v(n_1) = R^v(0) - (n_1/\tilde{N}_1)J_1[R(0)] \tag{26}$$

in which

$$\tilde{N}_1 = 1/KS_1{}^b, \qquad J_1[R(0)] = [R^v(0) - \sigma_1{}^v]/[R^c(0) - \sigma_1{}^c] \tag{27}$$

After the second fatigue loading has been applied, the residual strength $R(n_1 + n_2)$ at $n_1 + n_2$ cycles can be expressed in terms of the previous strength $R(n_1)$ using Eq 13 as

$$R^v(n_1 + n_2) = R^v(n_1) - (n_2/\tilde{N}_2)J_2[R(0)] \tag{28}$$

in which

$$\tilde{N}_2 = 1/KS_2{}^b, \qquad J_2[R(0)] = [R^v(0) - \sigma_2{}^v]/[R^c(0) - \sigma_2{}^c] \tag{29}$$

Summation of Eqs 26 and 28 leads to the expression for the residual strength $R(n_1 + n_2)$

$$R^v(n_1 + n_2) = R^v(0) - (n_1/\tilde{N}_1)J_1[R(0)] - (n_2/\tilde{N}_2)J_2[R(0)] \tag{30}$$

The distribution function, $F_{R(n_1+n_2)}(x)$, of the residual strength, $R(n_1 + n_2)$, can be obtained from that of $R(0)$ similar to the derivation for Eq 25, with the result,

$$y = (-\ln[1 - F_{R(n_1+n_2)}(x)]^{1/\alpha},$$

$$x^{\nu} = y^{\nu} - (n_1/\tilde{N}_1)J_1(y) - (n_2/\tilde{N}_2)J_2(y) \quad \text{for} \quad x \geq \sigma_2 \tag{31}$$

The probability of fatigue failure before $n_1 + n_2$ cycles is $F_{R(n_1+n_2)}(\sigma_2)$, and it can be obtained from Eq 31 by setting $x = \sigma_2$.

If the number of load cycles under the second cyclic loading is increased until fatigue fracture, then the fatigue life, N_{12}, under the second cyclic loading is a statistical variable. N_{12} can be obtained by substituting the condition of fatigue failure, that is, $R(n_1 + n_2) = \sigma_2$, $n_2 = N_{12}$, into Eq 30 as

$$N_{12} = \tilde{N}_2[R^c(0) - \sigma_2{}^c] - n_1\{[\tilde{N}_2 J_1[R(0)]/\tilde{N}_1 J_2[R(0)]\} \tag{32}$$

The distribution function, $F_{N_{12}}(n_{12}) = P[N_{12} \leq n_{12}]$, of the fatigue life under the second fatigue loading is obtained as

$$y = \{-\ln[1 - F_{N_{12}}(n_{12})]\}^{1/\alpha},$$

$$n_{12} = \tilde{N}_2(y^c - \sigma_2{}^c) - n_1[\tilde{N}_2 J_1(y)/\tilde{N}_1 J_2(y)] \tag{33}$$

For instance, the median fatigue life, $\hat{n}_{12}$, is obtained from Eq 33 by replacing $F_{N_{12}}(n_{12})$ and n_{12}, respectively, by 0.5 and $\hat{n}_{12}$.

It is interesting to observe from Eq 30 that the residual strength, $R(n_1 + n_2)$, does not depend on the load sequence. However, it is also observed from Eqs 32 and 33 that the fatigue life and its distribution under the second fatigue loading depend on the sequence of cyclic loading. Such a dependence will be examined in terms of Miner's damage sum.

Miner's Damage Sum

Let N_1 and N_2 be the fatigue lives under the cyclic loadings, σ_1 (stress range, S_1) and σ_2 (stress range, S_2), respectively, in which σ_1 may be greater or smaller than σ_2. Let the specimen be subjected to the first fatigue loading, σ_1, for n_1 load cycles and then to the second fatigue loading, σ_2, until fracture. Then, Miner's damage sum at fatigue failure, denoted by D, is

$$D = \frac{n_1}{N_1} + \frac{N_{12}}{N_2} \tag{34}$$

in which N_{12} is given by Eq 32.

The fatigue life, N_1, due to σ_1 alone is obtained from Eq 19 as

$$N_1 = \tilde{N}_1[R^c(0) - \sigma_1{}^c] \tag{35}$$

and the fatigue life, N_2, under σ_2 alone is

$$N_2 = \tilde{N}_2[R^c(0) - \sigma_2{}^c] \tag{36}$$

Substituting Eqs 32, 35, and 36 into Eq 34 yields

$$D = 1 + \frac{n_1}{\tilde{N}_1[R^c(0) - \sigma_1{}^c]}\left[1 - \frac{R^v(0) - \sigma_1{}^v}{R^v(0) - \sigma_2{}^v}\right] \tag{37}$$

It is observed from Eq 37 that Miner's damage sum, D, is a statistical variable since $R(0)$ is a statistical variable. Furthermore, the fatigue problem is associated with the situation where the ultimate strength, $R(0)$, is greater than σ_1 or σ_2. The following observations can plausibly be obtained from Eq 37. (1) Under the high-low loading sequence, where $\sigma_1 > \sigma_2$, D is always greater than unity and the lower bound of D is unity when $\sigma_1 = \sigma_2$, that is, $D \geq 1$. Furthermore, D increases as the difference between σ_1 and σ_2 increases or as v decreases. (2) Under the low-high loading sequence, where $\sigma_1 < \sigma_2$, D is always smaller than unity and the upper bound of D is unity when $\sigma_1 = \sigma_2$, that is, $D < 1$. Furthermore, D decreases as the difference between σ_1 and σ_2 increases or as v decreases. (3) The deviation of Miner's sum from unity increases as the damage, $n_1/\tilde{N}_1$, accumulated under the first fatigue loading increases.

From the observations just made, the low-high load sequence is more damaging. The probability density function of Miner's sum is schematically displayed in Fig. 3 for the low-high and the high-low load sequences. Moreover, the median Miner's sum, $\hat{D}$, can be obtained easily from Eq 37 by replacing D and $R(0)$ by $\hat{D}$ and $\hat{R}(0)$, respectively, where $\hat{R}(0) = (-\ln 0.5)^{1/\alpha}$ is the median value of $R(0)$.

Statistical Fatigue Under Spectrum Loading

Let a specimen be subjected to a spectrum loading in which the maximum stress of each load cycle is $\sigma_1, \sigma_2, \ldots \sigma_m$ (with a sequence of stress range, $S_1, S_2, \ldots S_m$), where some of the stress levels may be identical. Then, using the strength degradation model given by Eq 13, one can express the residual strength cycle-by-cycle as follows

$$\begin{aligned} R^v(1) &= R^v(0) - (J_1[R(0)]/\tilde{N}_1) \\ R^v(2) &= R^v(1) - (J_2[R(0)]/\tilde{N}_2) \\ &\text{-----------------------------} \\ R^v(m) &= R^v(m-1) - (J_m[R(0)]/\tilde{N}_m) \end{aligned} \tag{38}$$

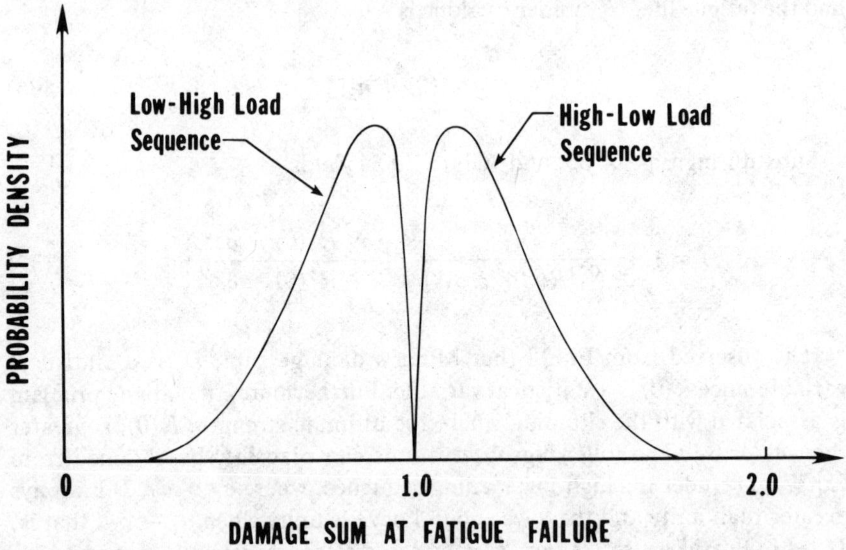

FIG. 3—*Probability density functions of Miner's sum for low-high and high-low load sequence.*

Summation of Eq 38 yields the residual strength, $R(m)$, after m load cycles

$$R^{v}(m) = R^{v}(0) - \sum_{i=1}^{m} J_i[R(0)]/\tilde{N}_i \tag{39}$$

in which

$$\begin{aligned} J_i[R(0)] &= [R^{v}(0) - \sigma_i^{v}]/[R^{c}(0) - \sigma_i^{c}] \\ \tilde{N}_i &= 1/KS_i^{b}; \qquad i = 1, 2, \ldots m \end{aligned} \tag{40}$$

Then, the distribution of $R(m)$, denoted by $F_{R(m)}(x) = P[R(m) \leq x]$, under the spectrum loading is obtained from that of $R(0)$ as

$$y = \{-\ln[1 - F_{R(m)}(x)]\}^{1/\alpha}, \qquad x = \left[y^{v} - \sum_{i=1}^{m} J_i(y)/\tilde{N}_i\right]^{1/v} \tag{41}$$

The fatigue life distribution under spectrum loadings can similarly be obtained [*10*].

Determination of Model Parameters

It follows from the previous discussions that after the model parameters α, β, b, c, K, and v have been determined, the theoretical model is capable of predicting the load sequence effect as well as the distributions of the residual strength and the fatigue life under variable amplitude or spectrum loadings. A minimum amount of baseline data [*7-10*] should be generated for the determination of these parameter values: (1) one set of the static strength data (approximately 15 specimens) for determining α and β, and (2) one set of the fatigue scan data with a total of approximately 40 to 45 specimens for determining b, c, K, and v. The fatigue scan data should consist of (*a*) the fatigue life data under various stress levels (approximately 25 to 30 specimens) and (*b*) residual strength data under three different stress levels up to certain number of cycles (approximately 15 specimens). The fatigue life data will be used to estimate the values of b, c, and K, while the residual strength data will be used to estimate the value of v.

When fatigue failure occurs, the model reduces to Eq 19 (the fatigue life model), that is, $N = [R^c(0) - \sigma^c]/KS^b$

$$R^c(0) = \sigma^c + KS^bN \tag{42}$$

Then, the fatigue life data in the fatigue scan set can be used to determine b, c, and K using Eq 42 following the procedures described in Refs *8* and *9*.

After determining the values of α, β, b, c, and K, $R(n)$ can be plotted against $KS^bn = n/\tilde{N}$ for different values of v using Eq 14 with $R(0) = 1.0$. Such a plot is shown in Fig. 2. Then, the residual strength data in the fatigue scan set can be plotted on the figure to determine v or a least squares fit to the data can be used.

The accuracy of the values of b, c, K, and v can further be improved if one wishes by (1) using the previous estimated values of b, c, K, and v as the initial trial values; (2) converting all the fatigue scan data into the equivalent static strength data [*8,9*] using Eq 14; and (3) matching the first three central moments as described in Refs *8* and *9*.

The procedure just described requires a minimum amount of baseline data for the determination of the model parameters. More accurate procedures can be used although more baseline data will be required. For instance, when a set of 30 fatigue life data is generated at a single stress level, α_f can be estimated and hence $c = \alpha/\alpha_f$. If m sets of fatigue life data are generated at m different stress levels, then $\tilde{N}_i (i = 1, 2, \ldots m)$ can be estimated directly and used in the model (note that $\tilde{N}_i = 1/KS_i^b$).

Experimental Verification

Test data on coupon specimens of cross-ply E glass/epoxy laminates (Scotchply Type 1002) have been published in Refs *11-13*. All of these fatigue tests were performed at 10 Hz with a stress ratio of 0.05. The average value of the ultimate strength was given as 448 MPa (65 ksi) [*11-13*]. From the ultimate strength data of glass/epoxy presented in Refs *1* and *4*, α is approximately 15.6 (7.7 percent dispersion) and hence $\beta = 464$ MPa (67.3 ksi).

More than 30 specimens were fatigue tested at each stress level 386, 338, 290, and 241 MPa (56, 49, 42, and 35 ksi). Fatigue scan data were selected uniformly from these four data sets as shown in Table 1. Using the procedure described before, as well as the data in Table 1, the values of b, c, and K were found to be

$$c = 6.0, \qquad b = 11.413, \qquad K = 6.083 \times 10^{-33}(22.6 \times 10^{-24}) \qquad (43)$$

which indicate that $\alpha_f = \alpha/c = 2.6$.

A set of residual strength data subjected to different stress levels up to various numbers of load cycles is also available in Ref *13*. From this set of residual strength data, it is determined that $v = 1.0$.

After determining the parameter values of α, β, b, c, K, and v from this limited amount of baseline data (that is, fatigue scan data), it is possible to predict: (1) the statistical distributions of the fatigue life and the residual strength, (2) the Miner's damage sum, and (3) the load sequence effect.

Four sets of fatigue life data [*13*] under a single stress level are plotted as circles in Fig. 4. Also plotted in Fig. 4 as solid curves are the predictions based on Eq 20 with the parameter values just determined. It is observed that the correlation between the theoretical predictions and the test results is satisfactory.

Fatigue life data under dual stress level cyclic loading are also available [*13*]. The median values (50 percent point) of the Miner's damage sum, $\hat{D}$, for various loading conditions have been computed using Eq 37. The results obtained herein are very close to those obtained by Broutman and Sahu

TABLE 1—*Fatigue scan results for cross-ply glass/epoxy composite laminates (fatigue life in cycles).*

Maximum Stress, MPa (ksi)			
386(56)	338(49)	290(42)	241(35)
202	1366	4 710	70 435
250	1670	7 050	87 045
291	2352	8 008	98 593
365	2412	9 371	119 005
406	2638	11 364	157 410
470	2766	12 877	199 429
601	2990	13 859	209 000
675	3250	17 731	236 952
723	3759	19 456	251 590
834		24 554	511 668
970		25 578	
		33 214	

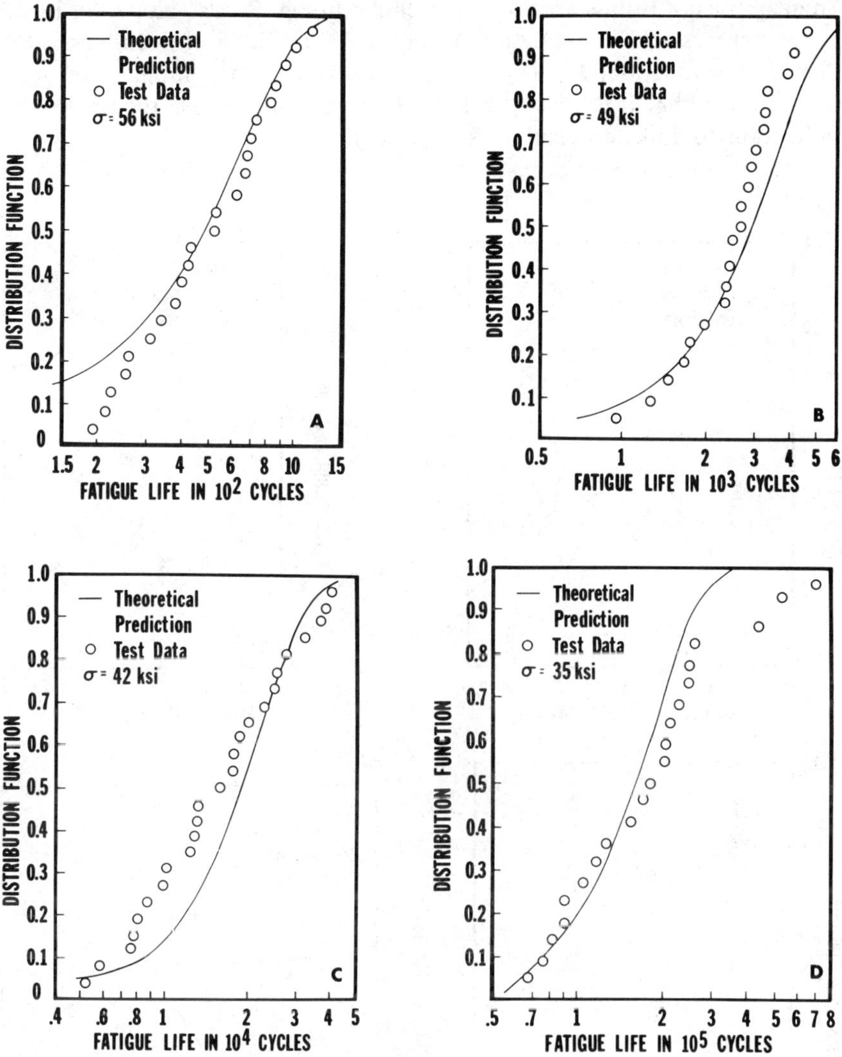

FIG. 4—*Distribution function of the fatigue life of glass/epoxy laminates at four different stress levels:* (a) *386 MPa (56 ksi),* (b) *338 MPa (49 ksi),* (c) *290 MPa (42 ksi), and* (d) *241 MPa (35 ksi).*

[*11-13*] and hence they are not repeated. Likewise, the load sequence effect is important since $\hat{D}$ varies from 0.6 to 1.3 depending on the loading conditions [*11-13*].

In Refs *11-13,* specimens were subjected to σ_1 stress level for n_1 load cycles and were then subjected to σ_2 stress level until fatigue failure. Fatigue life data N_{12} under the second fatigue loading σ_2 are also available [*11-13*].

Some of the test results are plotted as circles in Fig. 5. Also plotted in Fig. 5 as solid curves are the predictions using Eq 33. In Fig. 5, the ordinate is the distribution function, $F_{N_{12}}(n_{12})$, and the abscissa is the fatigue life, N_{12}, under σ_2 loading. It is observed that the correlation between the theoretical predictions and the test results is satisfactory.

Three additional sets of test results are plotted in Fig. 6 as circles along

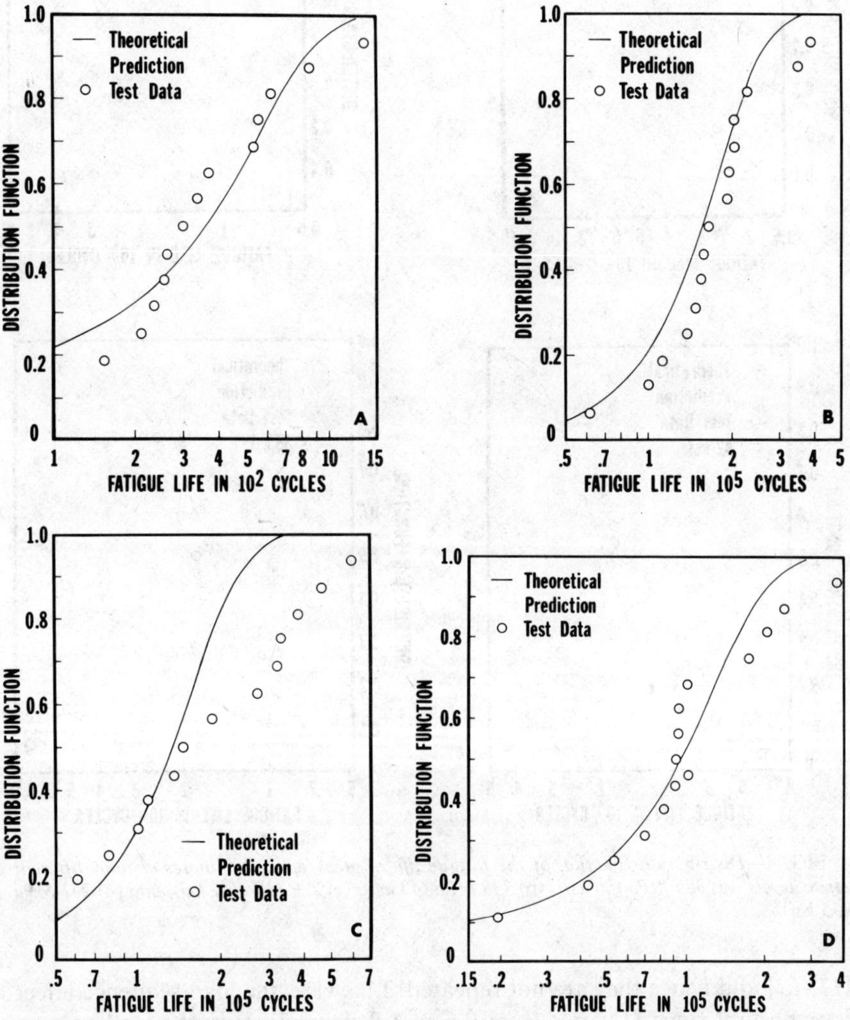

FIG. 5—*Distribution function for the fatigue life of glass/epoxy laminates under various low-high and high-low cyclic loadings:* (a) σ_1 = *290 MPa (42 ksi),* n_1 = *2000,* σ_2 = *386 MPa (56 ksi);* (b) σ_1 = *338 MPa (49 ksi),* n_1 = *250,* σ_2 = *241 MPa (35 ksi);* (c) σ_1 = *386 MPa (56 ksi),* n_1 = *250,* σ_2 = *241 MPa (35 ksi); and* (d) σ_1 = *290 MPa (42 ksi),* n_1 = *10 000,* σ_2 = *241 MPa (35 ksi).*

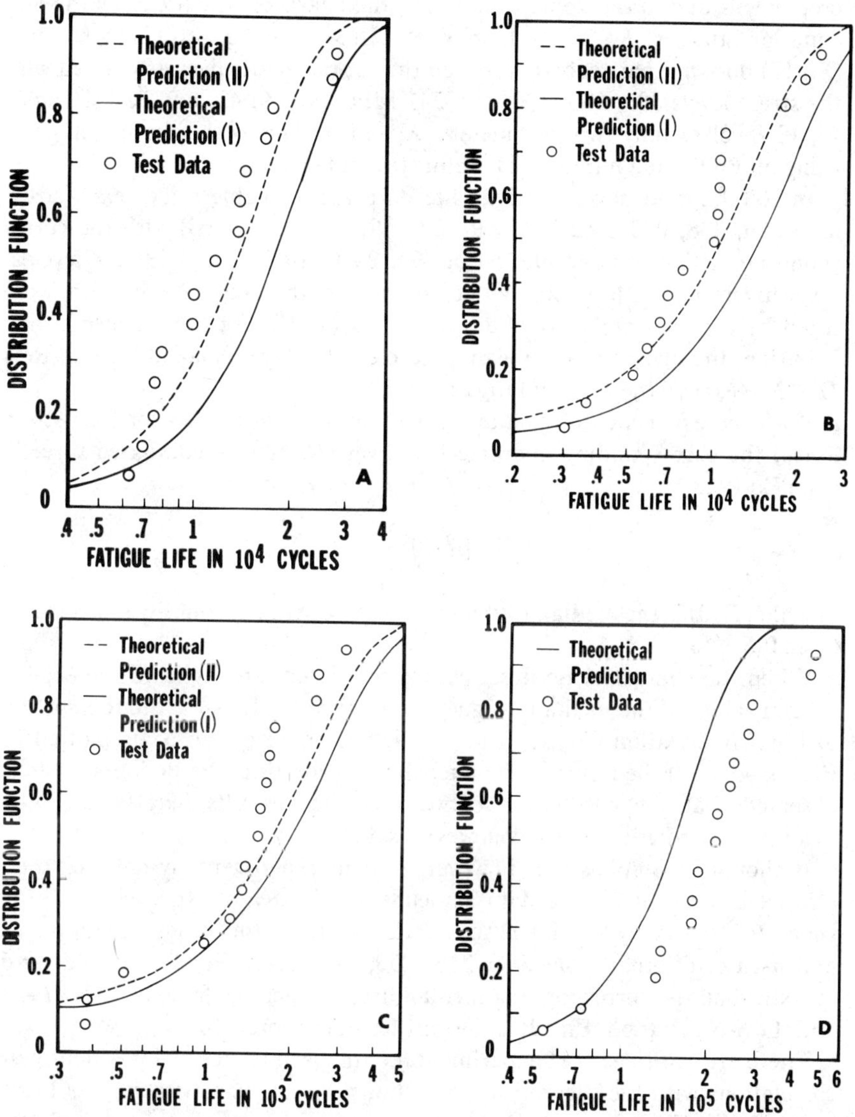

FIG. 6—*Distribution function of the fatigue life of glass/epoxy laminates under different high-low cyclic loadings.* (a) σ_1 = *386 MPa (56 ksi),* n_1 = *100,* σ_2 = *290 MPa (42 ksi);* (b) σ_1 = *338 MPa (49 ksi),* n_1 = *1000,* σ_2 = *290 MPa (42 ksi);* (c) σ_1 = *386 MPa (56 ksi),* n_1 = *250,* σ_2 = *338 MPa (49 ksi);* (d) σ_1 = *386 MPa (56 ksi),* n_1 = *100,* σ_2 = *241 MPa (35 ksi).*

with the solid curves representing the theoretical predictions. The correlation is not as good as that of Fig. 5. A careful examination reveals that the discrepancy is due to the approximation of the characteristic life $\tilde{N}$ by the *S-N* equation, that is, $KS^b\tilde{N} = 1$ or $\tilde{N} = 1/KS^b$, see Eq 12. The *S-N* curve (Eq 12) does not pass exactly through the characteristic lives associated with the stress levels 386, 338, 290, and 241 MPa (56, 49, 42, and 35 ksi), since Eq 12 involves only two parameters, K and b. This is the disadvantage of using only a limited number of baseline test data.

In Ref *13,* four sets of fatigue life data are available under each stress level 386, 338, 290, and 241 MPa (56, 49, 42, and 35 ksi) with the corresponding median fatigue life, $\hat{N}$, of 493, 2470, 14 700, and 172 200 cycles, respectively [*11-13*]. It should be mentioned that the fatigue data were fitted by the lognormal distribution in Refs *11-13.* For the lognormal distribution, the inverse of the mean log cycles to failure, denoted by N in Refs *11-13,* is exactly the median fatigue life, $\hat{N}$.

With the experimental median fatigue lives, $\hat{N}$, just given for four stress levels, the corresponding characteristic lives, $\tilde{N}$, can be computed directly from Eq 19 as

$$\tilde{N} = \hat{N}/[\hat{R}^c(0) - \sigma^c], \tag{44}$$

in which $\hat{R}(0)$ is the median ultimate strength obtained from Eq 1 as $\hat{R}(0) = (-\ln 0.5)^{1/\alpha}$.

When the characteristic lives, $\tilde{N}$, are computed directly from the experimental values of the median fatigue life, $\hat{N}$, through Eq 44 without resorting to the *S-N* equation (Eq 12), the prediction based on Eq 33 is plotted in Fig. 6*a-c* as dashed curves, denoted by "Theoretical Prediction II." It is observed that the correlation between the test results (circles) and the theoretical prediction (dashed curves) is satisfactory.

It should be emphasized, however, that to experimentally determine $\tilde{N}$, a large amount of fatigue data associated with each stress level has to be generated that may not be practical. Likewise, if the fatigue loadings σ_1 and σ_2 are different from 386, 338, 290, and 241 MPa (56, 49, 42, and 35 ksi), then the corresponding median lives are not available in Refs *11-13* and the *S-N* equation, Eq 12, is the only feasible approach.

There are several sets of experimental data given in Ref *13* for which their correlations with the theoretical predictions are not as satisfactory as those presented in Figs. 5 and 6*a-c.* The discrepancy exists whether $\tilde{N}$ is determined from the *S-N* equation using only a limited amount of the baseline data or from the experimental values of the median fatigue life. A typical example is presented in Fig. 6*d* where the correlation in the lower tail portion of the distribution is good, but at the 0.5 probability level, the predicted life was approximately 140 000 cycles with the actual value closer to 200 000 cycles.

Finally, the experimental data of graphite/epoxy $[\pm 45]_{2s}$ laminates have been reported in Ref *10* for high-low and low-high load sequences. For this particular material where $v = c = 10.0$, it has been shown (see Ref *10*) that the correlation between the test data and the theoretical prediction based on the simplified model (where $v = c$, see Eq 15) is satisfactory.

Conclusions

A comprehensive version of a fatigue and residual strength degradation model developed earlier by the authors is proposed to predict the effect of load sequence on the statistical fatigue behavior of composite laminates. The statistical distributions of the residual strength and the fatigue life under dual stress levels and spectrum loadings have been presented, and the statistical characteristics of Miner's damage sum have been discussed. The present model reduces to various fatigue models proposed in the literature when further approximations are made.

The model proposed herein has been verified by existing experiments on glass/epoxy laminates. It is shown that the correlation between the model and the test results under dual stress levels is reasonable. Furthermore, a simplified version of the present model in which $v = c$ has been verified by experiments on graphite/epoxy laminates [*10*], in which the correlation between the experimental results and the theoretical predictions under dual stress levels is satisfactory. The present model is also capable of predicting the effect of proof loads on the fatigue behavior of composite materials [*7,17*].

Acknowledgments

This research was supported by NASA Langley Research Center through Grant NSG 1604. The authors would like to thank Drs. J. R. Davidson, G. L. Roderick, and T. K. O'Brien of NASA Langley Research Center for their valuable discussions.

References

[*1*] Halpin, J. C., et al, "Characterization of Composites for the Purpose of Reliability Evaluation," AFML-TR-77-289, Wright-Patterson Air Force Base, Ohio, 1972.

[*2*] Halpin, J. C., Johnson, T. A., and Waddoups, M. E., *International Journal of Fracture Mechanics,* Vol. 8, 1972, pp. 465–472.

[*3*] Salkind, M. J. in *Composites Materials: Testing and Design, ASTM STP 497,* American Society for Testing and Materials, 1972, pp. 143–169.

[*4*] Hahn, H. T., "Fatigue Behavior and Life Prediction of Composite Materials," AFML-TR-78-43, Wright-Patterson Air Force Base, Ohio, 1978.

[*5*] Hahn, H. T. and Kim, R. Y., *Journal of Composite Materials,* Vol. 9, 1975, pp. 297–311.

[*6*] Ryder, J. T. and Walker, E. K., "Ascertainment of the Effect of Compressive Loading on the Fatigue Life Time of Graphite/Epoxy Laminates for Structural Application," AFML-TR-76-241, Wright-Patterson Air Force Base, Dec. 1976.

[7] Yang, J. N. and Liu, M. D., *Journal of Composite Materials,* Vol. 11, 1977, pp. 176–203.
[8] Yang, J. N., *Journal of Composite Materials,* Vol. 12, Jan. 1978, pp. 19–29.
[9] Yang, J. N. and Jones, D. L., *Journal of Composite Materials,* Oct. 1978, pp. 371–389.
[10] Yang, J. N. and Jones, D. L., *Proceedings,* AIAA/ASME/ASCE/ASH 20th Structures, Structural Dynamics and Materials Conference, April 1979, pp. 232–239.
[11] Broutman, L. J. and Sahu, S., *Composite Materials: Testing and Design* (Second Conference), *ASTM STP 497,* American Society for Testing and Materials, 1972, pp. 170–188.
[12] Broutman, L. J. and Sahu, S., "Progressive Damage of Glass Reinforced Plastics During Fatigue," *Proceedings,* 24th SPI Conference, Feb. 1969, p. 11-D.
[13] Sahu, S., "Fatigue and Fracture of Glass Filled Polymer Composites," Ph.D. thesis, Department of Metallurgical Engineering, Illinois Institute of Technology, 1970.
[14] Chou, P. C. and Wang, A. S. D., "Statistical Analysis of Fatigue of Composite Materials," AFML-TR-78-96, Wright-Patterson Air Force Base, Ohio, July 1978.
[15] Manning, S. D., Lemon, G. H., Waddoups, M. E., and Achord, R. T., "Composite Wing for Transonic Improvement, Vol. III—Structural Reliability Study," Air Force Flight Dynamic Laboratory Technical Report, AFFDL-TR-71-24, 1973.
[16] Wolff, R. V. and Lemon, G. H., "Reliability Prediction for Composite Joints," Air Force Materials Laboratory Technical Report, AFML-TR-74-197, 1974.
[17] Yang, J. N. and Sun, C. T., *Journal of Composite Materials,* Vol. 14, April 1980, pp. 168–176.
[18] McLaughlin, P. V., Jr., Kulkarni, S. V., Huang, S. N., and Rosen, B. W., "Fatigue of Notched Fiber Composite Laminates Part I: Analytical Model," Materials Science Corporation, MSC/TFR/7501, March 1975.
[19] Kulkarni, S. V., McLaughlin, P. V., Jr., and Pipes, R. B., "Fatigue of Notched Fiber Composite Laminates Part II: Analytical and Experimental Evaluation," Material Science Corporation, MSC/TFR/7601, March 1976.

C. T. Sun[1] *and E. S. Chim*[1]

Fatigue Retardation Due to Creep in a Fibrous Composite

REFERENCE: Sun, C. T. and Chim, E. S., **"Fatigue Retardation Due to Creep in a Fibrous Composite,"** *Fatigue of Fibrous Composite Materials, ASTM STP 723,* American Society for Testing and Materials, 1981, pp. 233–242.

ABSTRACT: A series of experiments were performed in an attempt to investigate the effect of creep strain and temperature on the fatigue life of a $[\pm 45]_{2s}$ graphite/epoxy composite laminate. The experimental results indicate that sustained periods of static loads have significant retardation effects and that cyclic loading with a lower frequency preceeding one with a higher frequency also prolongs the fatigue life.

KEY WORDS: composite materials, fatigue (materials), retardation, crack propagation, creep, hold time

The load frequency has been found to have definite effects on the fatigue life of laminated composites [*1,2*].[2] Recently, Sun and Chan [*1*] have conducted a sequence of tests on a $[\pm 45]_{2s}$ graphite/epoxy laminated composite and found that both temperature and creep had significant effects on the fatigue life. Based upon their test results, it was found that higher frequencies tended to increase the fatigue life while the rise in temperature is frequency-dependent, that is, at the same stress level, higher frequencies cause higher temperature rises in the material.

Figure 1 presents the test data obtained in Ref *1* in terms of time to fatigue failure. In Ref *1*, the numbers of cycles to failure corresponding to the four loading frequencies were presented. It was noted that for the first two frequencies (1 and 3 Hz) the fatigue life (in number of cycles) is in proportion to the loading frequency. As indicated by Fig. 1, the times to failure for both frequencies are practically the same. A similar frequency effect on the crack growth rate has been observed in both metals [*3*] and polymers [*4*]. An implication of this finding is that the life of a composite could be more meaningfully measured in terms of the time under load (hold time) rather than number of cycles.

[1] Professor and graduate student, respectively, School of Aeronautics and Astronautics, Purdue University, West Lafayette, Ind. 47907.

[2] The italic numbers in brackets refer to the list of references appended to this paper.

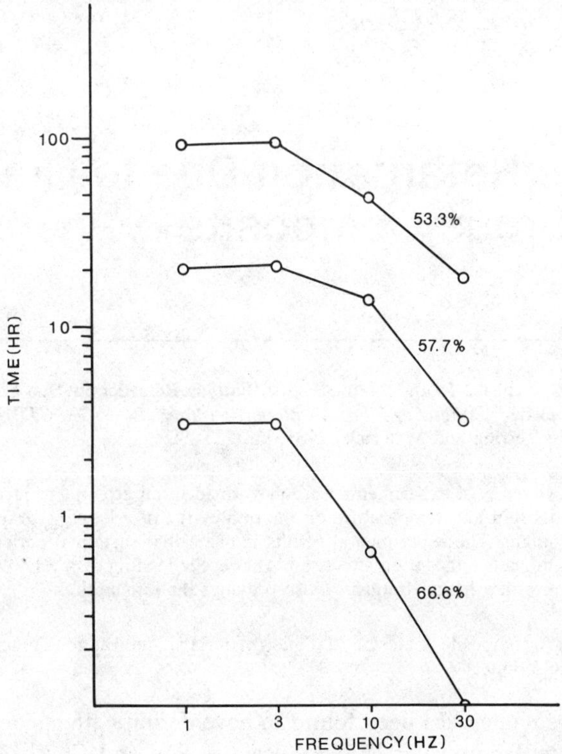

FIG. 1—*Average fatigue lives (hours) at three loading levels (53.3, 57.7, and 66.6 percent of the ultimate stress) and four different loading frequencies.*

The objective of this study is to examine the aforementioned implication both from theoretical arguments and experimental results.

Specimen

The material used in this study was T300/5208 graphite/epoxy composite that was the same as that used by Sun and Chan [*1*]. All specimens were flat 8-ply $[\pm 45]_{2s}$ laminates with a 6.25 mm (0.25 in.) diameter center hole. The specimen size was 254 mm (10 in.) long, 38.1 mm (1.5 in.) wide and 1.19 mm (0.047 in.) thick. Glass/epoxy end tabs of 38.1 mm (1.5 in.) in length were used in the testing.

The static strength for the laminate was obtained in Ref *1*. Five specimens from the same batch of material used for the fatigue tests were tested. The average net-section stress was found to be 155.41 MPa (22.54 ksi) with 160.65 MPa (23.3 ksi) and 151.69 MPa (22.0 ksi) being the maximum and minimum, respectively.

Effect of Static Creep

Test Procedures and Results

Eight specimens were used. In general, each specimen underwent a number of fatigue cycles under a constant amplitude tension-tension sinusoidal loading at intermittent intervals separated by periods during which the specimen was held statically under the peak load of the cyclic loading. The histories of the cyclic loading and static loading for each specimen are presented in Figs. 2 and 3. The peak stress for the sinusoidal loadings was 66.6 percent of the average ultimate static stress as determined in Ref *1*. The R ratio was 1/15.

Figure 2 shows the complete loading history for Specimen G-78. A shaded block represents the time for the sinusoidal loading at 10 Hz. The space between two shaded blocks is the time for the static loading. As shown in Fig. 2, the specimen was first cyclically loaded for 3000 cycles followed by static loading for 1 h. After 212 h of such loading, the specimen did not show any damage that was visible on the surface. Consequently, the static loading was discontinued and the specimen was put under the continuous cyclic loading until it accumulated 2.547×10^6 cycles. At that point, the test was stopped and the specimen was then loaded in tension to failure. The residual strength (net-section stress) was 149.21 MPa (21.64 ksi). Little degradation in the static strength was found.

The three specimens, G-79, G-80, and G-81, underwent a different loading pattern as shown in Fig. 3. The first cyclic loading lasted 3000 cycles and was followed by ½ h under static load. The subsequent cyclic loadings were 5000, 10 000, 20 000, 30 000, 40 000 cycles and so forth until failure. Specimens G-82, G-83, and G-86 were tested with different combinations of cyclic loadings and sustained static loads as shown in Fig. 3. A summary of the results is presented in Table 1. Specimen G-81 did not fail at 200 000 cycles. Hence, residual strength was determined and found to be 151.69 MPa (22.00 ksi). A trend was noted that the fatigue life in general increased as the time under static load (the creep time) increased. In fact, the fatigue life was significantly lengthened due to the static loading. The lives obtained in this series of experiments do not fall into the range of scatter in the fatigue life obtained with continuous cyclic loading (the maximum life of which was 34 090 cycles). The results of these experiments indicate that the previous surmise that hold time could be the determining factor in fatigue life needs to be reexamined carefully.

An Interpretation of the Results

There is wide agreement that the fatigue damage propagation in laminated composites is controlled by the matrix, and that degradation of the laminate under fatigue loading occurs in the matrix first. Damage in the

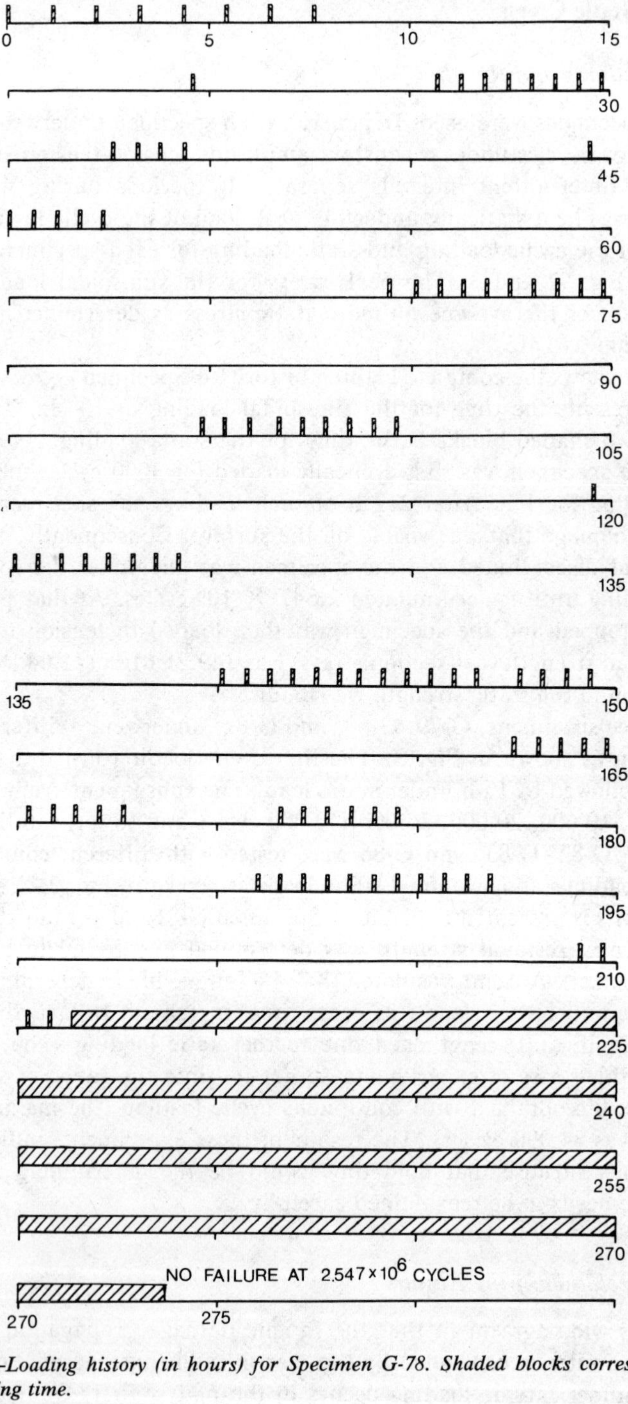

FIG. 2—*Loading history (in hours) for Specimen G-78. Shaded blocks correspond to the cyclic loading time.*

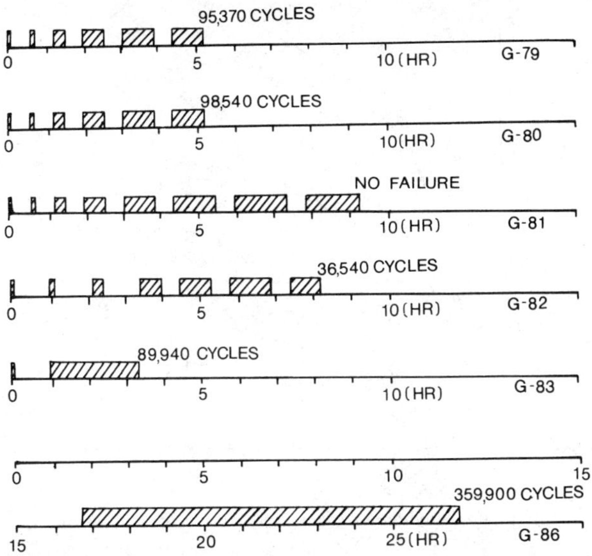

FIG. 3—*Loading histories for Specimens G-79, G-80, G-81, G-82, G-83, and G-86.*

TABLE 1—*Effects of sustained static loadings on fatigue life.*

Specimen Number	Total Creep Time, h	Fatigue Life, cycles
G-83	1	84 940
G-79	2.5	95 370
G-80	2.5	98 540
G-82	4.5	136 540
G-81	3.5	Loaded to failure after 200 000 cycles
G-86	16.75	359 900

fibers is the result of the matrix degradation. With this assumption, it is reasonable to say that the fatigue damage in laminated composites is in the form of a large number of cracks forming and propagating in the matrix and that the rate of degradation of the composite is related to the crack growth rate in the matrix.

To explain the hold time effect on the fatigue life of the composite, the crack growth rate in the composite is assumed to be a function of the hold time as shown in Fig. 4. According to this crack growth model, the crack extension per cycle of load is linearly proportional to the hold time when the hold time is small. The results found in Ref *1* could be interpreted with this linear relationship. As indicated by the sketch in Fig. 4, the amount of crack extension might reach a plateau and become independent of the

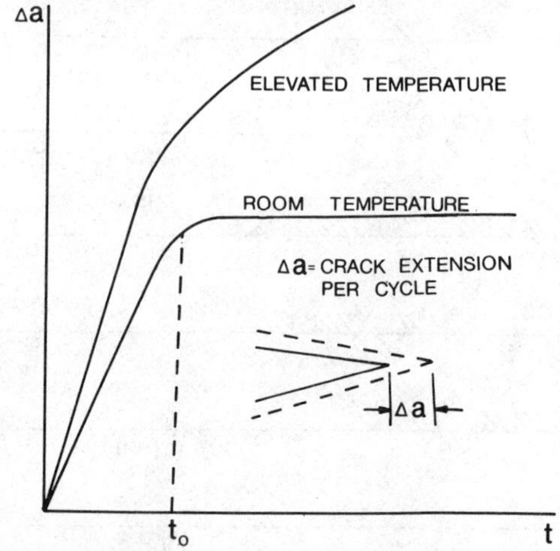

FIG. 4—*A model for crack extension in the composite.*

loading time after a certain period of time, t_0. Such a plateau might not exist when the composite creeps appreciably. For such a composite, static fatigue should be possible.

The static hold times used in this study, in most cases, were longer than ½ h. It is hypothesized that these hold times must be longer than t_0 and, consequently, the crack growth was no longer proportional to the hold time.

The Retardation Effect

As revealed by the results of the experiment, the continuous application of static load seems to have a retarding effect on fatigue degradation in the laminated composite. This means that the crack growth rate under the cyclic loading following each static loading period seems to slow down significantly. Such an effect may be viewed as the consequence of the permanent deformation buildup at the crack tips. In metals, the permanent strain (plastic strain) is produced by overloads. In the present case, with the loading level kept constant, the permanent strain might be the result of creep in the matrix. A longer hold time then could produce a larger permanent strain that reduces the stress intensity at the crack tip during the subsequent cyclic loading, and, as a result, the rate of crack growth could be reduced even though the far field stress might remain the same.

Frequency Interaction

The magnitude of the creep strain at the crack tip increases as the hold time increases. Then, we may expect lower loading frequencies to produce larger creep strains than higher frequencies at the same stress level. Thus, a cyclic loading with a lower frequency preceeding one with a higher frequency should have a retardation effect on the subsequent fatigue degradation rate. The effect should not exist if the order of loading is reversed.

With the foregoing in mind, we conducted two experiments using two specimens each. The first experiment involved switching from a high frequency to a low frequency loading (10 Hz → 1 Hz), while the second experiment was conducted with a low frequency followed by a high frequency (1 Hz → 10 Hz). Tension-tension sinusoidal loadings with a peak stress at 66.6 percent of the ultimate stress were used. Under the first loading frequency, the specimen was cycled to the mean life as obtained from continuous cyclic loading. The second loading was applied until the specimen failed.

Tables 2 and 3 present the results for both cases. The results seem to support our earlier reasoning.

The Temperature Effect

From Ref *1*, it was noted that higher temperatures tended to reduce the fatigue life. In general, the rise in temperature is either due to the rise in ambient temperature or due to the heat generated by the cyclic loading. The temperature rise due to the loading is usually localized and depends on the stress distribution as well as the damage level.

An interesting phenomenon in the temperature rise near the center hole was observed in our tests. A pair of thermocouples was placed about 1.5 mm (0.06 in.) from the hole on both sides of the specimen where the stress concentration was maximum. The temperature rise was recorded during the fatigue loading. A typical temperature-rise history for the specimens is

TABLE 2—*Fatigue life for low frequency to high frequency loading.*

Specimen Number	Number of Cycles at 1 Hz	Number of cycles at 10 Hz	Total Number of Cycles
G-85	5500	59 500	65 000
G-87	5500	62 500	68 000

TABLE 3—*Fatigue life for high frequency to low frequency loading.*

Specimen Number	Number of Cycles at 10 Hz	Number of Cycles at 1 Hz	Total Number of Cycles
G-84	12 000	9950	21 950
G-88	12 000	9190	21 190

presented in Fig. 5. In every case, the temperature rise was more than 5.5°C (10°F) above the room temperature in the first 3000 cycles. Following the static load application of either ½ or 1 h, the steady-state temperature rise dropped slightly in the subsequent fatigue loading. In Ref *1,* it was found that the near-hole temperature never dropped under continuous cyclic loading. If the heat generation rate was to be used as a measurement of the crack propogation rate in the laminate, then a drop in temperature might indicate a reduction in the fatigue degradation rate.

To quantify the temperature effect on the fatigue life of laminated composites, experiments should be conducted at different temperature levels while fixing the loading frequency. For a quick examination of such an effect, we performed an experiment with two specimens that underwent a fatigue test at 10 Hz using tension-tension sinusoidal loading with the peak stress 66.6 percent of the ultimate stress. After each 5000 cycles of loading, the specimen was unloaded and kept at room temperature for 6 h. The near-hole temperature rise for one of the specimens is shown in Fig. 6. After each 6-h rest, the temperature rose from the room temperature to a steady-state temperature during the renewed cyclic loading. Thus, the specimens did not experience continuous higher temperatures as in the case of the continuous cyclic loading. In view of this, the effect of temperature should be less severe as compared with that in the continuous loading. This effect was certainly revealed by the limited data obtained. From Table

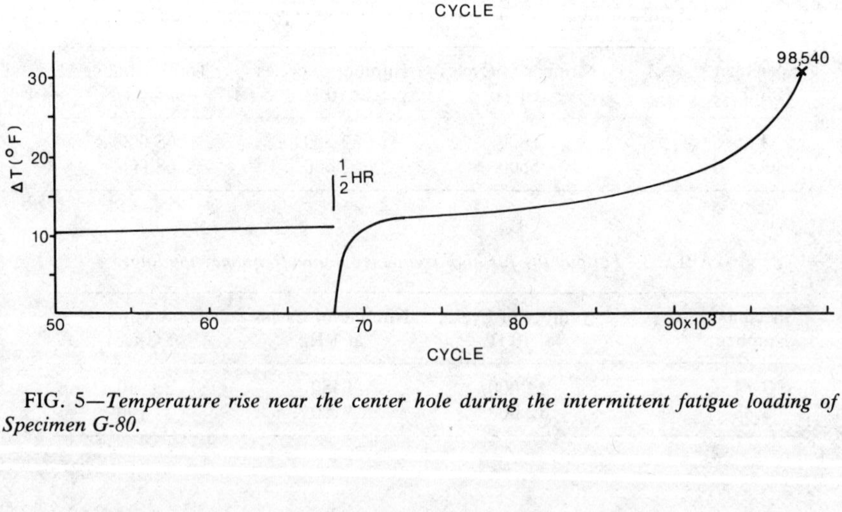

FIG. 5—*Temperature rise near the center hole during the intermittent fatigue loading of Specimen G-80.*

4, we note that the fatigue lives for these two specimens under the aforementioned intermittent cyclic loading are higher than the average life with continuous loading. In fact, these data fall near the upper bound of the data obtained in Ref *1*.

It is of interest to point up that the peak temperature after each rest never dropped below the previous one as clearly shown in Fig. 6. This may be taken as an indication that the rest does not retard the subsequent fatigue crack growth mechanism as the creep strain does. The slight increase in the fatigue life in this case is likely due to the initial reduction in the temperature in each period of cyclic loading.

TABLE 4—*Fatigue life for intermittent cyclic loadings with 6-h rest in between.*

Specimen Number	Total Relaxation Time, h	Fatigue Life, cycles
G-76	36	29 090
G-77	48	32 760

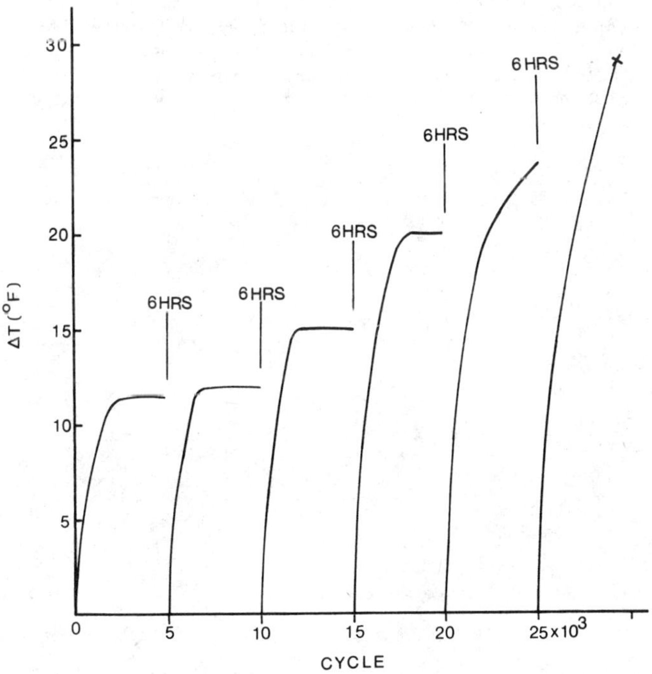

FIG. 6—*Temperature rise near the center hole for Specimen G-76 under the sinusoidal loading at 66.6 percent of the ultimate stress level and 10 Hz with 6-h rest after each 5000 cycles.*

Summary

A series of experiments have been performed to study the effect of creep strain and temperature on the fatigue life of the $[\pm 45]_{2s}$ laminated composite specimens. Sustained periods of static loadings appear to have significant retardation effects on the fatigue degradation rate of the laminated composite and that a cyclic loading with a lower frequency preceeding one with a higher frequency also prolongs the fatigue life. A creep crack hypothesis has been proposed that can provide reasonable interpretations for the experimental results.

Acknowledgment

This research was supported by the National Aeronautics and Space Administration, Langley Research Center under Grant No. NSG1278.

References

[*1*] Sun, S. T. and Chan, W. S. in *Composite Materials: Testing and Design* (Fifth Conference), *ASTM STP 674*, American Society for Testing and Materials, 1979, pp. 418–430.
[*2*] Reifsnider, K. L., Stinchcomb, W. W., and O'Brien, T. K. in *Fatigue of Filamentary Composite Materials, ASTM STP 636*, American Society for Testing and Materials, 1977, pp. 171–184.
[*3*] Yokobori, T. and Sato, K., *Engineering Fracture Mechanics*, Vol. 8, 1976, pp. 81–88.
[*4*] Mukherjee, B. and Burns, D. J., *Experimental Mechanics*, Vol. 11, 1971, pp. 433–439.

Jonathan Awerbuch[1] *and H. T. Hahn*[2]

Off-Axis Fatigue of Graphite/Epoxy Composite*

REFERENCE: Awerbuch, Jonathan and Hahn, H. T., **"Off-Axis Fatigue of Graphite/Epoxy Composite,"** *Fatigue of Fibrous Composite Materials, ASTM STP 723,* American Society for Testing and Materials, 1981, pp. 243-273.

ABSTRACT: Off-axis static and fatigue behavior of AS/3501-5A graphite/epoxy was studied in an effort to characterize the matrix/interface-controlled failure. Seven different off-axis angles were tested: 0, 10, 20, 30, 45, 60, and 90 deg. Initial (static) and post-fatigue residual strength were obtained together with *S-N* relationships. Fracture surfaces were examined through photomicrographs and stereo (three-dimensional) scanning electron microscope (SEM) photographs, in order to delineate failure modes, and the results of these inspections are discussed. The off-axis static strength, including scatter, was fully characterized by a polynomial and a nondimensional strength parameter. Essentially, no strength or modulus degradation was observed in the specimens surviving fatigue loading of 10^6 cycles regardless of the off-axis angle or fatigue stress level. When fatigue stress level is normalized with respect to static strength, all data seem to fall on the same *S-N* curve. Fatigue failure occurred without any warning or visible damage. Matrix failure characteristics vary with off-axis angle and appear in the form of serrations and axial and transverse cracks. Large scatter in life was observed at all off-axis angles; however, since the number of specimens employed in the present study is not sufficient to provide meaningful statistical *S-N* data, a more detailed investigation of the off-axis (and angle ply) behavior of graphite/epoxy composites is warranted.

KEY WORDS: composite materials, graphite/epoxy composites, fatigue tests, residual strength, acoustic emission, failure modes, off-axis fatigue, fatigue (materials)

Numerous studies on the fatigue behavior of composite materials have been performed with laminated composites. Two approaches are applied: the study of the statistical modeling of the failure process and the deterministic approach. In the latter, most of the attention is given to the understanding of damage growth during fatigue loading. This damage appears in the form of delamination, matrix cracking and crazing, fiber failure, and

*The experimental part of this work was performed at the Air Force Materials Laboratory, Wright-Patterson Air Force Base, Ohio.

[1]Department of Mechanical Engineering and Mechanics, College of Engineering, Drexel University, Philadelphia, Pa. 19104.

[2]Department of Mechanical Engineering, Washington University, St. Louis, Mo. 63130.

so on, all of which depend on the basic lamina behavior as well as the interaction between adjacent laminae within the laminated composite.

However, as composites technology approaches its maturity, more attention is directed toward the so-called matrix-controlled failure modes such as ply failure and delamination. Thus, it becomes extremely important to better understand the failure of matrix and interface in unidirectional composites. A general review on the various parameters influencing the fatigue behavior of composites has been recently published [*1*].[3]

It is well established by now that the failure in multidirectional composite laminates initiates in the ply whose fibers run normal to the applied tensile load. This ply failure manifests itself as cracks in the matrix and along the matrix/fiber interface [*2*]. The ply failure occurs because unidirectional composites are much weaker in the transverse direction, that is, normal to the fibers [*1–5*]. Cracks form parallel to the fibers and multiply as fatigue proceeds. The number of cracks reaches an equilibrium only when stresses in the ply are sufficiently reduced by the formation of the cracks themselves. Furthermore, these cracks frequently grow along the interfaces between plies, leading to delamination prior to failure. In all fibrous composites, the ply failure occurs long before the final failure, which usually coincides with final fracture of the fibers.

The ratio between transverse failure strain and longitudinal failure strain affects the ply failure. For example, glass/epoxy composites, for which the ratio is in the order of 0.1, fail at an earlier stage of loading than graphite/epoxy composites for which the ratio is higher, about 0.6. If ply failure could be avoided, the excellent longitudinal strength of graphite/epoxy could be used even more efficiently.

Although ply failure in fatigue has been receiving more attention in recent years, the basic mechanisms of matrix/interface-controlled failure are not well understood at present. Therefore, the purpose of this paper is to study the matrix/interface-controlled failure under combined-stress fatigue loadings using off-axis coupons. The data obtained from this study will be helpful in predicting the ply failure in composite laminates.

Relatively few works on the failure mechanism of unidirectional composites are available. A basic study on the static strength of unidirectional off-axis glass/epoxy [*6*] shows that the static strength can be predicted using the tensor polynomial failure criteria [*7,8*]. A follow-up study [*9*] shows an excellent agreement between the off-axis static strength data and the tensor polynomial failure criteria for boron/epoxy. The nonlinear behavior of off-axis boron/epoxy and unidirectional on-axis graphite/epoxy has been discussed in Ref *10*. The fatigue behavior of unidirectional on-axis graphite/epoxy has been studied intensively in recent years. Statistical studies on the static and residual strength and life have been published [*11–16*]. However,

[3]The italic numbers in brackets refer to the list of references appended to this paper.

studies on the fatigue behavior of glass/epoxy [*17*] and boron/aluminum [*18*] are among the few which attempted to understand the fatigue behavior of unidirectional off-axis composites. Surprisingly, no information on the fatigue behavior of off-axis graphite/epoxy is available despite the recognized importance of understanding the fatigue properties of the constituent laminae in a laminated composite. This point has been very well argued in Ref *17*.

Therefore, we have chosen graphite/epoxy AS/3501-5A to study the off-axis fatigue behavior. Seven off-axis angles were chosen: 0, 10, 20, 30, 45, 60, and 90 deg. Static and post-fatigue residual strength (after 10^6 cycles) data are presented, and the matrix/interface-controlled failure is characterized in terms of a polynomial. Fatigue tests were performed to obtain the *S-N* curves for the various off-axis angles at different dynamic stress levels ranging from 50 to 75 percent of the average static strengths. Acoustic emission counts were taken during all static and residual strength tests. In addition, static and fatigue fracture surfaces were examined through photomicrographs and through stereo (three-dimensional) scanning electron microscope (SEM) photographs. Finally, the mechanical properties are correlated with the failure modes observed by the SEM.

Experimental Procedure

Specimens

The material chosen for this study is unidirectional graphite/epoxy AS/3501-5A with a nominal fiber content of 70 percent by volume. Specimens were 20.0 cm long by 1.27 cm wide (8 in. by 0.5 in.) and consist of eight plies. Unidirectional glass/epoxy tabs were 3.8 cm (1.5 in.) long, and the gage length was 12.4 cm (5 in.). Two 60-cm by 60-cm panels were fabricated following the manufacturer's recommended cure procedure, and a total of 120 specimens were prepared from these panels. The number of specimens tested (102) in each loading mode are given as follows: 21 specimens for static tests to provide baseline data; 81 specimens for fatigue tests including 21 specimens tested for residual strength after 10^6 cycles.

Static Tests

All static tests were performed in tension on an MTS machine of either 50 000 or 20 000 lb capacity under a load control mode with the loading rate of 4.5 kg/s (10 lb/s). An extensometer of 2.54 cm (1 in.) gage length was used to obtain the load deflection curves for all static test specimens. Three specimens were tested at each of the seven off-axis angles.

Fatigue Tests

All fatigue tests were carried out on the same MTS machine as the static tests, under a load control mode. The constant amplitude fatigue tests were performed at about 18 Hz until failure or until 10^6 cycles, whichever occurred first. The stress ratio was $R = 0.1$, and the maximum fatigue stress applied was 50, 55, 60, 65, 70, and 75 percent of the average static strength given for each off-axis angle; one to four tests were performed at each stress level.

Residual Strength Tests

Those specimens which had survived 10^6 cycles were tested in static tension up to failure to obtain residual strength data. Load deflection curves were obtained using 2.54 cm (1 in.) gage length extensometer. The test procedure for the residual strength tests was the same as for the static tests.

Acoustic Emission

Acoustic emission (AE) was monitored during all static and residual strength tests by using a Dunegan/Endevco 300 series AE system with Model 801P preamplifier. The transducer, Model D9201, which has a flat frequency response in the range of 100 KHz to 1 MHz, was attached to the test specimen with vacuum grease. The total system gain was set at 41 dB with threshold voltage of 1 V peak. The filter in the system was set to 100 to 300 KHz. The AE rate was monitored as a function of time.

Results and Analyses

Static Behavior

Stress strain curves for the seven different off-axis angles are shown in Fig. 1. Nonlinearity and a sharp decrease in both ultimate strength and axial modulus appear at the very small off-axis angles, as expected. A detailed discussion on the nonlinear behavior of unidirectional off-axis composites is given in Ref *10*. For large off-axis angles, for example, 60 and 90 deg, the stress-strain curves are almost linear. The residual stress-strain curves of specimens fatigued for 10^6 cycles (not shown here) are very similar to the static stress-strain curves; data of the residual ultimate strength, axial modulus, and maximum strains to failure do fall within the experimental scatter of the static test data.

The static ultimate strength data as a function of off-axis angle are shown in Fig. 2 together with predictions (solid line) based on the tensor polynomial failure criteria formulated by Tsai and Wu [*7,8*]. The numbers in parentheses

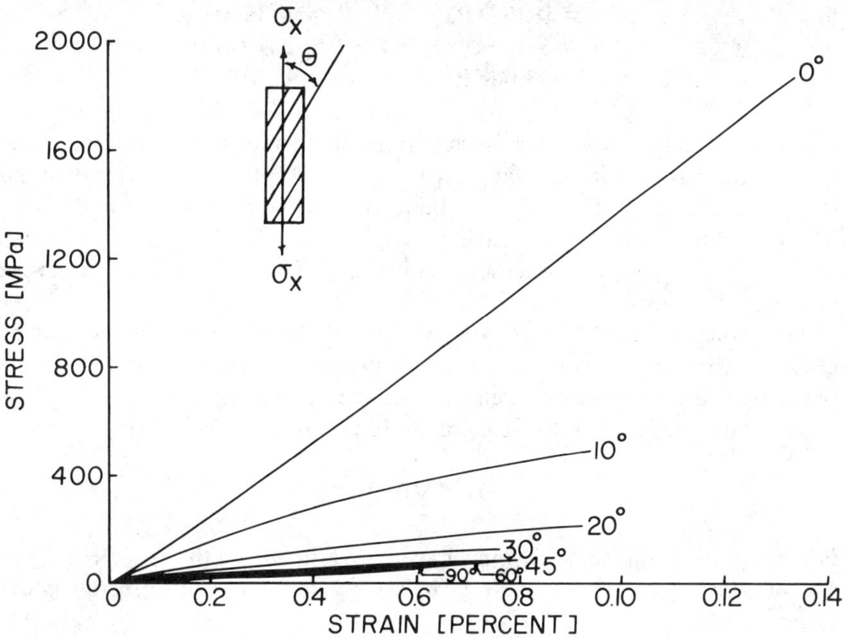

FIG. 1—*Stress-strain relationships of off-axis graphite/epoxy laminae for various fiber orientations.*

in Fig. 2 indicate the number of identical data points. The assumed quadratic failure surface in the stress space [*7,8*] takes the form

$$f(\sigma_k) = F_i\sigma_i + F_{ij}\,\sigma_i\,\sigma_j = 1\ (i, j, k = 1 - 6) \tag{1}$$

where F_i and F_{ij} are the second and fourth rank strength tensors, respectively. For the state of plane stress and for orthotropic materials, Eq 1 in expanded form can be rewritten as follows

$$F_1\,\sigma_1 + F_2\,\sigma_2 + F_6\,\sigma_6 + F_{11}\,\sigma_1^2 + F_{22}\,\sigma_2^2 + F_{66}\,\sigma_6^2 + 2F_{12}\,\sigma_1\,\sigma_2 = 1 \tag{2}$$

Introducing the stress transformation equations for the case of off-axis loading, the maximum failure stress X_θ can be calculated from

$$f(\theta)\,X_\theta^2 + g(\theta)\,X_\theta - 1 = 0 \tag{3}$$

where $f(\theta)$ and $g(\theta)$ include the strength tensors F_i and F_{ij} and the off-axis angle, θ. The strength tensors were calculated using the following material strength data

$$
\begin{array}{ll}
X^{(a)} = 1836 \text{ MPa} & X'^{(a)} = 1836 \text{ MPa} \\
Y^{(a)} = 56.9 \text{ MPa} & Y'^{(b)} = 207 \text{ MPa} \\
S^{(b)} = 93.0 \text{ MPa} & S'^{(b)} = 93 \text{ MPa}
\end{array}
\tag{4}
$$

where X, Y, and S are the material strengths in the fiber direction, transverse to the fibers, and in shear, respectively, and (a) = present results, and (b) = data from Ref *19*. The "prime" represents negative strength data. The values for F_i and F_{ij} appearing in Eq 2 (with the exception of F_{12}) can be calculated [*8*] using the data given in Eq 4. The value of F_{12} is assumed to be equal to zero.

The uniaxial strength, X_θ, as a function of the off-axis angle, θ, can be obtained from Eq 3, and the results are plotted in Fig. 2. As expected, the prediction, Eq 3, agrees very well with the experimental results.

The choice of $F_{12} = 0$ can be justified if the stability criteria [*8*]

$$F_{12} \leq \sqrt{F_{11}F_{22}} \tag{5}$$

is considered. From Eq 5, F_{12} may have any value within the range -36.5×10^{-6} (MPa)$^{-2}$ to $+36.5 \times 10^{-6}$ (MPa)$^{-2}$. However, the different values

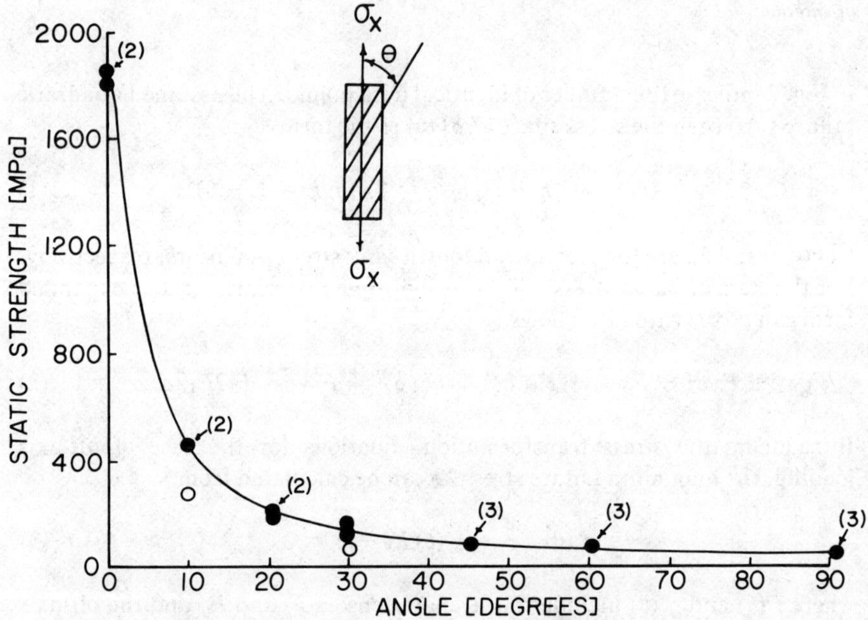

FIG. 2—*Off-axis tensile strength: comparison of the tensor polynomial strength criteria (Eq 3) with experimental data. Numbers in parentheses indicate number of identical data points, and open circles denote data to be discarded in the subsequent analysis.*

of F_{12} have little if any effect on the final predicted axial failure stress. This justifies the choice of $F_{12} = 0$. A similar conclusion has been given in Ref *9* for boron/epoxy. In fact, by the nature of Eq 2, the effect of F_{12} term on the predicted failure stress is negligible for the case of off-axis loading. Therefore, no attempt has been made in this work to determine experimentally the value of F_{12}. A detailed discussion of the experimental determination of F_{12} is given in Refs *7* and *8*.

A different approach can be taken assuming that there is no interaction between fiber-controlled failure and matrix/interface-controlled failure. As suggested in Ref *20*, the failure surface for the latter case is described using the equation

$$F_2 \sigma_2 + F_{22} \sigma_2^2 + F_{66} \sigma_6^2 = 1 \tag{6}$$

Here σ_2 and σ_6 are the transverse and shear stresses, respectively, in the material symmetry axes.

To determine the F's in Eq 6, the off-axis strength X_θ in Fig. 2 is first converted to the stresses σ_2 and σ_6

$$\sigma_2 = \frac{X_\theta}{2} (1 - \cos 2\theta), \quad \sigma_6 = \frac{X_\theta}{2} \sin 2\theta \tag{7}$$

Substitution of Eq 7 into Eq 6 for each failure stress $X_\theta^{(i)}$ leads to an algebraic equation for the F's

$$\begin{bmatrix} \sigma_2^{(1)} & \sigma_2^{(1)2} & \sigma_6^{(1)2} \\ \sigma_2^{(2)} & \sigma_2^{(2)2} & \sigma_6^{(2)2} \\ \cdots & \cdots & \cdots \\ \sigma_2^{(n)} & \sigma_2^{(n)2} & \sigma_6^{(n)2} \end{bmatrix} \begin{Bmatrix} F_2 \\ F_{22} \\ F_{66} \end{Bmatrix} = \begin{Bmatrix} 1 \\ 1 \\ \cdots \\ 1 \end{Bmatrix} \tag{8}$$

Here n is the total number of the data, which is 18 in the present case (three test data for each off-axis angle). Rewriting, for convenience, Eq 8 in the following abridged form

$$[\sigma] \{F\} = \{1\} \tag{9}$$

the F's are determined from

$$\{F\} = ([\sigma]^T [\sigma])^{-1} [\sigma]^T \{1\} \tag{10}$$

where the superscripts T and -1 denote the matrix transpose and matrix inversion, respectively.

Once the F's are known, the scatter in the off-axis strengths can be characterized by using a nondimensional strength parameter, s. Suppose (σ_2, σ_6) is a pair of stress components at failure, Fig. 3. Another pair (σ_{02}, σ_{06}) is then defined in such a way that

$$\frac{\sigma_2}{\sigma_{02}} = \frac{\sigma_6}{\sigma_{06}} = s > 0 \tag{11}$$

where (σ_{02}, σ_{06}) satisfies the failure criteria

$$F_2\sigma_{02} + F_{22}\,\sigma_{02}^2 + F_{66}\,\sigma_{06}^2 = 1 \tag{12}$$

That is, (σ_{02}, σ_{06}) defines in the stress space the point of intersection between the failure surface and the line connecting the origin and the failure point (σ_2, σ_6), Fig. 3. Note that by substituting Eq 11 into Eq 12, s is simply obtained from

$$s = \frac{1}{2}\,[F_2\sigma_2 + (F_2^2\sigma_2^2 + 4\,F_{22}\sigma_2^2 + 4\,F_{66}\sigma_6^2)^{1/2} \tag{13}$$

The scatter in combined strength can now be characterized in terms of s [*20*]. To this end, a two-parameter Weibull distribution is chosen

$$R\,(s) = \exp\left[-\left(\frac{s}{\hat{s}}\right)^{\alpha}\right] \tag{14}$$

where α is the shape parameter and $\hat{s}$ the characteristic nondimensional strength. The parameters α and $\hat{s}$ are then determined by using the method of maximum likelihood (see Ref *21*).

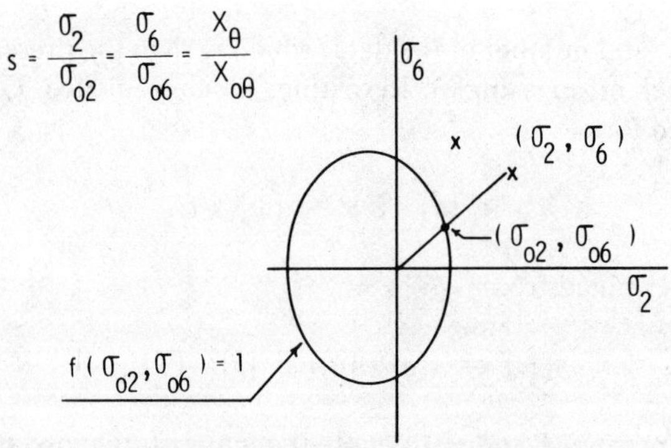

FIG. 3—*Definition of the nondimensional strength parameter,* s.

For an off-axis strength, X_θ Eqs 11 and 13 become

$$\frac{X_\theta}{X_{o\theta}} = s \tag{15}$$

where

$$X_{o\theta} = \frac{-F_2(1-\cos 2\theta) + \{F_2{}^2(1-\cos 2\theta)^2 + 4\,[F_{22}(1-\cos 2\theta)^2 + F_{66}\sin^2 2\theta]\}^{1/2}}{F_{22}(1-\cos 2\theta)^2 + F_{66}\sin^2 2\theta} \tag{16}$$

Thus, the off-axis strength distribution is described by

$$R(X_\theta) = \exp\left[-\left(\frac{X_\theta}{\hat{X}_\theta}\right)^\alpha\right] \tag{17}$$

where the characteristic off-axis strength $\hat{X}_\theta$ is determined by

$$\hat{X}_\theta = \hat{s}\,X_{o\theta} \tag{18}$$

Several problems arose in analyzing the data as previously described. First, using all the strength data in Fig. 2 resulted in a negative value for F_{22}. Since infinite compressive strength is not possible in reality, a transverse compressive strength of 207 MPa [*19*] was used to remedy this situation. Second, the feasibility of using a Weibull distribution for s was checked by linearizing Eq 14,

$$\ln(-\ln R) = \alpha \ln s - \alpha \ln \hat{s} \tag{19}$$

The two data points represented by open circles, Fig. 2, deviated too much from the line represented by Eq 19. Therefore, although the exact causes of deviation are not known, these two points were deemed extraneous and were deleted from the analyses. Thus, the final results, discarding these two data points, are obtained from Eq 10 as

$$\begin{aligned} F_2 &= 1.1777 \times 10^{-2}\ (\text{MPa})^{-1} & \alpha &= 16.319 \\ F_{22} &= 8.0195 \times 10^{-5}\ (\text{MPa})^{-2} & \hat{s} &= 1.0408 \\ F_{66} &= 1.3411 \times 10^{-4}\ (\text{MPa})^{-2} \end{aligned} \tag{20}$$

The variation of the tensile strength with off-axis angle is shown in Fig. 4 and compares favorably with Eq 16 using the strength values of Eq 20.

Figure 5 shows both the experimental data and Eq 12. The open circles again represent those two data that were excluded in the analysis. It is of in-

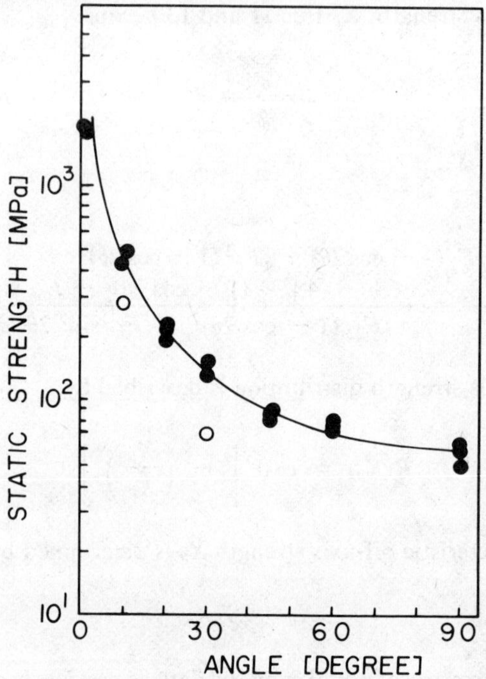

FIG. 4—*Off-axis tensile strength: comparison of the second order polynomial of* σ_2 *and* σ_6 *with experimental data. Open circles denote discarded data in the analysis (Eq 16).*

terest to note that according to Eq 6 the true shear strength X_{LT} is 86.4 MPa, whereas the shear strength obtainable from the 10-deg off-axis test [22] is only 75.8 MPa.

The calculated values of s are plotted against R in the linearized format of Eq 19, Fig. 6. The data follow the assumed linear relationship fairly closely, indicating that the Weibull distribution is an adequate representation of the strength distribution. The final determined strength distribution is shown in Fig. 7.

The initial axial moduli for the various off-axis are shown in Fig. 8. The variation of the axial modulus, E_θ, with off-axis angle is described by the equation

$$\frac{1}{E_\theta} = \frac{\cos^4\theta}{E_L} + \left(\frac{1}{G_{LT}} - \frac{2\nu_{LT}}{E_L}\right)\cos^2\theta\,\sin^2\theta + \frac{\sin^4\theta}{E_T} \qquad (21)$$

where E_L, E_T, and G_{LT} are the longitudinal, transverse, and shear moduli, respectively, and ν_{LT} is the major Poisson's ratio. The moduli E_L, and E_T have been measured to be

$$E_L = 137 \text{ GPa}, \qquad E_T = 9.6 \text{ GPa}$$

Using a typical value of 0.3 for ν_{LT} and the measured value of E_θ for $\theta = 10$ deg, Eq 21 can be solved for the shear modulus, G_{LT}, to obtain: $G_{LT} = 5.24$ MPa.

The resulting prediction for E_θ, Eq 21, is represented by the curve in Fig. 8. An excellent agreement is seen between the data and the prediction at the remaining off-axis angles.

Fatigue Behavior

An *S-N* (dynamic stress versus number of cycles to failure) scan is shown in Fig. 9 for six of the seven off-axis angles. The dynamic stress level for the 0-deg specimens is much higher than for all other off-axis angles and is therefore excluded from the expanded *S-N* data. The effect of the off-axis angle on the *S-N* relationship can be easily identified. Generally, the fatigue life for all off-axis angles is random, extending at least over three decades, from 10^3 to 10^6 cycles. It should be remembered that all fatigue tests were terminated at 10^6 cycles for the purposes of obtaining residual strength data. These run-outs, obtained for fatigue stress level in the range of $0.5 - 0.7\overline{X}_\theta$, are not shown here for the sake of clarity.

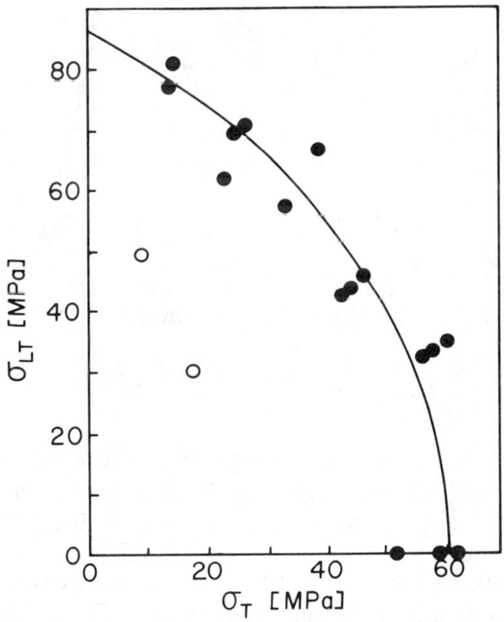

FIG. 5—*Predicted failure envelope and experimental data for off-axis coupons (Eq 12).*

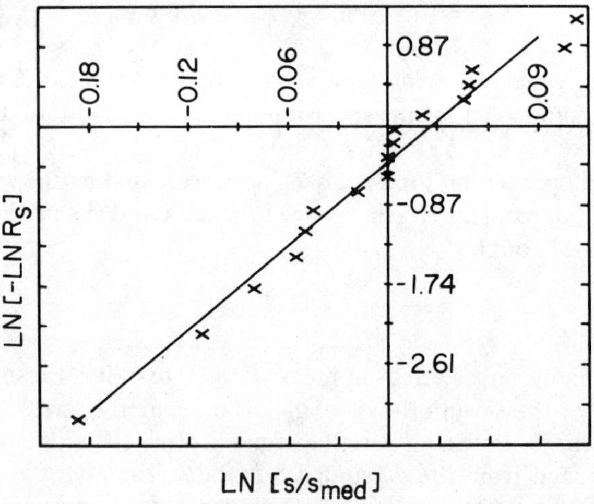

FIG. 6—*Weibull plot, in a linear format, for the nondimensional strength parameter* s (S_{med} *is the median value).*

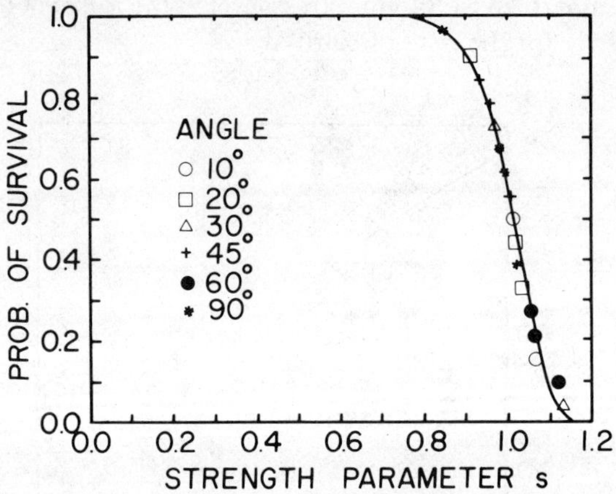

FIG. 7—*Axial static strength distribution (Eq 17).*

It is customary to present the *S-N* relationship with the maximum fatigue stress being normalized with respect to the average off-axis static strength $\overline{X}_\theta$. Such a normalized *S-N* scan is shown in Fig. 10 including all seven off-axis angles. It should be noted that in calculating the average static strength for the 10 and 30-deg specimens, the two lowest static strength data points (see Figs. 2 and 4) have been censored. However, even with the censored data, the normalized *S-N* relationship does not give any indication of the effect of off-axis angle on life, that is, all data seem to fall on the same *S-N*

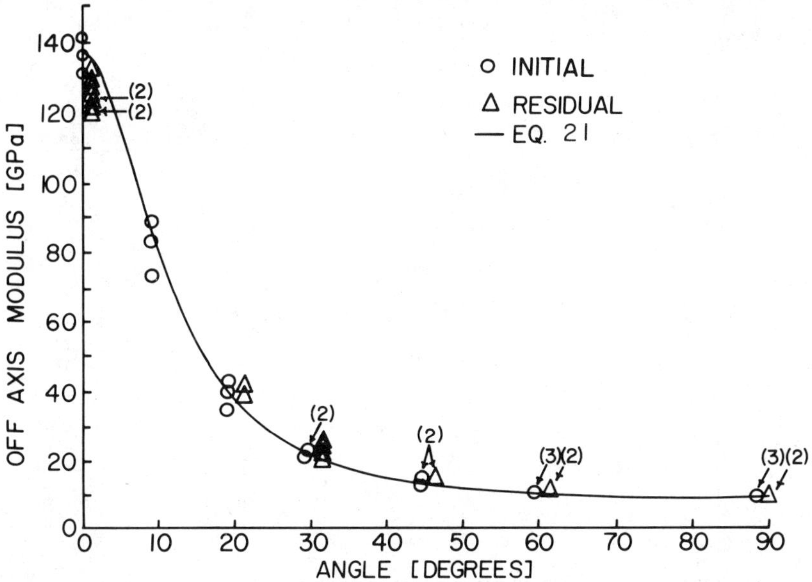

FIG. 8—*Axial modulus versus off-axis angle (Eq 21). Numbers in parentheses indicate identical data points.*

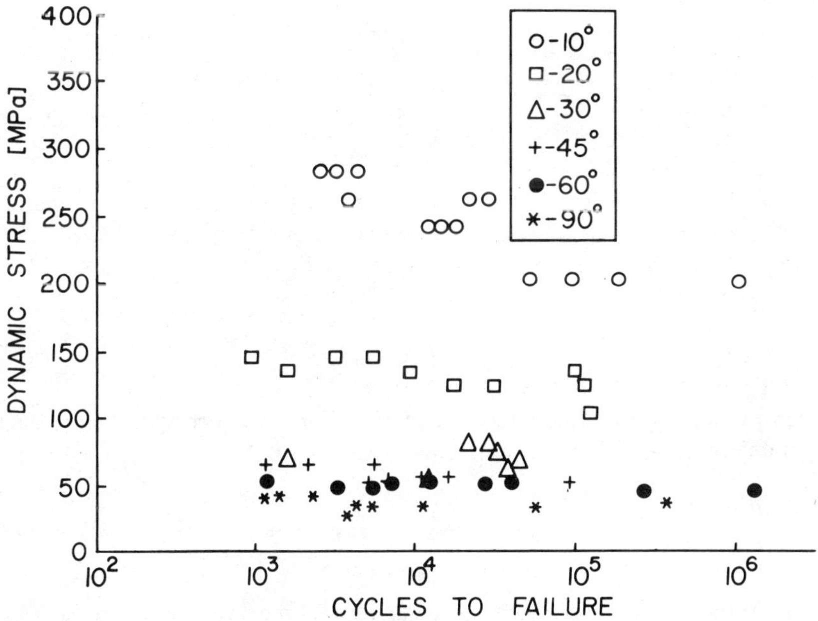

FIG. 9—*Expanded form of* S-N *relationship for various off-axis angles. The fatigue strength of 0-deg specimens is about 1200 MPa.*

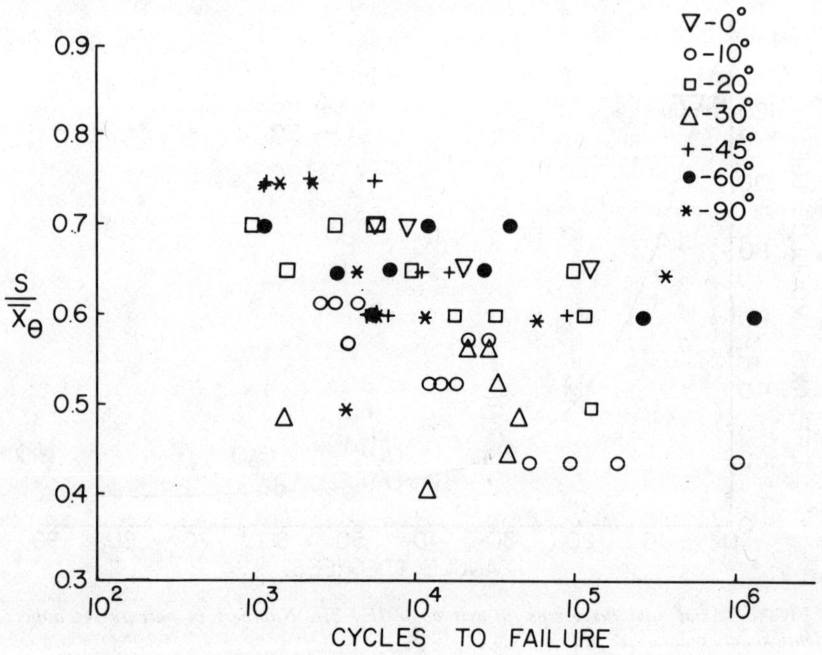

FIG. 10—*Normalized* S-N *relationship for various off-axis angles. All data points regardless of off-axis angle fall on the same curves.*

curve. This is of interest since once the normalized *S-N* relationship is known for any angle, it is known for every other off-axis angle. The same may be true in defining the endurance limit, but more testing is needed to substantiate this.

Since the off-axis tensile failure is very likely to be controlled by inherent flaws [*20*], the homologous stress ratio can be employed to analyze the *S-N* data [*11, 12, 23*]. To this end, fatigue data are ordered $[N_i \mid i = 1, 2, \ldots, n]$ at a fatigue stress, S

$$N_1 \leq N_2 \leq \cdots \leq N_n \tag{22}$$

Then, assuming a unique stress-life relationship, X_θ, corresponding to each n_i is determined from the equation

$$R = 1 - \frac{i - 0.3}{n + 0.4} = \exp\left[-\left(\frac{X_\theta}{\hat{X}_\theta}\right)^{\alpha_s}\right] \tag{23}$$

Note that the median rank has been used for R. Once X_θ is determined, S/X_θ versus $\log N_i$ is plotted. This way, all the data at different fatigue stresses are combined and finally fit by a power law

$$\left(\frac{S}{X_\theta}\right)^\alpha N_i = b \tag{24}$$

Figure 11 shows the results of the foregoing analyses for each off-axis angle. The parameters a and b are listed in Table 1 together with the coefficient of correlation. Pictorially, Fig. 12 shows the best consistency of the data and Fig. 13 shows the worst. The inconsistency in the data is due to insufficient sample size for the amount of scatter. Therefore, the S-N curves in Fig. 11 should be considered tentative. Yet, it is plausible to conclude that the off-axis fatigue strength at 10^6 cycles is slightly lower than half the static strength regardless of off-axis angle, Fig. 11.

Residual Strength and Modulus

All specimens surviving 10^6 cycles were tested for residual strength and modulus. The residual strength data shown in Fig. 14*a* indicate that all those

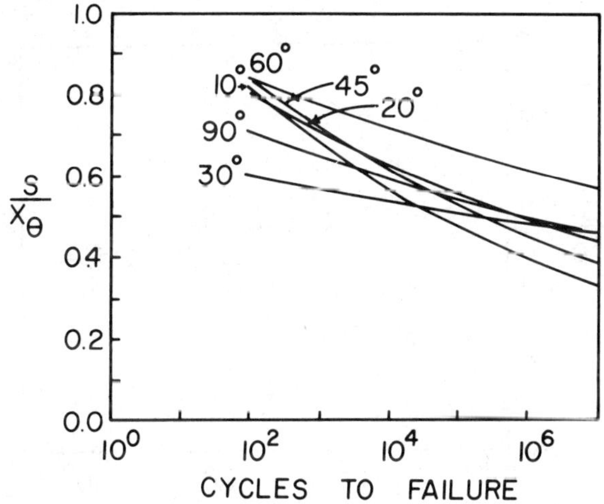

FIG. 11—*Best fit of power law (Eq 24) with experimental data for all off-axis angles.*

TABLE 1—*Parameters* a *and* b.

Off-Axis Angle, deg	Number of Data	a	b	Coefficient of Correlation
10	13	12.738	8.151	0.9479
20	10	18.762	1.759	0.9396
30	8	42.715	3.227×10^{-8}	0.4585
45	10	14.724	7.619	0.8445
60	9	29.514	5.170×10^{-1}	0.8281
90	10	26.419	1.044×10^{-2}	0.5753

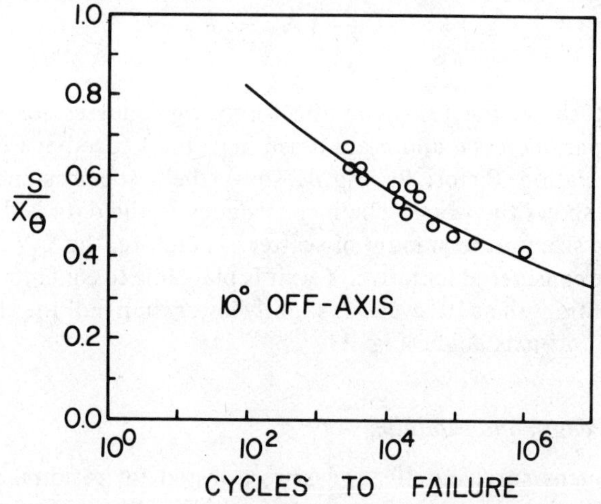

FIG. 12—*Best fit of power law (Eq 24) with experimental data for 10-deg coupons.*

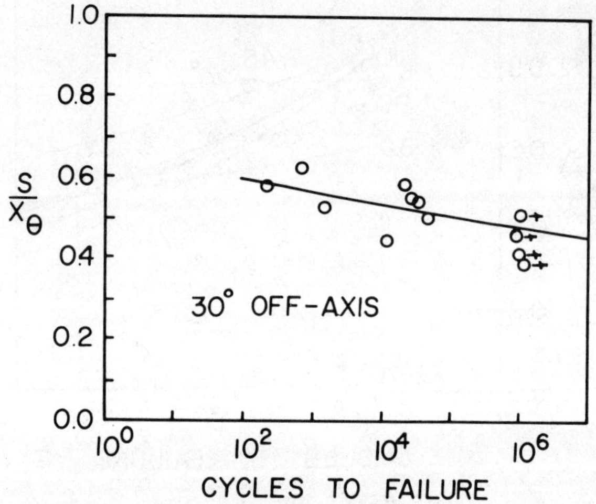

FIG. 13—*Best fit of power law (Eq 24) with experimental data for 30-deg coupons.*

specimens which survived 10^6 cycles exhibit practically no reduction in strength regardless of the off-axis angle or the fatigue stress level. With few exceptions, run-outs occurred when the fatigue stress level was about $0.5\overline{X}_\theta$. The $S/\overline{X}_\theta$ ratio is given in parentheses whenever it is different from 0.5. Only the 0-deg specimens show a strength reduction of about 10 percent, Fig. 14*b*. In an earlier study of unidirectional T300/5208 graphite/epoxy [*11*], the

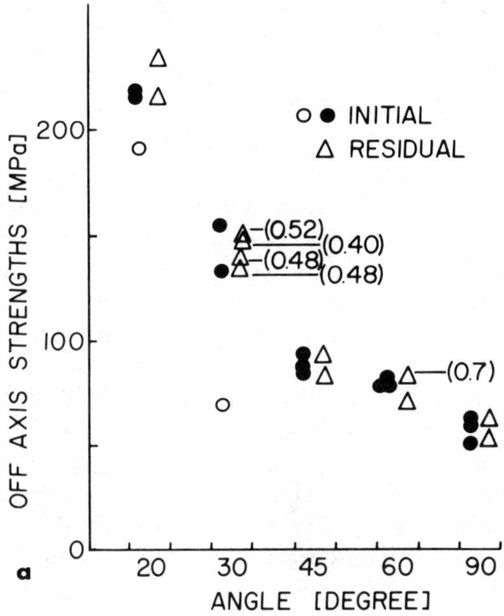

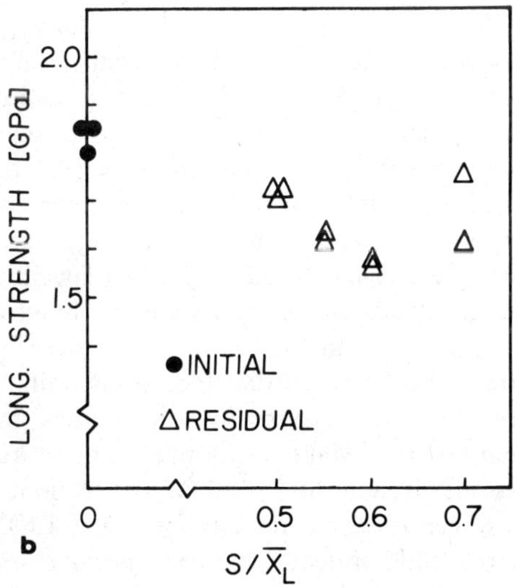

FIG. 14—*Comparison of initial (static) and residual strength (after 10^6 cycles) data;* (a) *off-axis coupons: essentially no strength degradation occurs during fatigue loading; and* (b) *unidirectional coupons: slight decrease in strength is observed.*

strength reduction was found to be negligible up to two million cycles. Such a discrepancy in behavior may be due to the difference in material. To obtain a definite answer would require further study. The reason for the slight increase in residual strength in the 0-deg specimens at $S/\overline{X}_\theta = 0.7$ is not clear; however, it should be pointed out that the number of specimens tested is not sufficient to draw any definite conclusions.

The axial moduli measured during the residual strength tests (for those specimens which survived 10^6 cycles) are compared with the initial (static) elastic moduli, Fig. 8. The comparison indicates that no modulus changes occur after fatigue loading, regardless of off-axis angle or fatigue stress level. The exception is again the 0-deg specimens for which a reduction in longitudinal modulus of about 10 percent is observed after fatigue loading. These results for the 0-deg specimens of the post-fatigue residual tests may seem to contradict the results obtained for unidirectional T300/5208 graphite/epoxy [*11*]. However, the slight increase in modulus observed in Ref *11* was for specimens which were proof tested prior to the fatigue loading. Besides, the data presented here are limited and, with the inherent statistical scatter in graphite/epoxy composites, any general conclusions may be premature at this stage. In any case, for all practical purposes it seems that no strength degradation or moduli changes occur due to the cyclic loading.

The large scatter in life as evidenced by the fatigue data shown in Figs. 9 or 10 together with the residual strength results, showing no strength degradation after 10^6 cycles, indicate that the underlying fatigue failure mechanism is one of sudden death; that is, a specimen being fatigue cycled at a certain dynamic stress level may fail randomly at any cycle number, as low as 10^3 cycles. However, if the specimen survives the fatigue loading, which can be as long as 10^6 cycles, no strength degradation is evident. A specific example is given as follows: a 30-deg specimen, being cycled at a dynamic stress level of about 60 percent of its average static ultimate stress may fail at 43 000 cycles. A supposedly identical specimen, being cycled at the same stress level will survive 10^6 cycles and retain a (residual) strength of 139 MPa, which is well within the scatter of the average static strength of 143 MPa. It should also be noted that the fatigue failure occurs without any visible evidence of failure. Apparently, stable crack growth is minimal and most life is spent before unstable crack propagation. Further discussion of failure mechanisms is given in the Failure Modes section. Such a "sudden-death" behavior has been observed for unidirectional T300/5208 graphite/epoxy [*11*]. Since the sudden-death behavior became evident in several studies, an attempt has been made to develop a statistical model based on the sudden-death assumption, for example Ref *14*. In the present study, no attempt has been made at statistical analysis due to the lack of sufficient data. However, the present results indicate that such a detailed statistical study is warranted.

Acoustic Emission (AE)

As previously mentioned, acoustic emission was also monitored for all static and residual tests. Two different AE characteristics were observed depending on whether the off-axis angle was greater or less than 20 deg. For off-axis angles larger than 20 deg, practically no AE events occurred before final fracture in either static or residual strength tests, independent of the transducer distance from the fracture zone. These results were based upon the gain and threshold levels previously described. This, combined with the earlier observation based on residual strength data that no strength degradation occurred during fatigue, points to the conclusion that there is no critical damage growth before final fracture in either static or fatigue loading. This is concluded in spite of the fact that many failed fibers and axial and transverse cracking of the matrix were observed at the fracture surface in both static and residual test specimens, which will be discussed in the next section.

When the off-axis angle was less than 20 deg, AE activity was substantial, increasing monotonically with applied stress [*11*]. In residual strength tests after fatigue, AE activity started only when the applied stress exceeded the dynamic stress level, which is known in the literature as the Kaiser effect [*24*]. This also indicates that no appreciable damage occurred during fatigue loading. Similar results were obtained in Ref *11* for unidirectional T300/5208 graphite/epoxy.

Failure Modes

Direct inspections of the failed specimens do not reveal any dependence of the failure modes on off-axis angle. Figure 15 shows a general view of six specimens for six different off-axis angles failed under fatigue loading. Similar macroscopic failure modes were observed for the statically loaded specimens (not shown here because the failed specimens look identical to those failed under fatigue load). The failure occurs along the fiber directions, the same as for glass/epoxy [*17*] and boron/epoxy [*9*]. The 0-deg specimens failed in a brush-like manner under both static and fatigue loading conditions as discussed in Ref *11*. All specimens, regardless of off-axis angle or dynamic stress level, failed abruptly without any early warning; nor do photomicrographs of the static, fatigue, and residual test specimens fracture surfaces reveal any additional information about the effect of the different loading conditions or the off-axis angles on the fracture surface.

Scanning electron microscope (SEM) photographs were also taken, inspecting the fracture surface after the static, fatigue, and residual strength tests. In order to obtain a better view of the fracture surfaces and better understand the different modes of failure, a successful attempt has been made to obtain stereo (three-dimensional) views of the fracture surfaces.

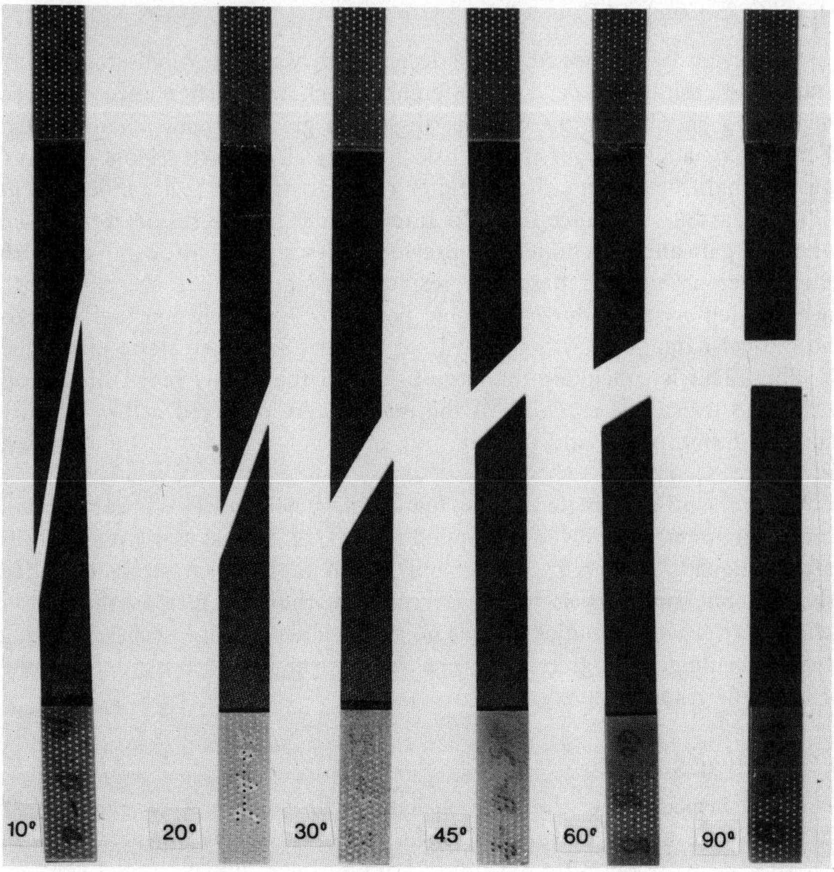

FIG. 15—*Off-axis graphite/epoxy tensile coupons after fatigue failure. Similar failure is obtained under static loading.*

The procedure for obtaining these stereo views is well known in the SEM laboratory and will not be presented here in detail. Briefly, the technique consists of taking two pictures of the same fracture surface at two different angles about 5 to 10 deg apart. These are obtained by rotating the specimens in the SEM vacuum chamber. Viewing the two resulting photographs through a stereo viewer gives a three-dimensional effect of the fracture surface. The "depth" of the inspected area depends on the specimen's angle of rotation, being "deeper" with the larger angle of rotation. The stereo view obtained is somewhat exaggerated; however, more details are revealed using this technique.

The photographs in Fig. 16 are paired to be viewed under a three-dimensional viewer. They show the fatigue fracture surfaces of three specimens of three different off-axis angles: 30, 60, and 90 deg. The reader is invited

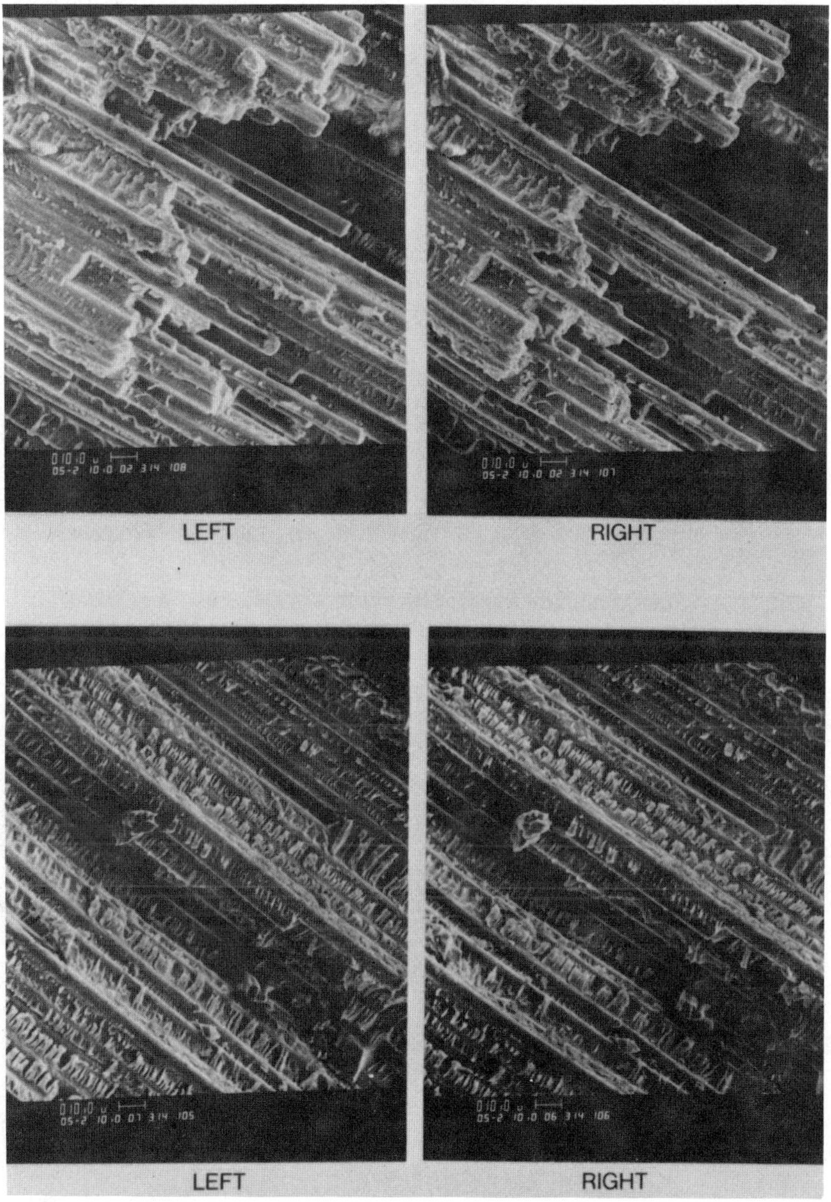

FIG. 16*a*—*Fatigue fracture surfaces of 30-deg off-axis specimens;* $S/\overline{X}_\theta$ *= 0.6,* N_f *= 43 430 cycles.*

FIG. 16b—*Static and fatigue fracture surfaces of 60-deg off-axis specimens: top-static failure, bottom-fatigue failure;* $S/\overline{X}_{\theta} = 0.6$, $N_f = 262\ 890$ *cycles.*

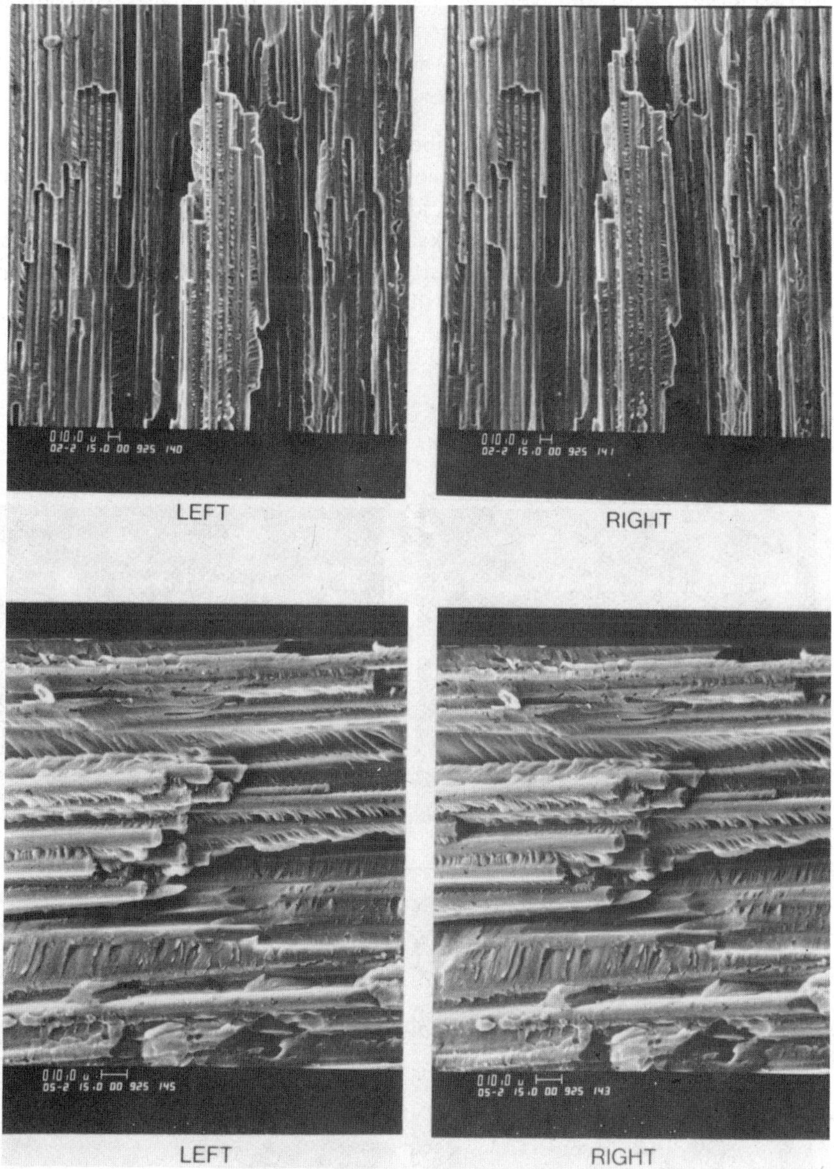

FIG. 16c—*Fatigue fracture surfaces of 90-deg off-axis specimens:* $S/\overline{X}_\theta$ = *0.65,* N_f = *4320 cycles.*

to study these three-dimensional effects using a stereo viewer. As the eyesight of each person is different, the reader may have to change the distance between the "right" and "left" photographs so that both images will coincide under the stereo viewer.

For the typical fatigue fracture surfaces shown in Figs. 16 through 21, the following failure modes are observed:

1. A relatively large amount of fiber fracture.
2. In many cases, failure of the complete tows and even of groups of tows such as is shown in Figs. 16, 19, 20, and 21.
3. In all specimens, axial cracks parallel to the fiber directions for all off-axis angles. These cracks, mostly pronounced in the 0-deg specimens, seem to appear between the tows of fibers, Fig. 17.
4. A rugged, rather than planar, fracture surface. This irregularity is in the direction of the specimen thickness and can be seen only by using the stereo viewer.

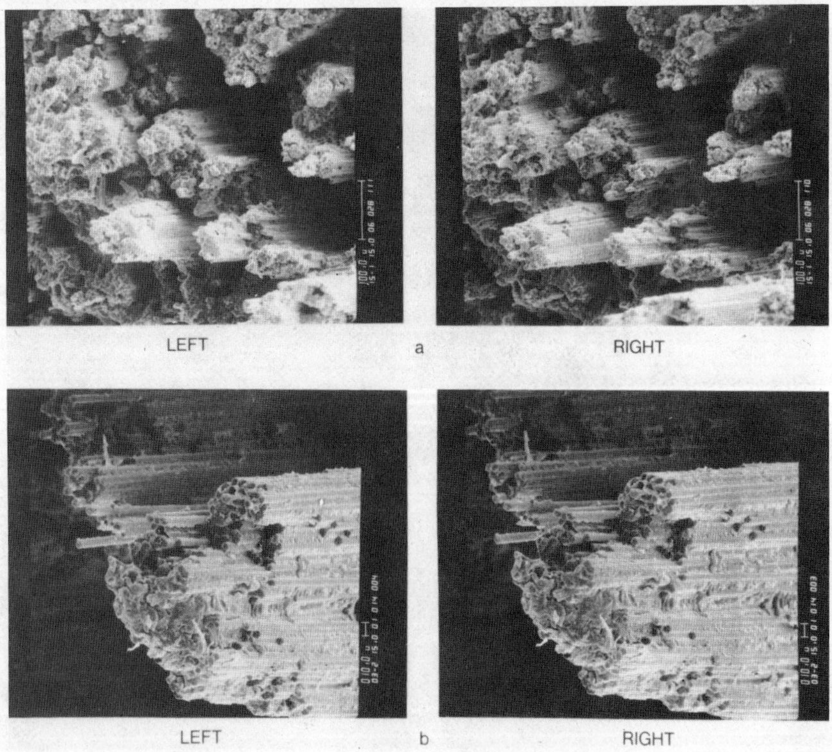

FIG. 17—*Static and fatigue fracture surfaces of 0-deg specimens:* (a) *static failure, and* (b) *fatigue failure:* $S/\overline{X}_\theta = 0.7$, $N_f = 5460$ *cycles.*

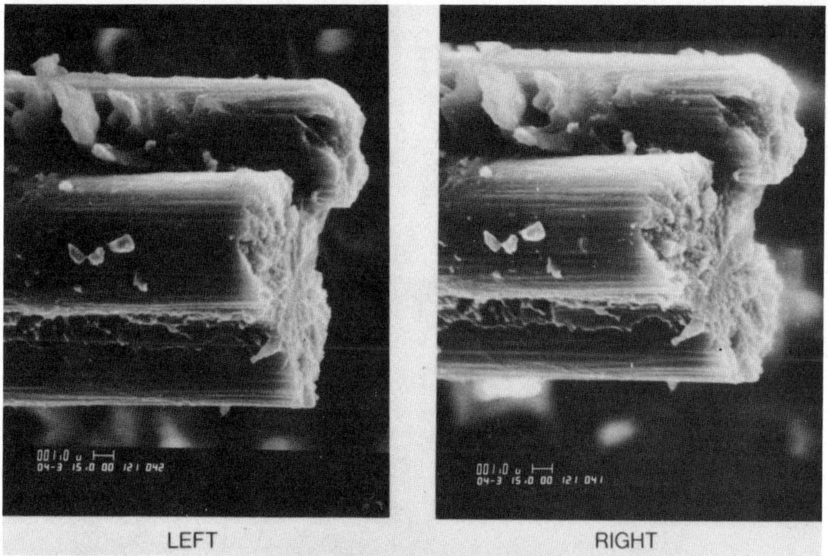

FIG. 18—*Fiber fracture under static loading for 10-deg off-axis specimens. Similar failure is observed for all off-axis angles under static and fatigue loadings.*

5. Fiber surfaces are clean of matrix residue, Fig. 16, indicating poor bonding between fibers and matrix.
6. Matrix failure in the form of serrations and cleavage, depending on the off-axis angle. The formation of serrations and cleavages appear mostly in resin rich areas.
7. Many transverse microcracks (about 0.1-μm) under detailed inspection in the resin rich areas and in the zones where the serrations are significant. These microcracks are in the direction of the specimen thickness (normal to the fibers) and indicate actual cracking of the matrix during loading.
8. In a scattered form, debris of matrix material clinging to the fiber surfaces and throughout the fracture surface, Figs. 16, 18, 19, and 20.

All of these failure modes have been observed at each fatigue fracture surface. Similar failure modes have also been observed on each of the static fracture surfaces. However, the matrix serrations and microcracks (Failure Modes 6 and 7 previously explained) are somewhat less pronounced in the static fracture surfaces.

Matrix serrations and microcracks are more pronounced in the matrix-dominated failure areas, that is, in the resin rich areas. In the interface-dominated fracture surfaces, that is, in the matrix between adjacent fibers, they appear to be much smaller with higher density. It should be noted that the appearance of a "resin rich area" is a result of whole groups of fibers or tows being pulled out during the final failure stage.

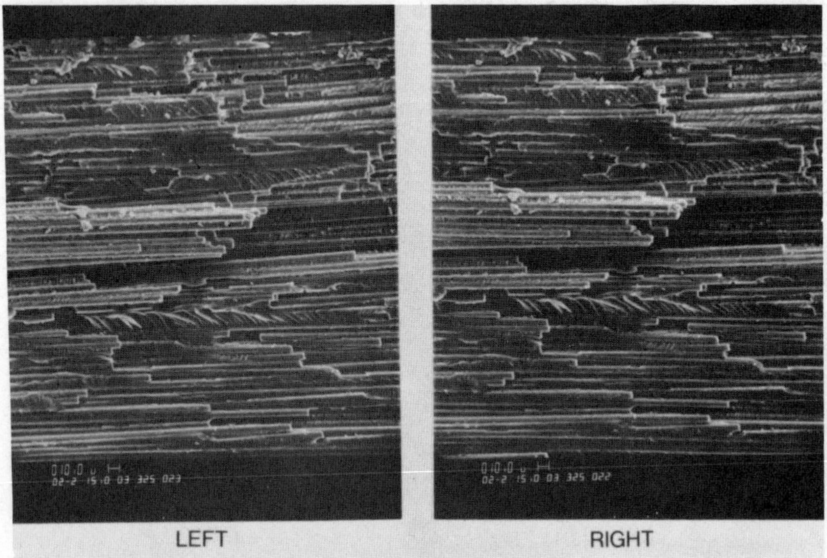

FIG. 19—*Static fracture surface of 30-deg off-axis specimens.*

The degree of serration in the resin rich areas can, to a certain degree, serve as an identification tool between the static and the fatigue failure surfaces as well as identifying approximately the off-axis angle. Comparing the static and fatigue fracture surfaces (for example, Fig. 16*a* with Fig. 19 or the static and fatigue fracture surfaces in Fig. 16*b*), it can be seen that larger serrations are observed on the fatigue fracture surfaces than on the static fracture surfaces. Moreover, it seems that the density of the serrations, that is, number of serrations per unit length, in the static fracture surfaces is higher than in the fatigue fracture surfaces. However, those differences between the static and fatigue fracture surfaces are not highly pronounced.

The degree of serration also changes to a certain extent with the off-axis angle, being more extensive for the smaller angles. This point was also discussed in Ref *25*. At 60 and 90 deg (see Figs. 16*b*, 16*c*, and 21), most of the matrix-dominated failure areas (resin rich areas) are showing the cleavage mode of failure which is typical to brittle tensile fracture. The serration seen in Figs. 16*b* 16*c* and 21 are less distinct and smeared over the fracture surface. A detailed inspection of those serrated areas through the stereo viewer reveals that those smeared serrations appear at angles to the fracture surface which resulted from the fracture surface propagating between the fiber tows. As no shear stress components are applied on the 90-deg specimens, the lack of distinct serrations is self explanatory. The shear stresses which are higher at the smaller off-axis angles are the main cause of those serrations. The fact that serrations appear also in the 0-deg

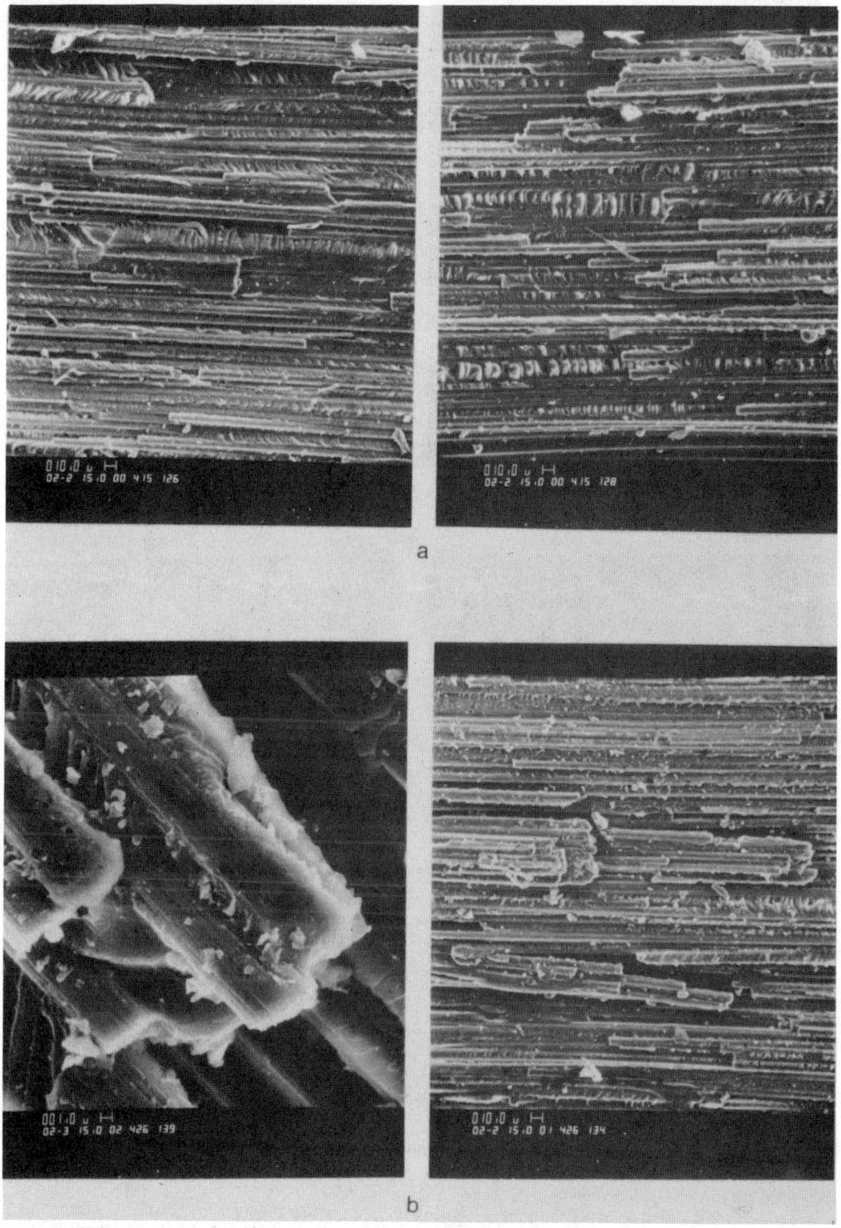

FIG. 20—*Fatigue fracture surface of 45-deg off-axis specimens;* (a) $S/\bar{X}_\theta = 0.6$, $N_f = 6650$ *cycles, and* (b) $S/\bar{X}_\theta = 0.75$, $N_f = 5570$ *cycles.*

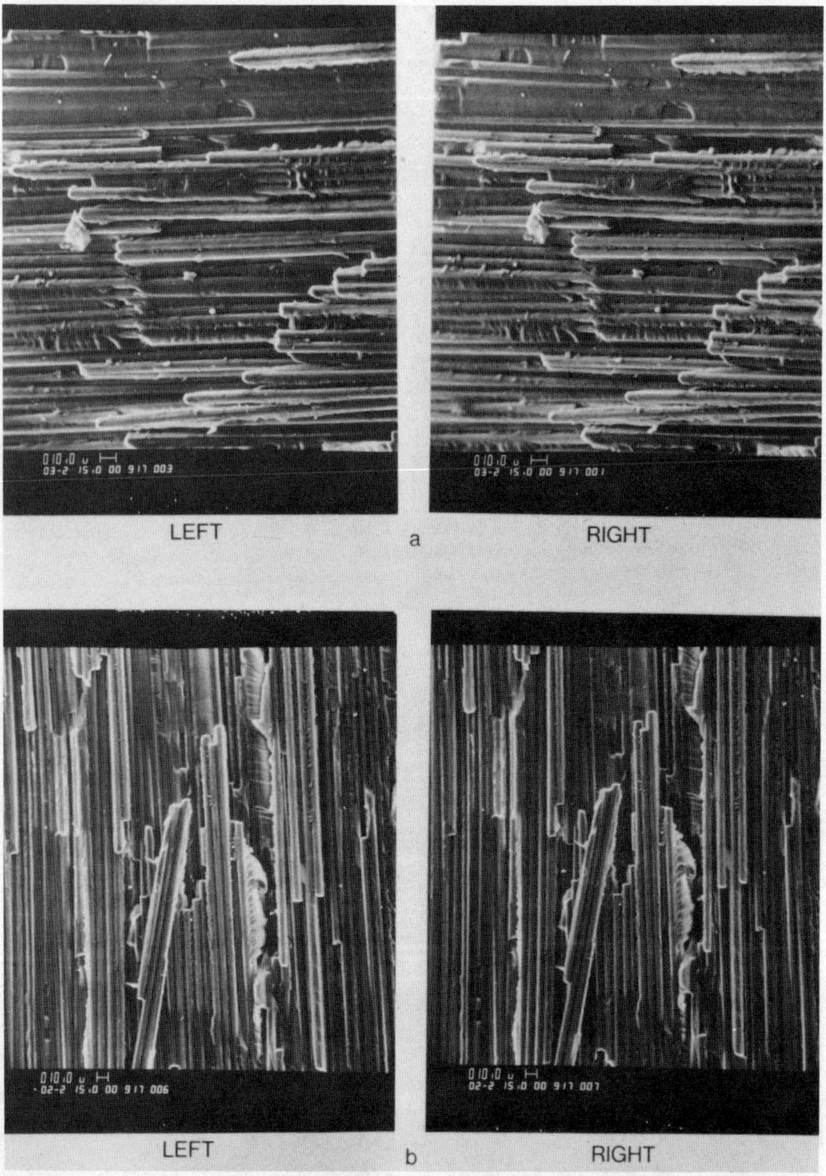

FIG. 21—*Static fracture surface of 90-deg off-axis specimens.*

specimens, Fig. 17, indicates that the axial cracks in on-axis unidirectional graphite/epoxy are a result of shear deformation.

Inspection of the fracture surfaces of the residual strength specimens through the stereo photographs reveals failure modes similar to those observed on the fatigue fracture surfaces, and therefore they are not shown here. It can be concluded that no noticeable damage occurs in fatigue before final failure.

An important question to be answered is whether the fractured fibers and tows observed on the fracture surfaces broke during fabrication or whether the fracture actually occurred during loading. Inspection of the fracture surface reveals that the latter is the case for two reasons. (1) In most broken fibers the serrations are observed also at the fiber ends. If those fibers had been broken prior to loading, such serrations would not have been possible. (2) Inspection of the "resin rich" areas reveals an imprint of the missing fibers, which would indicate that the fibers were intact at the time of fabrication.

The broken tows observed in Figs. 16 through 21 seem to be caused when the cracks branch out as the matrix cracking and interface failure spread to different locations along each specimen. It seems that the actual fracture of the fibers and tows occurs at the final stage of failure, considering that no strength or modulus degradation with life had been observed.

The sudden-death behavior of the material tested in this program, that is, no strength degradation after fatigue combined with large scatter in life, indicates that once a crack initiates it can grow to a critical size within a few cycles. As unidirectional graphite/epoxy is a highly notch-sensitive material especially in the transverse direction, a small crack is sufficient to abruptly reduce the specimen's strength by 30 to 40 percent which is well within the dynamic stress level, causing complete failure of the specimen. The fast crack propagation at the final fracture stage can also result in broken fibers (or fiber tows) even at large off-axis angles, 45 to 60 deg. Although the tensile stress component in the fiber direction is quite low for these cases, the appearance of serrations and microcracks in the matrix material can create high stress concentrations along the fibers causing them to fail along some weak spots.

Conclusions

Off-axis static and fatigue behavior of AS/3501-5A graphite/epoxy composite can be summarized as follows.

1. Off-axis strength has been characterized by a second-order polynomial of σ_2 and σ_6, and scatter in strength by a two-parameter Weibull distribution.
2. Practically no strength degradation or axial modulus changes occur during fatigue loading of off-axis graphite/epoxy laminae.
3. The relationship between normalized fatigue strength and life is only weakly dependent on off-axis angle.
4. Large scatter in life is observed for all off-axis angles.
5. Fatigue failure of off-axis graphite/epoxy is like a sudden death, that

is, strength decreases rapidly immediately before fatigue failure. Failure in fatigue occurs abruptly without early warning or prior visible damage.

6. The technique for obtaining stereo (three-dimensional) photographs of the fracture surface is helpful in obtaining more detailed information and better identification of the failure modes.

7. Fracture surfaces include a combination of several interdependent failure modes such as fracture of individual fibers and fiber tows, matrix serration (shear failure), matrix cleavage, and matrix/interface cracking parallel to fibers.

8. Matrix serrations increase with longitudinal shear stress. In the absence of the shear stress, a cleavage type of failure prevails.

9. Matrix serrations are larger on the fatigue fracture surfaces and smaller on the static fracture surfaces.

10. Fracture surfaces are not planar; they consist of matrix-dominated failure regions, interface-dominated failure regions, and combinations thereof.

11. No acoustic emission activity is observed before fracture in static tension for off-axis angles greater than 20 deg. These results are based upon the gain and threshold levels used for the 0-deg specimens.

12. The sudden-death behavior can be attributed to the rapid crack growth immediately before the final fracture. The appearance of serrations and microcracks in the matrix material can create high stress concentrations along the fibers causing them to fail along some weak spots.

The number of specimens tested in the present work is not sufficient to provide statistically meaningful life data. Furthermore, the results obtained show that the identification of failure modes through the examination of fracture surface is very promising. Therefore, a more detailed investigation is recommended.

Acknowledgments

This work has been performed in the Mechanics and Surface Interactions Branch, Nonmetallic Materials Division, Air Force Materials Laboratory, Wright-Patterson Air Force Base, Ohio. The support of S. W. Tsai in our program is acknowledged with appreciation. The authors also wish to express their appreciation to R. Y. Kim, R. Esterline, and R. Cornwell of the University of Dayton Research Institute for the preparation and testing of the composite specimens. The efforts of Mr. B. Strope of System Research Laboratory on the SEM are also acknowledged with appreciation.

References

[1] Hahn, H. T., "Fatigue Behavior and Life Prediction of Composite Laminates," Technical Report AFML-TR-78-43, Air Force Materials Laboratory, Wright-Patterson Air Force Base, Ohio, 1978.

[2] Tsai, S. W. and Hahn, H. T. in *Inelastic Behavior of Composite Materials*, C. T. Herakovich, Ed., AMD-Vol. 13, ASME, 1975, pp. 73–96.
[3] Broutman, L. J. and Sahu, S. in *Proceedings*, 24th Technical Conference, The Society of Plastics Industry, 1969, 11-D.
[4] Owen, M. J. in *Composite Materials, Vol. 5, Fracture and Fatigue*, L. J. Broutman, Ed., Academic Press, New York, 1974, pp. 313–340.
[5] Reifsneider, K. L., Henneke, E. G., and Stinchcomb, W. W., "Defect-Property Relationships in Composite Materials," Technical Report, AFML-TR-76-81, Part III, Air Force Materials Laboratory, Wright-Patterson Air Force Base, Ohio, 1978.
[6] Tsai, S. W. in *Fundamental Aspects of Fiber Reinforced Plastic Composites*, R. T. Schwartz and H. S. Schwartz, Eds., Wiley Interscience, New York, 1968.
[7] Tsai, S. W. and Wu, E. M., *Journal of Composite Materials*, Vol. 5, 1971, pp. 58–80; also Technical Report AFML-TR-71-12, Air Force Materials Laboratory, Wright-Patterson Air Force Base, Ohio, 1972.
[8] Wu, E. M., in *Composite Materials*, Vol. 2, G. P. Sendeckyj Ed., 1974, pp. 353–431.
[9] Pipes, R. B. and Cole, B. W., *Journal of Composite Materials*, Vol. 7, 1973, pp. 246–256.
[10] Hahn, H. T. and Tsai, S. W., *Journal of Composite Materials*, Vol. 7, 1973, pp. 102–118.
[11] Awerbuch, J. and Hahn, H. T. in *Fatigue of Filamentary Composite Materials, ASTM STP 536*, American Society for Testing and Materials, 1977, pp. 248–266.
[12] Hahn, H. T. and Kim, R. Y., *Journal of Composite Materials*, Vol. 9, 1975, pp. 297–311.
[13] Yang, J. N. and Liu, M. D., *Journal of Composite Materials*, Vol. 11, 1977, pp. 176–203.
[14] Chou, P. C. and Croman, R. in *Composite Materials: Testing and Design (Fifth Conference), ASTM STP 674*, American Society for Testing and Materials, 1979, pp. 431–454.
[15] Chou, P. C. and Croman, R., "Residual Strength in Fatigue Based on the Strength Life Equal Rank Assumptions," Technical Report, AFML-TR-78-96, Air Force Materials Laboratory, Wright-Patterson Air Force Base, Ohio, 1978.
[16] Yang, J. N., *Journal of Composite Materials*, Vol. 12, 1978, pp. 19–39.
[17] Hashin, Z. and Rotem, A., *Journal of Composite Materials*, Vol. 7, pp. 448–464.
[18] Toth, I. J., "An Exploratory Investigation of the Time Dependent Mechanical Behavior of Composite Materials," Technical Report AFML-TR-69-9, Air Force Materials Laboratory, Wright-Patterson Air Force Base, Ohio, 1969.
[19] Kim, R. Y., University of Dayton Research Institute, private communication.
[20] Hahn, H. T. and Eriksen, J., "Representation of Matrix/Interface-Controlled Strength of Unidirectional Composites," Technical Report, AFML-TR-78-85, Air Force Materials Laboratory, Wright-Patterson Air Force Base, Ohio, 1978.
[21] Mann, N. R., Schafer, R. E., and Singpurwalla, N. D., *Methods for Statistical Analysis of Reliability and Life Data*, Wiely, New York, 1974.
[22] Chamis, C. C. and Sinclair, J. M., *Experimental Mechanics*, Vol. 17, 1977, pp. 248–266.
[23] Jakus, J., Coyne, D. C., and Ritter, J. R., Jr., *Journal of Materials Science*, Vol. 13, 1978, pp. 2071–2080.
[24] Kaiser, J., "Untersuchungen Uber das Auftreten von Gerauschen beim Zugversuch," Dr.-Ing. thesis, Technische Hochschule, Munich, 1950.
[25] Sinclair, J. H. and Chamis, C. C., Fracture Surface Characteristics of Off-Axis Composite, NASA TN-73-700, 1979.

R. T. Bhatt[1] *and H. H. Grimes*[1]

Fatigue Behavior of Silicon-Carbide Reinforced Titanium Composites

REFERENCE: Bhatt, R. T. and Grimes, H. H., **"Fatigue Behavior of Silicon-Carbide Reinforced Titanium Composites,"** *Fatigue of Fibrous Composite Materials, ASTM STP 723,* American Society for Testing and Materials, 1981, pp. 274–290.

ABSTRACT: The low cycle axial fatigue properties of 25 and 44-fiber volume percent SiC/Ti(6Al-4V) composites were measured at room temperature and at 650°C. At room temperature, the *S-N* curves for the composites showed no anticipated improvement over bulk matrix behavior. Although axial and transverse tensile strength results suggest a degradation in silicon-carbide fiber strength during composite fabrication, it appears that the poor fatigue life of the composites was caused by a reduced fatigue resistance of the reinforced Ti(6Al-4V) matrix. Microstructural studies indicate that the reduced matrix behavior was due, in part, to the presence of flawed and fractured fibers created near the specimen surfaces by preparation techniques. Another possible contributing factor is the large residual tensile stresses that can exist in fiber-reinforced matrices. These effects as well as the effects of fatigue testing at high temperature are discussed.

KEY WORDS: metal matrix composite, SiC/Ti(6-4), low cycle fatigue, ultimate tensile strength, modulus, Poisson's ratio, microstructure, fiber flaws, residual stresses, mechanisms, composite materials, fatigue (materials)

Fiber reinforcement of a metal can often yield a composite with mechanical fatigue properties superior to those of the unreinforced metal. The predominant mechanism for this improvement is the fact that at the same fatigue stress level the cyclic stress on the reinforced matrix is reduced, which correspondingly reduces the initiation and propagation of matrix fatigue cracks. Abundant examples of composite fatigue superiority over unreinforced metal are available in the literature [*1*].[2] In many cases, however, the improvement is not attributed to matrix stress reduction but to other factors such as the deflecting and impeding role of the fibers and interfaces on the fatigue crack. Although these factors, as well as fiber flaws, reaction zones, residual stresses, etc., may have large effects on fatigue

[1] Materials scientist and head, Physics and Solids Section, National Aeronautics and Space Administration, Lewis Research Center, Cleveland, Ohio 44135.

[2] The italic numbers in brackets refer to the list of references appended to this paper.

properties, their significance is often difficult to evaluate due to the overwhelming effect of matrix stress reduction. Thus, the importance of these other factors can only be evaluated in experiments where stress reduction effects are accounted for by comparing the composite matrix and the unreinforced metal fatigue properties at the same matrix stress.

The purpose of this experimental study is to evaluate on a microstructural level those factors that control the fatigue resistance of Ti(6Al-4V) metal reinforced with silicon-carbide (SiC) fiber. This composite system was selected for study for several reasons. First, the titanium alloy matrix is considerably harder and more fatigue-failure prone than most of the composite metal matrices previously studied. Second, the fatigue properties of the unreinforced matrix have been studied extensively [*2*]. Based on these studies, it is anticipated that factors influencing matrix fatigue behavior such as notch sensitivity and residual stresses will be significantly modified in the composite to warrant investigation. Third, because both the fiber and the matrix in SiC/Ti(6Al-4V) composites deform elastically up to fracture of the composite at room temperature, it should be possible in the absence of stress concentrations from broken fibers to estimate the matrix stresses during the fatigue test from elastic theory. With this information, then, one can eliminate the stress reduction factor from the composite fatigue data and thus better evaluate the influence of other microstructural factors on matrix behavior. Finally, no prior fatigue study of this composite system was found despite its potential for aerospace application.

Experimental

Materials

The silicon-carbide reinforced titanium (SiC/Ti) composite used in this study was manufactured by TRW using AVCO 142 μm (5.6 mil) silicon carbide on carbon fiber and Ti(6Al-4V) alloy matrix. Unidirectional composite panels containing 14 or 18 plies sandwiched by 0.76-mm thick Ti(6Al-4V) cover skins were consolidated by foil/filament diffusion bonding methods described by Brentnall and Toth [*3*]. The panels were pressed at 0.10 GN/m^2 (15 000 psi) for 30 min at a temperature of 870°C. Two volume fractions of fibers were used, 25 percent and 44 percent by volume.

For reasons relating to faulty tooling dies used during diffusion bonding, the original material produced in this way was found to be less than fully dense, with some void areas appearing between the fibers in the plane of the lamina. Therefore, an additional hot isostatic pressing (HIP) was used to further consolidate the plates to the point where no voids were observable. In this "HIPing" process, the temperature was held at 843°C, below that used for diffusion bonding (870°C), and the time was limited to ½ h. Subsequent microstructural examination of the fiber/matrix interface reaction zone revealed no voids and no significant increase of the zone thickness. This interface zone, shown in Fig. 1, was 0.5 μm thick, typical of well-bonded SiC/Ti(6Al-4V) composites [*3*].

Fatigue and tensile specimens were cut from the plates to the dimensions shown in Fig. 2. Both axial (0-deg) and transverse (90-deg) tensile specimens were prepared.

Testing Procedures

Fatigue testing, both at room temperature and at elevated temperature, was done in a tension-tension mode, using an axial fatigue test facility developed at the NASA-Lewis Research Center. This facility that is described elsewhere [*4*] could be operated by a servocontrolled closed-loop hydraulic system in either a load- or strain-controlled mode. All fatigue testing was conducted at 0.5 Hz in a load-controlled mode. Conventional wedge-type grips were found to be unsuitable for these tests due to slipping within the grips. To overcome this problem, a new grip assembly was designed and is shown schematically in Fig. 3.

In the elevated temperature fatigue tests, the specimens were resistance heated with chopped a-c currents of 600 A. The grips were water cooled to prevent any creep or relaxation of the gripping during high temperature

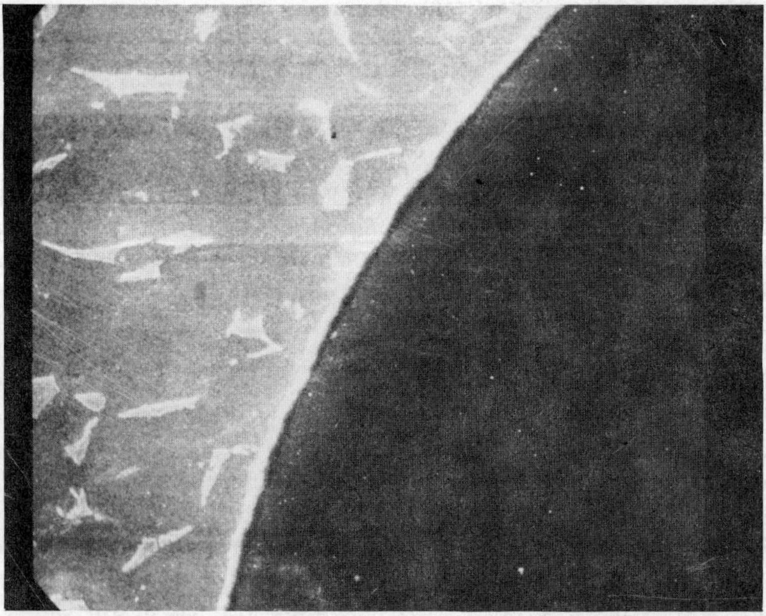

FIG. 1—*Interface region of SiC/Ti(6Al-4V) composite in as-fabricated and "HIPped" condition (×300).*

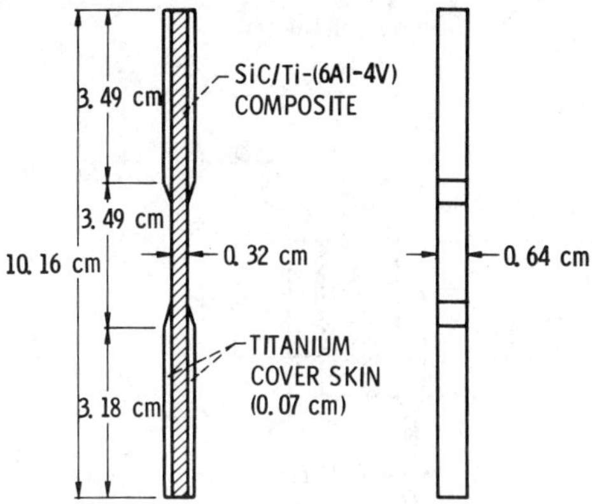

FIG. 2—*Tensile and fatigue specimen geometry.*

tests. The temperature at the center of the specimen was monitored by an attached chromel-alumel thermocouple and held constant within ±1°C.

Tension testing at room temperature was accomplished in a standard Instron machine using conventional wedge-type grips and a constant cross-head speed of 0.51 mm/s (0.02 in./min). Room temperature elastic modulus and Poisson's ratio measurements were made on the composite using the fatigue testing facility. The axial strain was measured with two resistance-type strain gages cemented on parallel sides of the specimen. The transverse strain was measured using a diametrical extensometer clamped onto the specimen. Load was changed in increments of 200 lb to trace the load-strain plots.

Measurement was made of the tensile strength of silicon-carbide fibers both before consolidation into a composite and after removal from the composite by acid dissolution of the matrix using a hot 50 percent hydrochloric acid 50 percent water solution. Fibers of each type were pulled to failure in an Instron testing machine using the test method described in Ref 5.

Results and Discussion

Tensile Data

The room temperature elastic modulus, Poisson's ratio, and tensile strength as a function of percent by volume fiber are shown in Figs. 4 to 6 for the SiC/Ti(6Al-4V) composites used in this study. The data points indicate the range and average value for each volume fraction. Typically, eight tests were performed for each data point. Also included for comparison purposes are the averaged data of Brentnall and Toth [*3*] and Tsareff et al [*6*]. Our measured values of axial modulus, E_{11}, and Poisson's

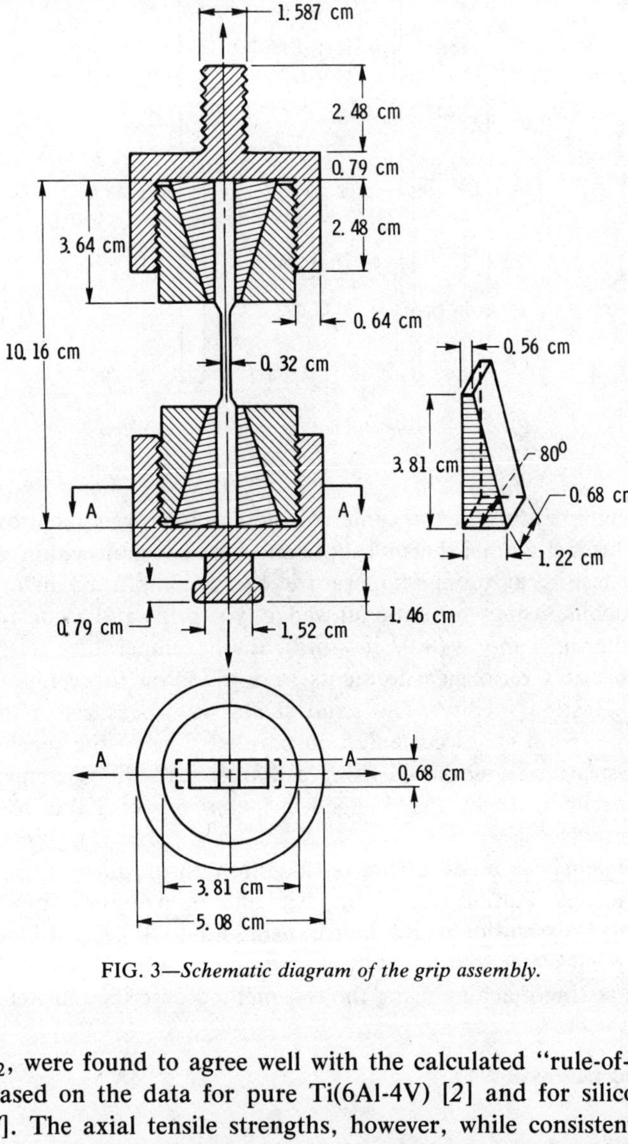

FIG. 3—*Schematic diagram of the grip assembly.*

ratio, ν_{12}, were found to agree well with the calculated "rule-of-mixtures" values based on the data for pure Ti(6Al-4V) [*2*] and for silicon-carbide fibers [*7*]. The axial tensile strengths, however, while consistent with the data of Refs *3* and *6*, show rather low values compared with the rule-of-mixtures predictions based on an average tensile strength of the as-received fiber of 3.4 GN/m^2 (500 000 psi) and a value of 0.85 GN/m^2 (124 000 psi) for the stress on the matrix strained to the fracture strain of the fiber (0.008). The average measured value of axial tensile strength was 1.03 GN/m^2 (150 000 psi) for the 44 percent by volume composite and 0.86 GN/m^2 (125 000 psi) for the 25 percent by volume composite. The corresponding calculated rule-of-mixtures values are 1.99 GN/m^2 (289 000 psi)

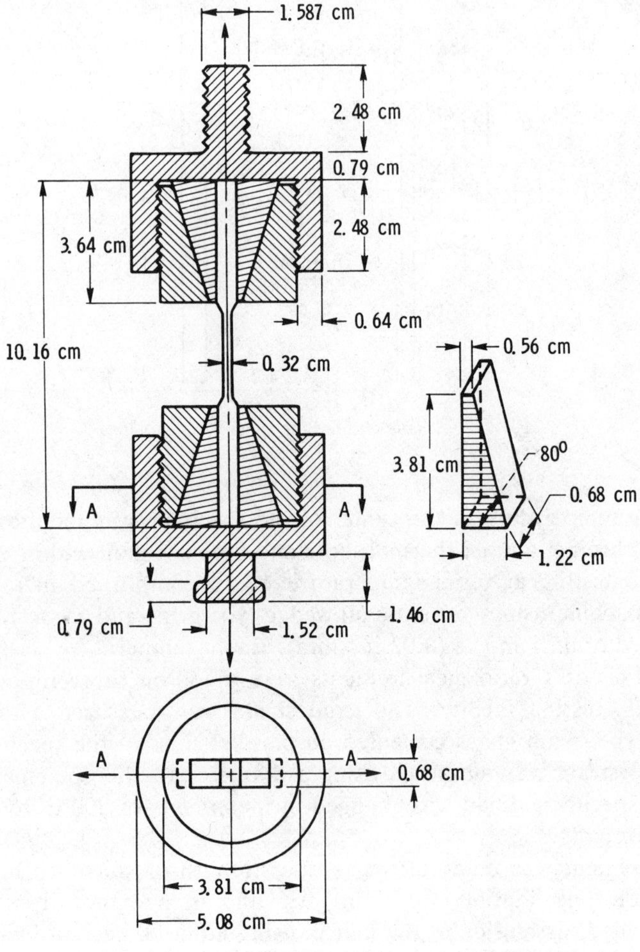

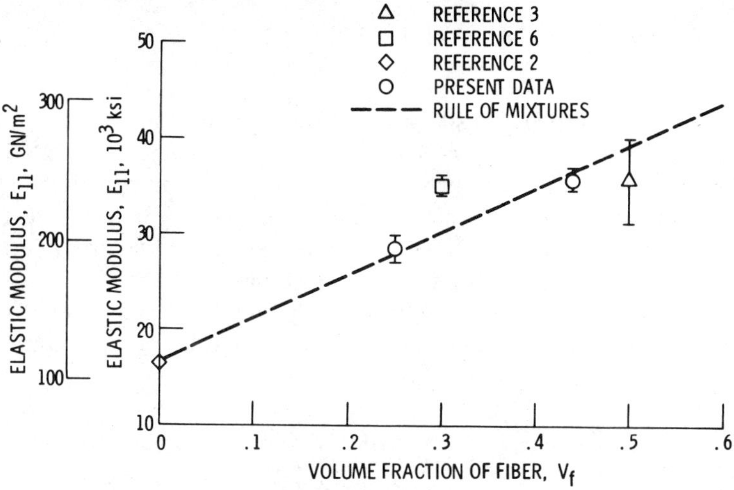

FIG. 4—*Axial elastic modulus* (E_{11}) *of SiC/Ti(6Al-4V) composite as a function of fiber content.*

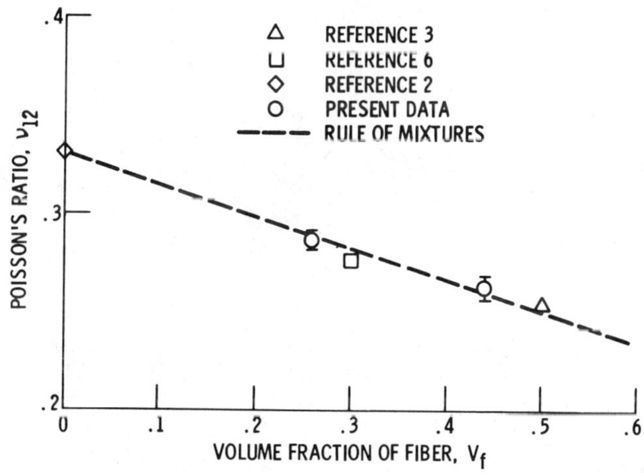

FIG. 5—*Poisson's ratio* (ν_{12}) *of SiC/Ti(6Al-4V) composite as a function of fiber content.*

for the 44 percent by volume composite and 1.50 GN/m^2 (218 000 psi) for the 25 percent by volume composites.

The transverse tensile strength for the 44 percent by volume composite, 0.27 GN/m^2 (40 000 psi), was also found to be low when compared with the ultimate tensile strength we measured for an annealed Ti(6Al-4V) alloy, 0.92 GN/m^2 (135 000 psi). For no fiber/matrix bonding, a lower bound can be estimated for the transverse tensile strength, σ_T, from the Eq 4*a* of Ref *8*

$$\sigma_T = \sigma_m \left[1 - 2\left(\frac{V_f}{\pi}\right)^{1/2}\right]$$

where σ_m is the ultimate tensile strength of the matrix and V_f the volume fraction of the fiber. For $V_f = 0.44$, the σ_T value obtained is 0.23 GN/m^2 (34 000 psi). Since our measured value of transverse strength was slightly higher than the calculated value for no bonding, it may be concluded that the fibers are only partially bonded to the matrix, or that the interfacial bond is weak, or that the fiber's transverse strength was low.

To understand the mechanisms responsible for the low composite strength values, SEM micrographs were taken. Typical tensile fracture surfaces of

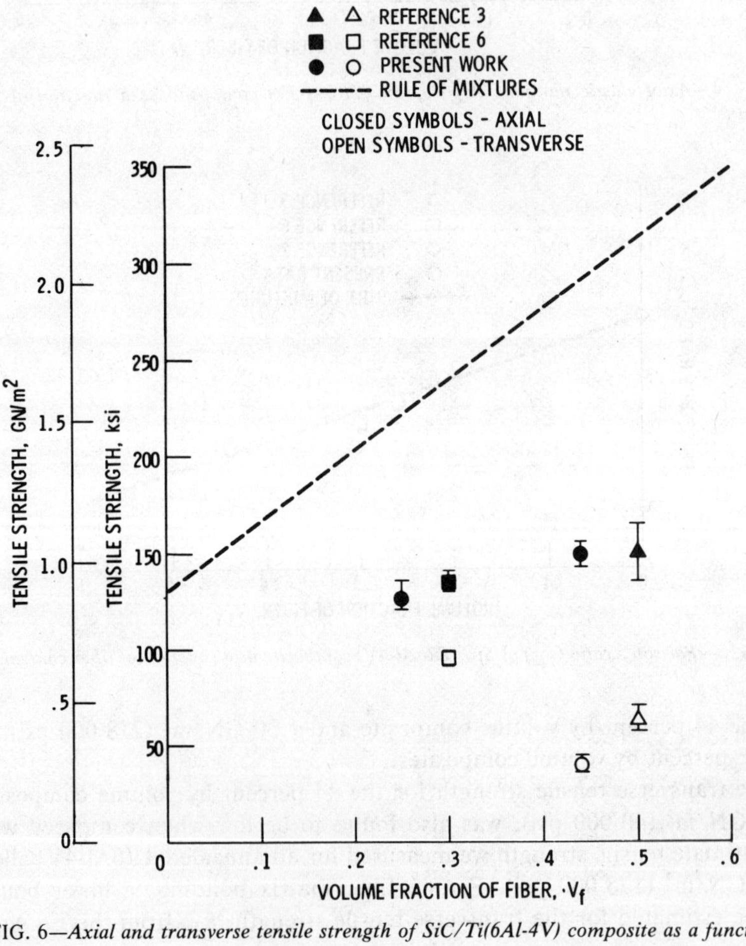

FIG. 6—*Axial and transverse tensile strength of SiC/Ti(6Al-4V) composite as a function of fiber content.*

specimens stressed in the axial and transverse directions are shown in Figs. 7 and 8, respectively. The fracture surface of Fig. 7 gives no evidence of extensive fiber pull-out, characteristic of poor bonding. This result, then, favors the argument for fiber strength degradation rather than weak or partial interface bonding. The fracture surface of Fig. 8 further supports this view in that fiber splitting occurs in preference to fiber debonding. As part of silicon carbide fiber manufacturing, a thin carbon coating is deposited on the fiber surface to improve handleability by decreasing surface flaw sensitivity. Any process that mechanically or chemically degrades this coating will result in fiber weakening. This effect was discussed by DeBolt and Krukonis [7] as causing silicon carbide fiber strength decreases of 0.69 to 1.38 GN/m^2 (100 000 to 200 000 psi).

To determine whether fiber degradation did occur, tensile tests were conducted on 2.54 cm gage lengths of as-received silicon carbide fibers and on fibers chemically removed from the 44 percent by volume composite after consolidation. The acid reagent used to extract the fibers does not attack the silicon carbide nor the protective carbon coating. Nevertheless, the average of the strengths measured for 65 fibers removed from the as-fabricated composite was 2.07 GN/m^2 (300 000 psi) compared with an average value of 3.45 GN/m^2 (500 000 psi) for 30 as-received silicon carbide fibers. Thus, while it is not completely understood what the degradation

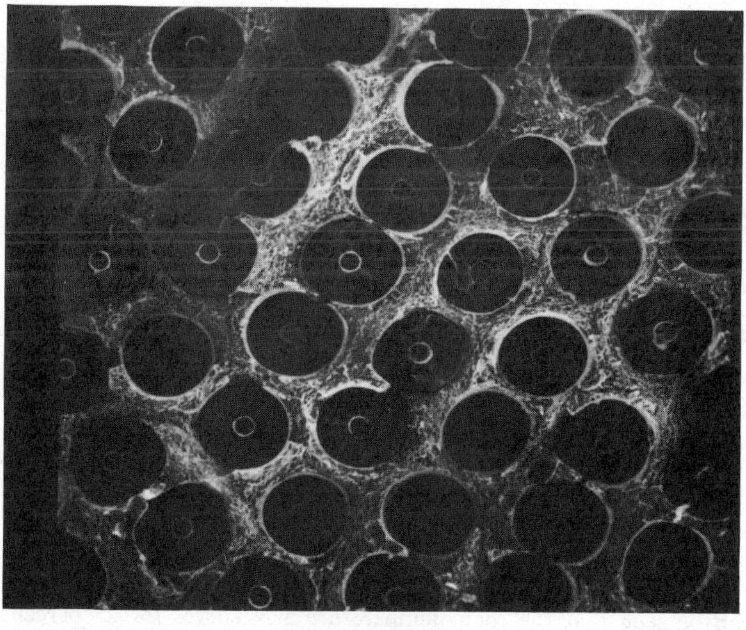

FIG. 7—*Fracture surface of a SiC/Ti(6Al-4V) composite stressed in axial direction (×100).*

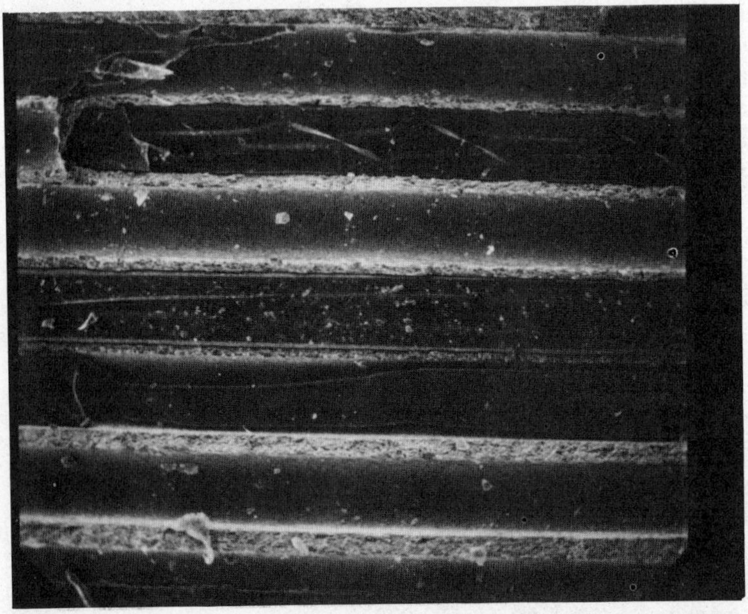

FIG. 8—*Fracture surface of a SiC/Ti(6Al-4V) composite stressed in transverse direction (×100).*

mechanism is, it appears that this coating may be rendered ineffective by composite fabrication with a titanium matrix. If we use our measured average value of 2.07 GN/m² (300 000 psi) for the degraded fiber strength in the rule-of-mixtures equation and the stress on the matrix for a strain equal to the fiber fracture strain (0.0045), we obtain an axial tensile strength of 1.20 GN/m² (175 000 psi) that is in better agreement with our measured value of 1.03 GN/m² (150 000 psi).

Fatigue Data

Axial low cycle fatigue testing of 25 and 44 percent by volume SiC/Ti-(6Al-4V) composites was done at room temperature and 650°C at a stress ratio, $R = 0.0$ (R = minimum stress/maximum stress). The data obtained are presented in Fig. 9 as applied maximum external stress versus number of cycles to failure (*S-N* curves). The data points of Fig. 9 usually represent the results from one experiment, except at the lower stresses where increased scatter necessitated several tests and the average value is reported. Matrix fatigue behavior at room temperature for annealed Ti(6Al-4V) bar stock of identical specimen configuration is also shown in this figure for comparison. It can be seen that the room temperature *S-N* curves for 25 percent and 44 percent by volume composites and the unreinforced matrix alloy are all

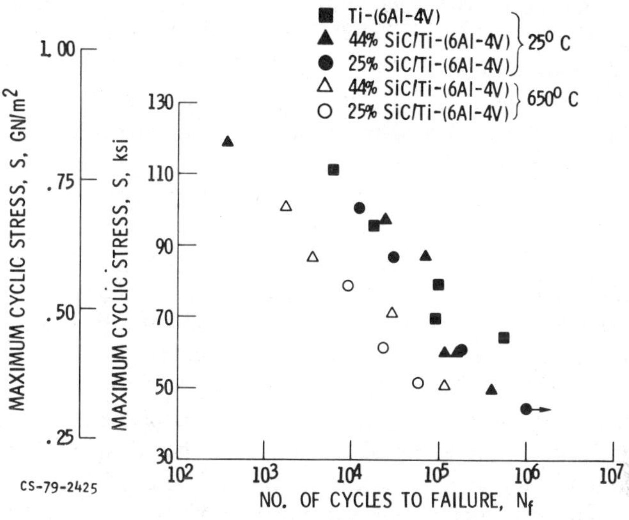

FIG. 9—*Room and elevated temperature fatigue curves of SiC/Ti(6Al-4V) system at a stress ratio of* R = *0.0.*

quite similar up to 10^6 cycles, the maximum number of cycles tested. No definite approach to a fatigue limit is indicated up to this cycle level for these composite materials, nor is any beneficial effect noted by the presence of fibers.

The *S-N* curves for composites fatigued at 650°C are also shown in Fig. 9. The curves, while similar to the room temperature *S-N* curves, have shorter cycle lifetimes for the same maximum stress values. At both temperatures, little difference can be observed between the 25 percent and the 44 percent by volume composites data.

In contrast to the fatigue data for many of the other metal-matrix composites studied [*1*], it appears that the SiC/Ti(6Al-4V) system does not exhibit the anticipated improvement in room temperature fatigue properties with fiber reinforcement. The reason for this and for the apparent lack of dependence on fiber loading was further investigated in this study.

Because there exists no indication that ceramic fibers such as silicon-carbide or boron are subject to fatigue damage at stresses well below the failure stresses of even weakened fiber, one can conclude that the lifetime-limiting processes in ceramic fiber reinforced metal matrix composites is the initiation or growth or both of the fatigue crack in the matrix [*9*]. In the composite, the matrix carries only a fraction of the load experienced by the monolithic matrix material alone in a comparable test. Thus, any effects in the matrix due to weak interfaces, stress concentrations or flaws, tensile residual stresses, work hardening or softening, etc., on crack initiation or propagation may be overwhelmed by the fatigue life improvement due to the reduced strain on the matrix. For this reason, the room temperature fatigue strength data of Fig. 9 were replotted in Fig. 10 as matrix strain amplitude versus cycles-to-failure in order to compare the matrix in the composite and the unreinforced matrix material under conditions of equal cyclic strain. This plot seems reasonable for SiC/Ti(6Al-4V) for two reasons. First, at room temperature, the fatigue and fracture strains observed for this composite were all within the elastic limit of both the fiber and the matrix. The load-deflection curves were indeed linear. Second, it appears that the amplitude of the matrix cyclic strain remains constant throughout the load-controlled fatigue test. For this to be so, the fatigue life of the matrix would have to be controlled almost entirely by fatigue crack initiation time rather than crack propagation time. Support for this can be found in the literature for this alloy [*10*] and for the composite in the results of the following experiment designed to determine the relative importance of matrix crack initiation and crack propagation on fatigue life.

Several fatigue specimens were instrumented within the gage length with probes to measure the axial d-c resistance throughout the room temperature fatigue test. A transverse matrix crack would reduce the specimen cross sectional area and consequently reduce the measured resistance. The resistance during cycling remained constant to within ±1 percent throughout over 90 percent of the test duration, after which it increased exponentially

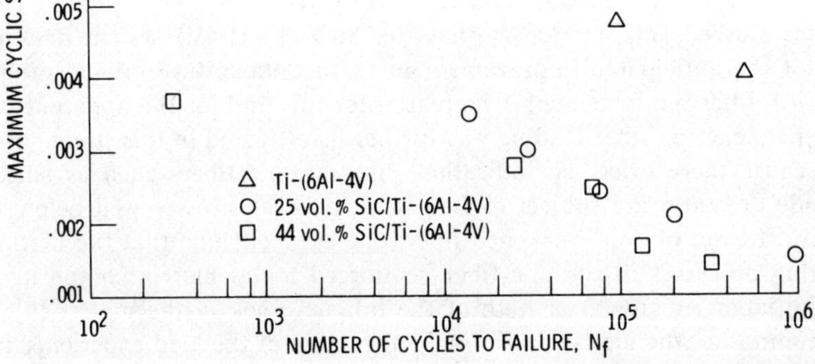

FIG. 10—*Room temperature fatigue data of Fig. 9 for SiC/Ti(6Al-4V) and Ti(6Al-4V) plotted as maximum cyclic matrix strain versus cycles to failure. The matrix strains in the composite are calculated using the composite moduli determined in this study.*

to the time of specimen failure. A visible fatigue crack was observed within a few thousand cycles after the resistance increase. The resistance and visual observation of the fatigue crack confirm a mechanism in which the greater part of the fatigue life is spent in crack initiation, followed by a rather rapid crack growth to failure.

In addition, metallography carried out on the fatigue fractured specimens shows very few secondary cracks. This would occur if, because of the rapid growth of the primary crack, little time is available for secondary crack growth. Again, this supports the conclusion that fatigue crack initiation time is longer than the propagation time.

Having thus established the validity of Fig. 10, let us now examine its two primary implications. Clearly, at a constant cyclic strain, the composite fatigue lifetime is reduced from that for the unreinforced matrix, suggesting that incorporation of the silicon carbide reinforcement serves to degrade the fatigue properties of Ti(6Al-4V). In addition, increasing the fiber content from 25 percent to 44 percent by volume appears to further reduce the lifetime, but only slightly.

One possible factor for the reduced fatigue behavior of the titanium composite is the presence of flawed and fractured fibers in the surface region of the specimens that could serve as initiation sites for the fatigue crack by raising the local stress in the proximate matrix. The role of such flaws in facilitating fatigue crack initiation has long been known [*1*]. To establish the presence of the flawed and fractured fibers in the surface of the specimen and their role in fatigue failure, metallographic investigation was carried out on the as-received and fatigue-fractured specimens. Figure 11*a*, a photomicrograph of the machined surface of the as-received composite, shows such fractured fibers. Removal of the first two layers of the fibers by careful grinding and polishing produced no evidence of the broken fibers in the interior of the specimen. Thus, we conclude that the surface fiber cracks occurred during specimen machining. Photomicrographs of the fatigue-fractured specimens indicate that the matrix fatigue cracks always initiated from these broken surface fibers (Fig. 11*b*). Metallographic analysis of the entire cross section of the fatigued specimen failed to produce any clear evidence of fatigue crack initiation at internal sites or even at the unmachined specimen surface. Thus, although fiber reinforcement reduces the overall matrix strain and thereby increases fatigue life, the presence of fractured fibers by their ability to act as localized stress raisers in the proximate matrix, can also shorten the matrix crack initiation time and thus shorten the fatigue life in SiC/Ti(6Al-4V).

Another possible factor contributing to the poor fatigue resistance of SiC/Ti(6Al-4V) composites compared to unreinforced Ti(6Al-4V) is the internal state of stress in the composite. Cooling of composites from the fabrication temperature can introduce residual stresses into the composite because of the difference in thermal expansion coefficients of the fibers

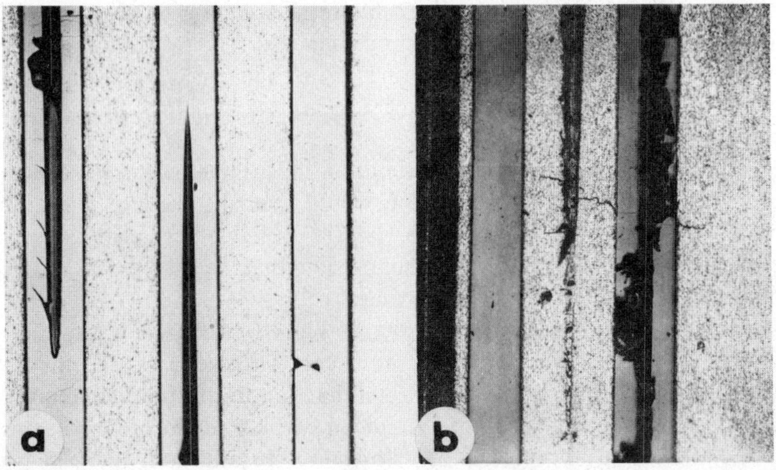

FIG. 11(*a*)—*Photomicrograph of as-received SiC/Ti(6Al-4V) composite specimen showing the fractured fibers on the machined surface (×100).* (*b*) *Photomicrograph of a fatigued specimen showing fatigue crack initiation at the cracked surface fiber (×100).*

and the matrix. Even though quantitative estimates of the magnitude of these residual stresses are rather difficult due to possible plastic relaxation of the matrix at high temperatures, one can assume elastic maxtrix behavior during cooling from the fabrication temperature and thus calculate the maximum residual stress in the matrix, σ_R, from the equations used by Schaefer et al [*11*].

$$\sigma_R = \left[\frac{E_m(\alpha_m - \alpha_f)(T_c - T_o)}{E_m V_m / E_f V_m + 1}\right]$$

where T_c is the composite consolidation temperature, T_o is room temperature, α is thermal expansion coefficient, E is elastic modulus, V is fiber volume fraction, and subscripts m and f refer to matrix and fiber, respectively. Using $T_c = 870°C$ and values for α_m, α_f, E_f, and E_m from Ref *12*, the residual matrix stress, α_R, was calculated for various volume fractions of fiber in the composite. The calculated values in Fig. 12 show that the matrix could be under a tensile stress as large as 0.38 GN/m^2 (56 000 psi) for the 44 percent by volume SiC/Ti-(6Al-4V) composite. Preliminary X-ray line broadening studies on our as-received composite do, in fact, indicate the presence of residual stresses, but to date we have not been able to quantify them sufficiently for comparison with the calculated values.

The presence of a tensile residual stress in the matrix can be an additional source of reduced fatigue behavior in the composite. For example, from Ref *2*, we learn that increasing the mean stress, at a constant cyclic stress, decreases the fatigue life of Ti(6Al-4V). This is indicated in the Goodman diagram of Ref *2* replotted here in Fig. 13. The fatigue behavior of the matrix in our composites can be compared with the fatigue data of Ref *2* for Ti(6Al-4V) assuming the calculated residual stresses of Fig. 12 to be present. As indicated by the insert of Fig. 13, the mean stress on the matrix in our composites is obtained by adding half of the component of the applied cyclic stress experienced by the matrix, σ_a, to the calculated tensile residual stress, $\sigma_r = \sigma_{min}$, for the two fiber volume fractions studied. The resulting alternating versus mean stresses calculated from the best fit data of Fig. 10 are plotted as the solid lines in Fig. 13 for 10^4, 3×10^4, and 10^5 cycle lives.

Several interesting conclusions result from the comparison of these plots with the Ti(6Al-4V) fatigue data of Ref *2*. First, the fatigue behavior of the matrix in our composite is similar to that of Ti(6Al-4V) in a notched condition. Thus, not only do the fibers provide no beneficial effect except to

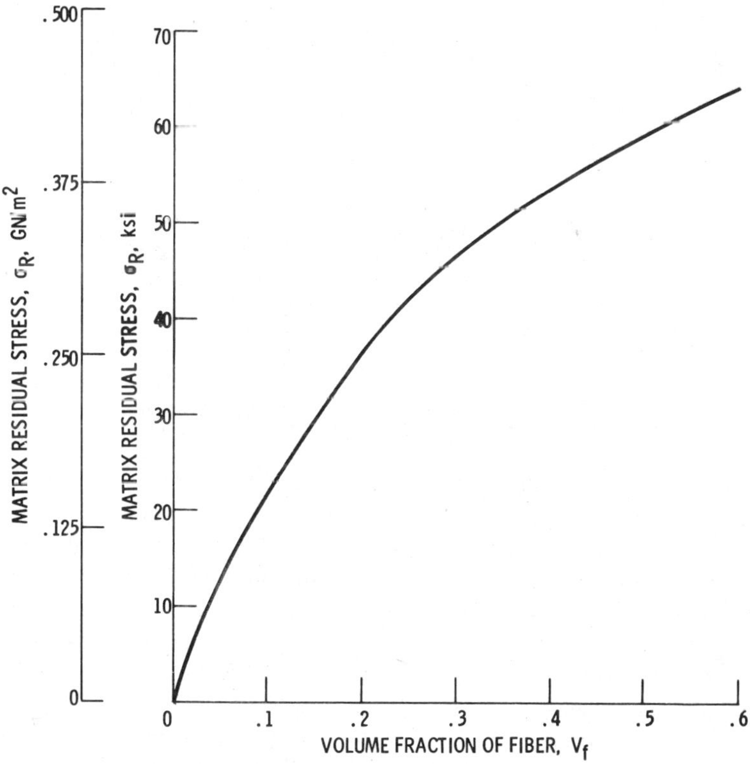

FIG. 12—*Calculated matrix tensile residual stress* (σ_R), *in SiC/Ti(6Al-4V) composites as a function of fiber content. Elastic matrix behavior is assumed during cooling from fabrication temperature of 870°C. Thermal expansion coefficients for the calculations obtained from Ref* 12.

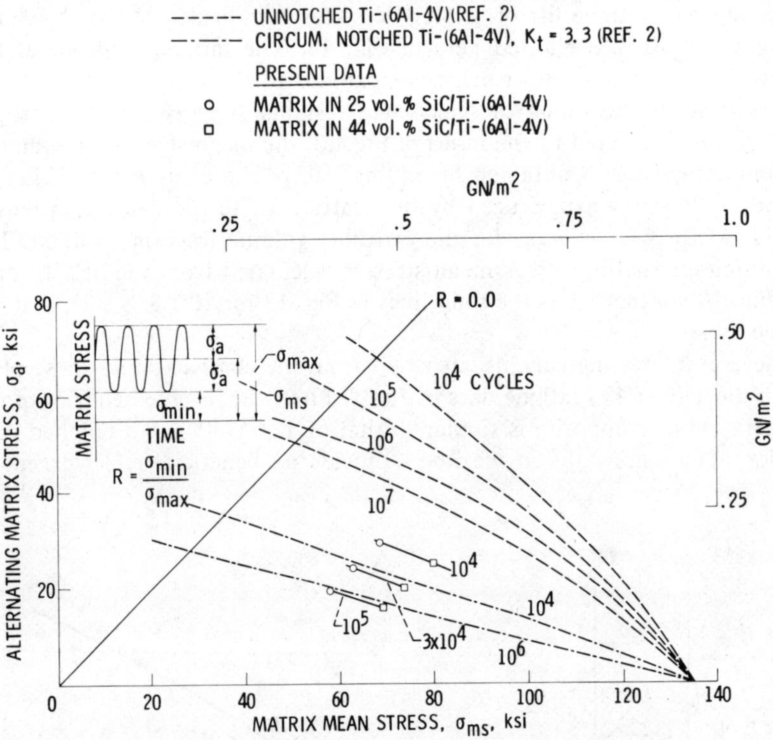

FIG. 13—*Comparison of the Goodman diagram for notched and unnotched annealed Ti(6Al-4V) with the properties of the Ti(6Al-4V) matrix in the composites. The mean stress on the matrix in the composite was obtained by adding half of the component of the applied cyclic stress experienced by the matrix,* σ_a*, to the tensile residual stresses,* σ_R*, calculated for two volume fractions studied.*

reduce the matrix stress, it is likely that their presence provides additional notch sensitivity to the matrix material, reducing its fatigue properties. Second, cooldown of the composite from the fabrication temperature to room temperature can produce residual stresses in the matrix that further reduce the matrix fatigue behavior. This reduction is proportional to the fiber volume fraction and can account for the observed decrease of fatigue properties with increased fiber content. Examined together, however, it appears that the fiber notch effect has a much more significant degrading effect on matrix fatigue than the residual stress effect.

The understanding of the high temperature fatigue results is complicated by several changes in the composite due to the temperature. These include: (1) a marked decrease in the strength and an increase in the ductility of the matrix, (2) relief of any tensile residual stress in the matrix, (3) relief of any compressive residual stress in the fibers, (4) an increase in the fraction of the load carried by the fibers, (5) a possible increase in the interface reaction layer, and (6) a general reduction in the stress intensity factor at the tip of a matrix crack. Item 1 is known to decrease matrix fatigue resistance [*2*]. Items 2, 4, and 6 might be expected to increase matrix fatigue resistance, Item 3 might be expected to have no effect, and Item 5 could go either way. In addition, there is no fatigue data available in the literature for the matrix beyond 500°C. Our own efforts to run fatigue tests at 650°C were unsuccessful because of excessive creep of the specimens. Without this comparison, interpretation of the results in terms of the preceding factors becomes rather speculative and was not attempted in this study.

Summary and Conclusions

The strength and fatigue behavior of SiC/Ti(6Al-4V) composites can be summarized as follows:

1. While measured axial elastic moduli and Poisson's ratios appear to agree with the rule-of-mixtures values, the axial and transverse tensile strengths are distinctly lower. This strength disparity can be resolved if the fiber strength used in the calculation is that measured for fibers removed from the composite after consolidation. The low fiber strengths thus measured probably result from the return of surface flaw sensitivity of the silicon carbide due to abrasion or surface reaction during fabrication.
2. The room temperature *S-N* curves for this composite system indicate no enhancement, in fact, even a reduction of fatigue life at a given stress amplitude when compared with unreinforced matrix fatigue life. The fact that the composite deforms elastically at room temperature coupled with the observation that fatigue crack initiation in the matrix accounts for the major portion of the fatigue life of this composite allows one to compare the room temperature fatigue data on the basis of matrix strain amplitude. Fractured and flawed fibers near the composite surfaces are found to be contributory to the reduction of matrix fatigue life through their effect on crack initiation time. Thermal residual stresses generated during composite fabrication are also considered to be a prime contributor to the degradation of matrix fatigue behavior and can explain the decrease in fatigue properties with increasing fiber content.
3. The explanation of the fatigue properties at high temperatures is complicated by unknown effects of matrix plasticity, modification of residual stresses, and interface reaction effects.

In summary, it appears that significant improvement in the fatigue properties of silicon carbide fiber reinforced Ti(6Al-4V) may be achieved with the elimination of broken or abraided fibers and the elimination or reduction of possible residual tensile strains in the matrix. Both problems might be alleviated to some extent by the use of a softer matrix material. The increased plasticity should reduce stress concentrations at flaws and also relieve residual stresses.

References

[*1*] Hancock, J. R. in *Composite Materials,* first ed., L. J. Broutman, Ed., Academic Press, New York, Vol. 5, Chapter 9, pp. 371–414.

[*2*] Mayakuth, D. J., Monroe, R. E., Favor, R. J., and Moon, D., "Titanium Base Alloys/6Al-4V, DMIC Processes and Properties Handbook," Battelle Memorial Institute, Feb. 1971.

[*3*] Brentnall, W. D. and Toth, I. J., "High Temperature Titanium Composites," Technical Report AFML-TR-73-223, Air Force Materials Laboratory, Dayton, Ohio, 1973.

[*4*] Hirschburg, H. H. in *Manual on Low Cycle Fatigue Tests, ASTM STP 465,* American Society for Testing and Materials, 1969, pp. 67–86.

[*5*] Smith, R. J., "Changes in Boron Fiber Strength Due to Surface Removal by Chemical Etching," NASA-TN-T-5219, National Aeronautics and Space Administration, Washington, D.C., 1976.

[*6*] Tsareff, T. C., Sippel, G. R., and Herman, M., "Metal-Matrix Composites," DMIC Memo 243, Defense Metals Information Center, Battelle Memorial Institute, May 1969.

[*7*] DeBolt, H. and Krukonis, V., "Improvement of Manufacturing Methods for the Production of Low Cost Silicon Carbide Filament," Technical Report, AFML-TR-73-140, Air Force Materials Laboratory, Dayton, Ohio, 1973.

[*8*] Cooper, G. A. and Kelly, A. in *Interfaces in Composites, ASTM STP 452,* American Society for Testing and Materials, 1969, p. 90.

[*9*] Salkind, M. and Patarini, V., *Transactions,* American Institute for Metallurgical Engineers, Vol. 239, P. 1268.

[*10*] Weinburg, J. G. and Hanna, I. E., "An Evaluation of Fatigue Properties of Titanium and Titanium Alloys," TML Report No. 77, Titanium Metallurgical Laboratory, July 1957.

[*11*] Schaefer, W. H. and Christian, J. L., "Evaluation of the Structure Behavior of Filament Reinforced Metal Matrix Composites," Technical Report, AFML-TR-69-36, Vol. 3, Air Force Materials Laboratory, Dayton, Ohio, 1969.

[*12*] DiCarlo, J. A. and Maisel, J. E., "High Temperature Dynamic Modulus and Damping of Aluminum and Titanium Matrix Composites," NASA TN-79080, National Aeronautics and Space Administration, Washington, D.C., 1979.

Ramesh Talreja[1]

Estimation of Weibull Parameters for Composite Material Strength and Fatigue Life Data

REFERENCE: Talreja, Ramesh, **"Estimation of Weibull Parameters for Composite Material Strength and Fatigue Life Data,"** *Fatigue of Fibrous Composite Materials, ASTM STP 723,* American Society for Testing and Materials, 1981, pp. 291-311.

ABSTRACT: Two classical methods for estimation of Weibull parameters, namely, the moment estimation and the maximum likelihood estimation are reviewed. It is demonstrated that, assuming the location parameter to be zero, the errors made in estimating both the scale and the shape parameters by these methods increase linearly with the true value of the location parameter.

An alternative method for Weibull parameter estimation is proposed. This method provides estimates for all three parameters when the shape parameter is close to one, and estimates the scale and the shape parameters more accurately than the other two methods for higher values of the shape parameter. Furthermore, unlike the other methods, this method is capable of separating the constituent components in a multi component sample and estimating the parameters of each component. This property makes the method suitable for statistical analysis of composite material strength and fatigue life data.

KEY WORDS: composite materials, Weibull distribution, statistical analysis, strength data, fatigue life data, reliability assessment, fatigue (materials)

In assessing the reliability of composite structures, Weibull distribution has proved to be a useful and versatile means of describing composite material strength and fatigue life data. However, proper use of this distribution requires that its parameters be accurately estimated. Furthermore, fracture of composite materials is complex, which often results in multicomponent strength and fatigue life data. This necessitates using an estimation procedure that is capable of detecting heterogeneity in given data and estimating parameters of each constituent component. The classical methods of parameter estimation, for example, the moment estimation and the maximum likelihood estimation, are incapable of this task. Both these methods

[1] Associate professor, Department of Solid Mechanics, Technical University of Denmark, Lyngby, Denmark; also, research scientist, Metallurgy Department, Risø National Laboratory, Roskilde, Denmark.

require homogeneous samples and are highly inefficient if all three Weibull parameters are to be estimated.

In statistical analyses of composite material strength and fatigue life data reported in the literature, a two-parameter Weibull distribution has often been assumed, implying that the location parameter is zero. The consequences of this assumption have not yet been examined. It is therefore demonstrated here that if given data, which actually belong to a three-parameter Weibull population, are assumed to be two-parameter Weibull data, then the estimates of these two parameters from these data will be higher than their true values. Furthermore, the errors made in so estimating the two parameters will increase linearly with the true value of the location parameter.

In the following, we shall describe a method that allows estimation of all three parameters except when the estimate of the location parameter becomes negative. In the latter case, requiring the location parameter to be zero or positive, good estimates of the other two parameters are obtained. Furthermore, with this method, the constituent components in heterogeneous data can be separated and parameters for each component can then be estimated.

The superiority of the proposed method is demonstrated by comparing its estimates from random samples with the estimates of the moment estimation and the maximum likelihood methods from the same samples. This is done for two values of the shape parameter using five samples of ten data points.

Finally, composite material strength and fatigue life data are taken from the literature to illustrate the use of the proposed method.

Estimation of Weibull Parameters

The Weibull distribution function is given by

$$F(x;\ a,\ b,\ c) = 1 - \exp\left\{-\left(\frac{x-a}{b}\right)^c\right\} \tag{1}$$

for $x > a$, $b > 0$, and $c > 0$, where a, b, and c are the location, scale, and shape parameters, respectively.

For strength distributions, $a < 0$ will not be acceptable for tensile strength. For fatigue life distributions, $c < 1.0$ will not be acceptable, as that would lead to a decreasing failure rate.

A set of observations, for example, test data, is called a sample and, when arranged in increasing order, an individual observation is called an order statistic denoted by x_i, such that $x_1 < x_2 < x_3 \ldots < x_N$. A complete statistical analysis of a given sample consists of the following steps [*1*]:[2]

[2]The italic numbers in brackets refer to the list of references appended to this paper.

1. testing the homogeneity of the sample,
2. testing the acceptability of one or more of the assumed distributions,
3. selecting one out of a set of acceptable distributions, and
4. estimating the unknown parameters of the accepted distribution.

Herein we shall discuss Step 4 with Step 1 forming part of one of the methods of estimation discussed.

Moment Estimation

This is a classical method introduced by K. Pearson in 1894. Essentially, it consists of equating as many of the population moments to the sample moments as the number of unknown parameters.

The sample moments are given by

$$A_k = \frac{\Sigma(x_i)^k}{N} \tag{2}$$

and

$$B_k = \frac{\Sigma(x_i - \bar{x})^k}{N} \tag{3}$$

where

$$\bar{x} = \frac{\Sigma x_i}{N} \tag{4}$$

and A_k are the moments about origin and B_k are the central moments.

The population moments are given by

$$\alpha_k = \int_{-\infty}^{\infty} x^k f(x) dx \tag{5}$$

and

$$\beta_k = \int_{-\infty}^{\infty} (x - \bar{x})^k f(x) dx \tag{6}$$

where $f(x)$ is the Weibull density function given by

$$f(x) = \frac{c}{b}\left(\frac{x-a}{b}\right)^{c-1} \exp\left\{-\left(\frac{x-a}{b}\right)^c\right\} \tag{7}$$

The first three population moments are thus given by

$$\alpha_1 = a + bg_1 \tag{8}$$

$$\alpha_2 = a^2 + b^2g_2 + 2abg_1 \tag{9}$$

$$\alpha_3 = a^2 + b^3g_3 + 3abg_2 + 3a^2bg_1 \tag{10}$$

$$\beta_1 = 0 \tag{11}$$

$$\beta_2 = b^2(g_2 - g_1{}^2) \tag{12}$$

$$\beta_3 = b^3(g_3 - 3g_1g_2 + 2g_1{}^3) \tag{13}$$

where

$$g_i = \left(\frac{i}{c}\right)! \tag{14}$$

For a two-parameter Weibull distribution, the parameters b and c may be estimated as follows.

Equating A_1 to α_1 and taking $a = 0$, we have

$$bg_1 = \bar{x} = \frac{\Sigma x_i}{N} \tag{15}$$

Equation 15 gives the estimate of b as

$$b^* = \frac{\Sigma x_i}{N} \tag{16}$$

To obtain an estimator for c, we form the ratio

$$V = \frac{(B_2)^{1/2}}{A_1} = \frac{[\Sigma(x_i - \bar{x})^2]^{1/2}}{\sqrt{N}\cdot\bar{x}} \tag{17}$$

The corresponding ratio for population moments is given by

$$K = \frac{(\beta_2)^{1/2}}{\alpha_1} = \frac{(g_2 - g_1{}^2)^{1/2}}{g_1} \tag{18}$$

which is only a function of c, according to Eq 14. Thus, the value of c satisfying

$$K = V \tag{19}$$

is taken as an estimate of c.

The values of K have been tabulated by Weibull and Weibull [*1*] for different values of c.

For three-parameter Weibull distribution, the estimators are obtained as follows.

The coefficients of skewness for sample and population are given by

$$G_s = \frac{B_3}{(B_2)^{3/2}} = \sqrt{N}\frac{\Sigma(x_i - \bar{x})^3}{[\Sigma(x_i - \bar{x})^2]^{3/2}} \tag{20}$$

and

$$G_p = \frac{\beta_3}{(\beta_2)^{3/2}} = \frac{g_3 - 3g_1g_2 + 2g_1^3}{(g_2 - g_1^2)^{3/2}} \tag{21}$$

The estimate of c is thus the value that satisfies the equality

$$G_s = G_p \tag{22}$$

The values of G_p have been tabulated by Gumbel [*2*] and by Weibull and Weibull [*1*] for different values of c.

Equating β_2 to B_2, we have

$$b^2(g_2 - g_1^2) = \frac{\Sigma(x_i - \bar{x})^2}{N} \tag{23}$$

that gives an estimate of b as

$$b^* = \frac{[\Sigma(x_i - \bar{x})^2]^{1/2}}{N^{1/2}(g_2 - g_1^2)^{1/2}} \tag{24}$$

in which g_i are calculated from the estimates of c.

The estimate of a is given by equating α_1 to A_1. Thus

$$a^* = \bar{x} - b^*\left(\frac{1}{c^*}\right)! \tag{25}$$

Maximum Likelihood Estimation

The likelihood function $L(a, b, c)$ for the first M order satistics from a sample of size N is given by

$$L_m = \ln L(a, b, c) = \ln N! - \ln(N - M)! + M(\ln c - c \ln b) + (c - 1) \sum_{i=1}^{M} \ln(x_i - a) - \sum_{i=1}^{M} [(x_i - a)/b]^c - (N - M)[(x_N - a)/b]^c \quad (26)$$

The maximum likelihood equations are obtained by equating to zero the partial derivatives of $\ln L$ with respect to each of the three parameters. Thus

$$\frac{\partial L_m}{\partial a} = cb^{-c}\left[\sum_{i=1}^{M} (x_i - a)^{c-1} + (N - M)(x_M - a)^{c-1}\right] - (c - 1) \sum_{i=1}^{M} (x_i - a)^{-1} = 0 \quad (27)$$

$$\frac{\partial L_m}{\partial b} = cb^{-(c+1)}\left[\sum_{i=1}^{M} (x_i - a)^c + (N - M)(x_M - a)^c\right] - Mcb^{-1} = 0 \quad (28)$$

$$\frac{\partial L_m}{\partial c} = \frac{M}{c} + \sum_{i=1}^{M} \ln\left[\frac{(x_i - a)}{b}\right] - \sum_{i=1}^{M} \left[\left(\frac{x_i - a}{b}\right)^c \ln\left(\frac{x_i - a}{b}\right)\right] - (N - M)\left[\frac{x_M - a}{b}\right]^c \ln\left[\frac{x_M - a}{b}\right] = 0 \quad (29)$$

Antle and Klimko [*3*] have discussed solutions of the likelihood equations. They have shown that if a solution (a^*, b^*, c^*) is a local maximum, then another solution (a', b', c') exists that corresponds to a saddle point. They have also demonstrated by a numerical example that even if a solution is a local maximum, it may not be the maximum likelihood estimate.

Harter and Moore [*4*] have developed an iterative procedure that involves estimating the three parameters, one at a time, in the cyclic order b, c, a. They claim that positive values of b^* and c^* can always be found in this way. In estimating a, however, they find that no value of a in the permissible interval $0 \leq a \leq x_1$ will satisfy the likelihood Eq 29. By censoring x_1 in such a case, they succeed in finding the estimates.

Antle and Klimko [*3*], however, do not recommend maximum likelihood estimation of all three parameters for sample size ≤ 10. Weibull and Weibull [*1*] used a computer program provided by Antle for maximum likelihood estimation of 300 random samples of size 10 and 20 each drawn from Weibull populations (0, 1, 1), (0, 1, 2), and (0, 1, 5). They found that for sample size 10 and for $c = 2$ and $c = 5$, about half the estimates were unacceptable, that is, they were not the local maximum solutions.

For a two-parameter ($a = 0$) complete sample ($N = M$), the likelihood function on eliminating b from Eqs 27 through 29 becomes

$$L_m(c) = \ln c - \ln \Sigma x_i^c + (1/N)[(c-1) \Sigma \ln x_i] \tag{30}$$

The maximum likelihood equation for estimating c is then given by

$$\frac{1}{c} - \frac{\Sigma x_i^c \ln x_i}{\Sigma x_i^c} + \frac{\Sigma \ln x_i}{N} = 0 \tag{31}$$

The maximum likelihood estimate of b is given by Eq 28 as

$$b^* = \left[\frac{\Sigma x_i^{c^*}}{N}\right]^{1/c^*} \tag{32}$$

For maximum likelihood estimation of two-parameter Weibull distribution, Cohen [*5*] has given variance-covariance matrices for censored and complete samples and Thoman, Bain, and Antle [*6*] and McCool [*7*] have given confidence intervals for estimates and unbiasing factors for shape parameter estimates.

Standardized Variable Estimation

The standardized variable z is defined by

$$z = \frac{x - a}{b} \tag{33}$$

The distribution function of the standardized variable is given by $F(z; 0, 1, c)$. Thus the order statistics, z_i, are independent of a and b, and depend only on the shape parameter, c.

The expected value, Ez_i, the median, Mz_i, and the variance, Vz_i, of the order statistics, z_i, were derived by Lieblein [*8*], and are given by

$$Ez_i = iC_i^N \left(\frac{1}{c}\right)! \sum_{j=0}^{i-1} (-1)^{i-1-j} C_j^{i-1} (N-j)^{-(1+1/c)} \tag{34}$$

$$Mz_i = [-\ln(1 - MP_i)]^{1/c} \tag{35}$$

$$Vz_i = E(z_i^2) - [E(z_i)]^2 \tag{36}$$

where

$$E(z_i^2) = iC_i^N \left(\frac{2}{c}\right)! \sum_{j=0}^{i-1} (-1)^{i-1-j} C_j^{i-1} (N - j)^{-(1+2/c)} \tag{37}$$

C_i^j are the binomial coefficients and MP_i are the median percentage points.

As seen in Eqs 34 through 37, all the characteristic values of the order statistics, z_i, depend only on the sample size, N, and the shape parameter, c. If, now, the order statistics, z_i, are estimated by their expected values, Ez_i, or by their median, Mz_i, we obtain from Eq 33

$$x_i = a + bEz_i \tag{38}$$

or

$$x_i = a + bMz_i \tag{39}$$

Thus, if the observations, x_i, are plotted against Ez_i or Mz_i for the true value of c, the data points (x_i, Ez_i) or (x_i, Mz_i) will, with due regard to the sampling scatter, fall along a straight line. The estimate of c may then be taken as the value for which the linear regression gives the maximum correlation coefficient. Once c has been estimated, the estimates of a and b will be given by the x_i-intercept and the slope, respectively, of the best-fit line.

The tables for Ez_i and Vz_i are given by Weibull [*9*] and for Mz_i by Weibull [*10*] for different values of c and samples sizes.

It was shown by Weibull and Weibull [*10*] that a good approximation to MP_i is given by the simple formula

$$MP_i = k(i)/[k(i) + k(N + 1 - i)] \tag{40}$$

where

$$k(i) = i - 0.334 + 0.0252/i \tag{41}$$

The computation of Mz_i is thus considerably simplified and preference may therefore be given to Eq 39 over Eq 38 for estimation of parameters.

It may be noted that this method of estimating parameters can be used also for censored or truncated samples.

The linear relationships in Eqs 38 and 39 indicate that if the observations,

x_i, belong to different populations, then in the (x_i, Ez_i) or (x_i, Mz_i) plots the data points will scatter about different straight lines. An inspection of these plots would then allow us to separate the different components in the sample and estimate parameters for each component by finding the shape parameter values that maximize correlation coefficients for the corresponding straight lines. This will be illustrated by analyzing actual composite strength data in the sequel.

Estimation from Random Samples

Five random samples of each size 10 were constructed for a chosen Weibull population using the relationship

$$x_i = a + b(-\ln r_i)^{1/c} \tag{42}$$

where r_i are random numbers drawn from a uniform distribution in the interval (0, 1). The order statistics, x_i, are shown in Table 1 for a Weibull population (0, 1, 1). Weibull parameters were estimated from these samples using each of the methods just described.

Moment Estimation

All three Weibull parameters were estimated for two populations (0, 1, 1) and (0, 1, 10) using Eqs 22, 24, and 25. The estimates are shown in Table 2. As seen here, negative estimates of location parameters are obtained for all samples of the first population and for two samples of the second population. These estimates can therefore not be accepted. The average errors made in estimating the scale and the shape parameters are higher for the higher value of the shape parameter. The variance of the estimates, the inverse of which defines the efficiency of the estimation method, is higher for the higher value

TABLE 1—*Random samples drawn from a Weibull distribution (0, 1, 1).*

i	x_i	x_i	x_i	x_i	x_i
1	0.315	0.013	0.200	0.024	0.027
2	0.342	0.229	0.236	0.272	0.221
3	0.589	0.233	0.333	0.574	0.265
4	0.590	0.244	0.401	0.772	0.277
5	1.117	0.365	0.911	0.920	0.419
6	1.470	0.489	0.966	0.989	0.665
7	1.693	0.606	1.219	1.157	0.980
8	2.041	0.909	1.773	2.264	1.390
9	2.679	1.413	1.789	3.040	1.509
10	3.667	1.773	1.997	3.299	2.647

of the shape parameter. It therefore appears that both the accuracy and the efficiency of this estimation method decrease with increasing values of the shape parameter.

Tables 3 and 4 illustrate the errors made in estimating the scale and the shape parameters when the location parameter is taken to be zero. The estimates in these tables are found by using Eqs 16 and 19 for Weibull populations (a, 1, 1) and (a, 1, 10) with a having values 0.1, 0.5, 1.0, and 2.0

TABLE 2—*Estimates of Weibull parameters by moment estimation.*

Population	(0, 1, 1)			(0, 1, 10)		
Sample	a^*	b^*/b	c^*/c	a^*	b^*/b	c^*/c
1	−0.29	2.07	1.81	0.72	0.32	0.39
2	−0.10	0.92	1.57	−1.81	2.97	3.23
3	−0.60	1.89	2.76	0.60	0.38	0.49
4	−0.52	2.21	1.88	−8.46	9.70	10.0
5	−0.23	1.25	1.49	0.12	0.86	0.87
Average	−0.35	1.67	1.90	−1.77	2.85	3.00
Variance	0.034	0.246	0.205	12.029	12.680	13.34

TABLE 3—*Estimates of Weibull parameters by moment estimation assuming* a = *0.*

Population	(0.1, 1, 1)		(0.5, 1, 1)		(1.0, 1, 1)		(2.0, 1, 1)	
Sample	b^*/b	c^*/c	b^*/b	c^*/c	b^*/b	c^*/c	b^*/b	c^*/c
1	1.72	1.51	2.20	1.94	2.76	2.51	3.82	3.66
2	0.80	1.36	1.27	2.20	1.81	3.31	2.84	5.62
3	1.21	1.70	1.67	2.41	2.21	3.33	3.24	5.24
4	1.57	1.35	2.07	1.76	2.64	2.30	3.71	3.42
5	1.00	1.23	1.51	1.80	2.07	2.56	3.13	4.15
Average	1.26	1.43	1.74	2.02	2.30	2.80	3.35	4.42
Variance	0.119	0.026	0.119	0.060	0.125	0.186	0.136	0.747

TABLE 4—*Estimates of Weibull parameters by moment estimation assuming* a = *0.*

Population	(0.1, 1, 10)		(0.5, 1, 10)		(1.0, 1, 10)		(2.0, 1, 10)	
Sample	b^*/b	c^*/c	b^*/b	c^*/c	b^*/b	c^*/c	b^*/b	c^*/c
1	1.15	1.67	1.55	2.27	2.05	3.08	3.04	4.76
2	1.06	1.11	1.45	1.67	1.96	2.16	2.96	3.33
3	1.11	1.67	1.50	2.50	2.01	3.12	3.00	4.76
4	1.14	1.07	1.54	1.43	2.04	2.00	3.04	3.03
5	1.08	1.13	1.48	1.67	1.98	2.17	2.99	3.33
Average	1.11	1.33	1.50	1.91	2.01	2.51	3.01	3.84
Variance	0.001	0.077	0.001	0.165	0.001	0.239	0.001	0.574

for each set of five samples. As seen here, the errors made in estimating the scale and the shape parameters increase with increasing value of the location parameter. The average ratios of the estimates to the true parameter values are plotted against the true location parameter values in Fig. 1 for $c = 1$ and in Fig. 2 for $c = 2$. As shown by these figures, the errors appear to increase linearly with the true location parameter value. Comparing these two figures, it can also be seen that the errors in estimating b and c are lower for higher value of c. Looking at the variance of estimates in Tables 3 and 4, it can be seen that the efficiency of estimating b decreases with increasing value of a for $c = 1$ and remains constant for $c = 10$, while the efficiency of estimating c decreases with increasing value of a for both $c = 1$ and $c = 10$.

In summarizing the Weibull parameter estimation by this method, it may be said that the method cannot be relied on for estimating all three parameters, especially not for higher values of the shape parameter. For two-parameter estimation, the errors in estimation increase linearly with the true value of the location parameter, which is assumed to be zero. The errors are more than 100 percent when the location parameter is equal to or greater than the scale parameter.

Maximum Likelihood Estimation

No attempt was made here to estimate all three parameters by this method, as the computations involved are tedious and as the estimates may not even be reliable (see section on Maximum Likelihood Estimation).

For two-parameter estimation, the shape parameter is estimated first by maximizing the function in Eq 30 or by solving Eq 31 for c. The former method was used here since it was found to involve simpler computations. After estimating c, b was estimated by using Eq 32 in which the estimate of c was used.

The results of estimation are shown in Table 5 for populations $(a, 1, 1)$ and in Table 6 for populations $(a, 1, 10)$ for $a = 0.1, 0.5, 1.0,$ and 2.0. As seen in these tables, the errors in estimating b and c increase with increasing value of the location parameter, which is taken to be zero in the estimation of b and c. The average ratios of the estimates to the true parameter values are plotted in Fig. 3 for $c = 1$ and in Fig. 4 for $c = 10$. Both these figures show that the average errors in estimating b and c increase linearly with increasing value of a. By comparing Figs. 3 and 4, it is apparent that the estimation errors are lower for higher value of c.

Looking at the variance values in Tables 5 and 6, it is seen that the efficiency of estimating b decreases with increasing value of a for $c = 1$ and remains constant for $c = 10$. This constant variance value is the same as that obtained in estimating b for $c = 10$ by the moment estimation method, as seen in Table 4. The efficiency in estimating c decreases with increasing value of a both for $c = 1$ and $c = 10$.

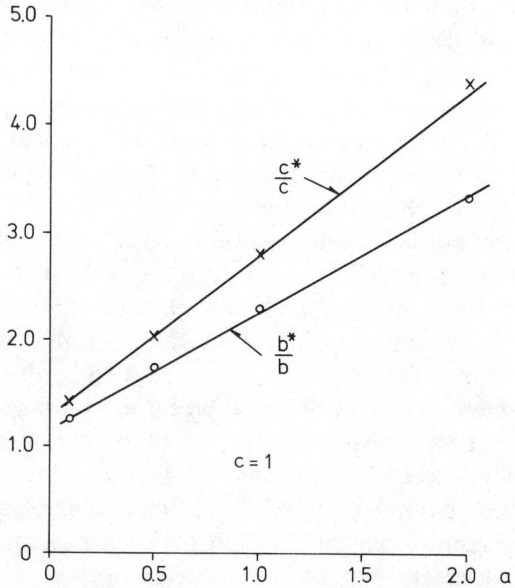

FIG. 1—*Ratios of the moment estimates to the true parameter values against the location parameter for* c = *1.*

The estimates of c can be improved slightly by multiplying these by the unbiasing factor, which, for $N = 10$, is given by Thoman, Bain, and Antle [*6*] as 0.859.

Standardized Variable Estimation

To start estimation by this method, one must plot the order statistics, x_i, against Ez_i or Mz_i for a trial value of c. It is advisable to start with $c = 1$. In Table 7, values of Ez_i and Mz_i are listed for $c = 1$ for a sample of size 10. For other values of c and sample sizes, see Refs *9* and *10*.

To illustrate the method, we first take a random sample from a Weibull population (0.1, 1, 1). The (x_i, Mz_i) plots are shown for $c = 0.5$, 1.0, and 2.0 in Fig. 5. The consecutive points are joined by a straight line. As seen here, the plot curves to the right for $c = 0.5$, which is less than the true value of c. For the true value of c (that is, $c = 1$), the data appear to scatter about a straight line. For a higher value of c, the scatter appears to increase. Thus, an inspection of the (x_i, Mz_i) plots gives an idea of the range in which a search for the estimate of c should be made. Once this is done, an estimate of c is found by making a linear regression analysis of the (x_i, Mz_i) data points for different values of c and choosing as estimate the value that maximizes the correlation coefficient. For the data plotted in Fig. 5, the following correlation coefficients r were obtained.

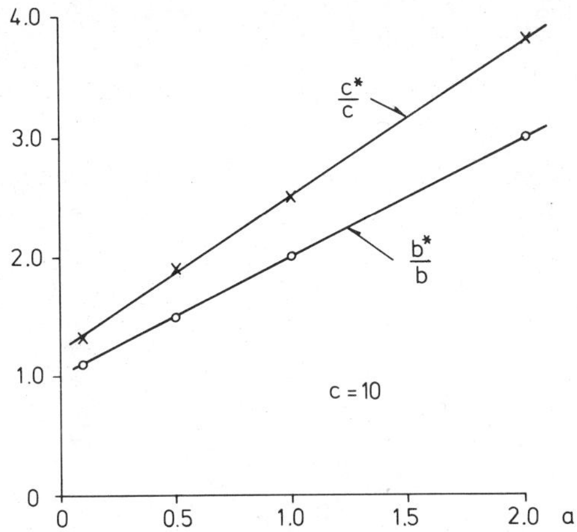

FIG. 2—*Ratios of the moment estimates to the true parameter values against the location parameter for* c = *10*.

TABLE 5—*Estimates of Weibull parameters by maximum likelihood estimation assuming* a = *0*.

Population	(0.1, 1, 1)		(0.5, 1, 1)		(1.0, 1, 1)		(2.0, 1, 1)	
Sample	*b*/b*	*c*/c*	*b*/b*	*c*/c*	*b*/b*	*c*/c*	*b*/b*	*c*/c*
1	1.75	1.60	2.21	2.00	2.77	2.50	3.84	3.50
2	0.80	1.40	1.28	2.20	1.82	3.10	2.85	4.80
3	1.22	1.70	1.68	2.50	2.22	3.40	3.25	5.10
4	1.56	1.30	2.08	1.80	2.65	2.40	3.72	3.40
5	1.03	1.30	1.52	1.90	2.07	2.50	3.13	3.70
Average	1.47	1.46	1.75	2.08	2.31	2.78	3.36	4.10
Variance	0.091	0.026	0.120	0.062	0.127	0.158	0.137	0.500

TABLE 6—*Estimates of Weibull parameters by maximum likelihood estimation assuming* a = *0*.

Population	(0.1, 1, 10)		(0.5, 1, 10)		(1.0, 1, 10)		(2.0, 1, 10)	
Sample	*b*/b*	*c*/c*	*b*/b*	*c*/c*	*b*/b*	*c*/c*	*b*/b*	*c*/c*
1	1.15	1.60	1.55	2.10	2.05	2.80	3.05	4.20
2	1.06	1.20	1.46	1.70	1.96	2.20	2.96	3.40
3	1.11	1.60	1.51	2.30	2.01	3.00	3.01	4.50
4	1.13	1.20	1.54	1.70	2.03	2.20	3.04	3.30
5	1.08	1.10	1.49	1.60	1.99	2.10	2.99	3.20
Average	1.11	1.34	1.51	1.88	2.01	2.46	3.01	3.72
Variance	0.001	0.046	0.001	0.074	0.001	0.134	0.001	0.278

$c = 1.0$	1.1	1.2	1.3
$r = 0.99359$	0.99527	0.99561	0.99509

Thus, the estimate $c^* = 1.2$ is taken. For further accuracy, interpolation between the c-values may be made. The estimates of a and b are now given by the x_i-intercept and the slope of the best fit line.

Table 8 shows the estimates of all three parameters for five samples from

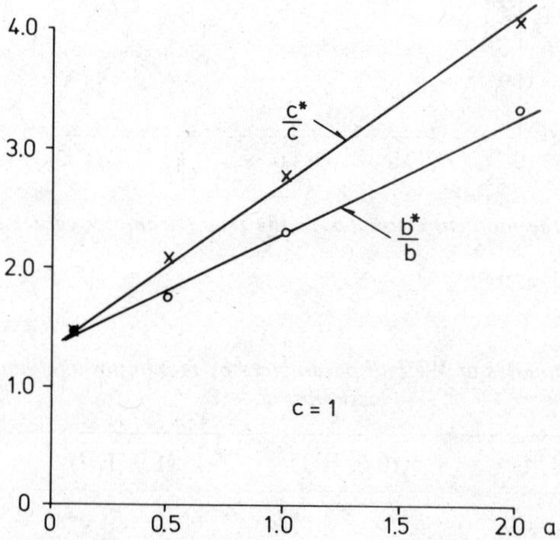

FIG. 3—*Ratios of the maximum likelihood estimates to the true parameter values against the location parameter for* c = *1.*

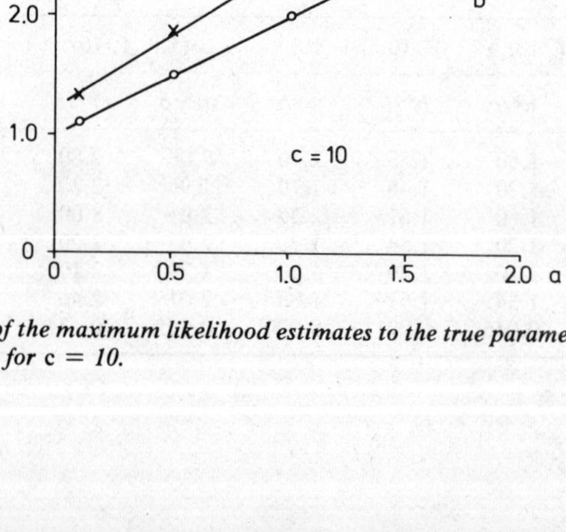

FIG. 4—*Ratios of the maximum likelihood estimates to the true parameter values against the location parameter for* c = *10.*

TABLE 7—*The expected value, median value, and variance of the order statistic z_i for a sample of size 10.*

i	$Ez_i, c = 1$	$Mz_i, c = 1$	$Vz_i, c = 1$	$Vz_i, c = 10$
1	0.10000	0.06905	0.01000	0.00827
2	0.21111	0.17700	0.02235	0.00426
3	0.33611	0.29916	0.03797	0.00298
4	0.47897	0.43864	0.05838	0.00236
5	0.64564	0.60092	0.08616	0.00200
6	0.84564	0.79476	0.12616	0.00178
7	1.09564	1.03539	0.18866	0.00164
8	1.42897	1.35264	0.29977	0.00159
9	1.92897	1.81878	0.54977	0.00166
10	2.92897	2.70725	1.54977	0.00209

Weibull population (0, 1, 1). For Samples 3 and 4, small negative estimates of a are obtained. To avoid these unacceptable estimates, we could look for the c-values that maximize the correlation coefficient within the constraint $a \geq 0$. Doing this, revised estimates for Samples 3 and 4 are obtained as (0, 1.22, 1.4) and (0, 1.46, 1.1), respectively. These estimates are indeed better than the corresponding estimates in Table 8.

If the estimates of a given by this method are positive, then the estimates of b and c will remain unchanged for higher values of a. However, if a is estimated to be negative, then the constraint $a \geq 0$ for estimation would change (improve) the estimates of b and c for higher values of a. This is illustrated for a set of samples taken from Weibull populations (a, 1, 10) with $a = 0.1$, 0.5, 1.0, and 2.0 below.

The (x_i, Mz_i) plots for one sample from population (0.1, 1, 10) are shown in Fig. 6 for $c = 1$ and $c = 10$. For $c = 1$, the plot curves to the right indicating that a higher value of c should be taken. For $c = 10$, the data points, except the lowest one, scatter about a straight line. This behavior of the lowest data point is typical for high values of c and is explained by the variance of the lowest order statistic (see Table 7). As seen in this table, Vz_i for $c = 10$ is much higher than the remaining Vz_i. It is therefore advisable to censor the first order statistic when the (x_i, Mz_i) plots indicate a high value of c. This was done for all samples from Weibull populations (a, 1, 10). Furthermore, the constraint $a \geq 0$ was employed in estimation as a^* otherwise became negative. The estimates are shown in Table 9.

As seen in Table 9, accurate estimates of b and c are obtained at the lowest value of a, and the accuracy diminishes at higher values of a. Beyond $a = 1$, however, the estimation errors increase at a lower rate WRT a. The efficiency of estimating both b and c decreases with increasing value of a.

In summarizing the standardized variable estimation method, it may be said that this provides accurate and efficient estimates of all three Weibull parameters for low values of c. For high values of c, however, it tends to give

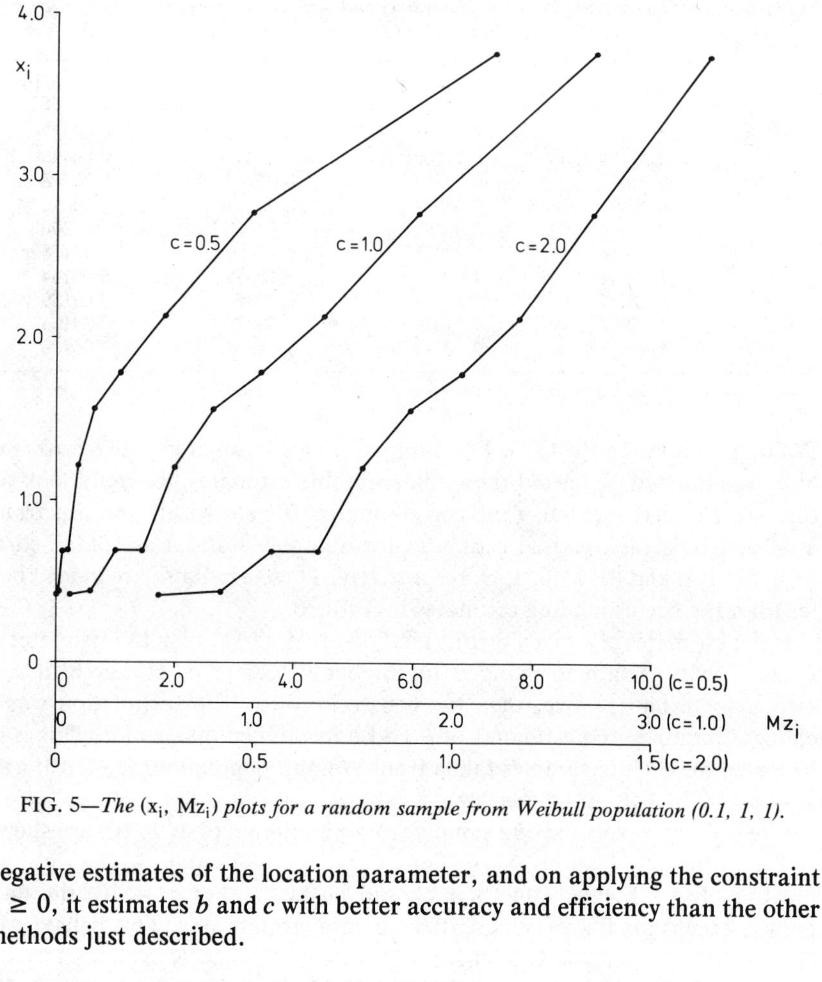

FIG. 5—*The* (x_i, Mz_i) *plots for a random sample from Weibull population (0.1, 1, 1).*

negative estimates of the location parameter, and on applying the constraint $a \geq 0$, it estimates b and c with better accuracy and efficiency than the other methods just described.

Estimation from Composite Material Strength and Fatigue Life Data

To illustrate the use of the standardized variable estimation method further, we take as examples some recently published data on composite material strength and fatigue life.

As the first example, we take the compression test results for two graphite/epoxy laminates reported by Ryder and Black [*11*]. Their test results, given in their Table 1 (Laminate 1) and Table 2 (Laminate 2), are plotted as (x_i, Mz_i) plots for a trial value of $c = 1$ in Figs. 7 and 8. These two figures show that the data points scatter about two straight lines in each plot. The highest point in each case, however, appears to be an outlier. It is therefore apparent that, except for the highest point, the data belong to a two-component population. According to the authors, the stress-strain plots indicated that

TABLE 8—*Estimates of Weibull parameters by standardized variable estimation.*

Population	(0, 1, 1)		
Sample	a^*	b^*/b	c^*/c
1	0.02	1.60	1.20
2	0.0	0.68	1.00
3	−0.37	1.55	2.00
4	−0.20	1.74	1.30
5	0.01	0.86	0.90
Average	−0.11	1.29	1.28
Variance	0.024	0.186	0.150

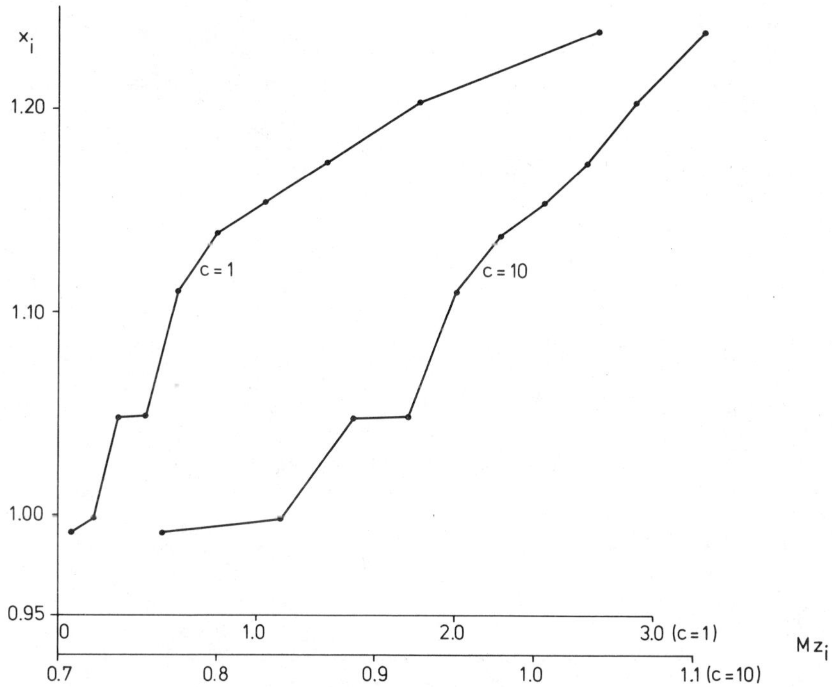

FIG. 6—*The* (x_i, Mz_i) *plots for a random sample from Weibull population (0.1, 1, 10).*

two different modes of fracture existed. The (x_i, Mz_i) plots therefore appear to confirm this observation. The Laminate 1 data in Fig. 7 indicate that the seven highest points (except perhaps the highest point) belong to one component and the remainder to another. According to the authors' observation of the stress-strain plots, however, there should be 11 highest points in a component. A linear regression analysis of the data points in the lower compo-

TABLE 9—*Estimates of Weibull parameters by standardized variable estimation with the condition* a ≥ *0.*

Population	(0.1, 1, 10)		(0.5, 1, 10)		(1.0, 1, 10)		(2.0, 1, 10)	
Sample	b^*/b	c^*/c	b^*/b	c^*/c	b^*/b	c^*/c	b^*/b	c^*/c
1	1.11	1.20	1.54	1.84	2.01	2.20	2.19	2.40
2	1.08	1.00	1.46	1.35	1.95	1.80	2.16	2.00
3	1.13	1.30	1.56	1.80	2.07	2.40	3.00	3.5
4	1.13	1.20	1.50	1.60	1.68	1.80	1.68	1.80
5	1.10	1.00	1.52	1.40	2.02	1.90	2.95	2.90
Average	1.11	1.14	1.52	1.60	1.95	2.02	2.40	2.52
Variance	0.0004	0.014	0.001	0.040	0.019	0.058	0.256	0.382

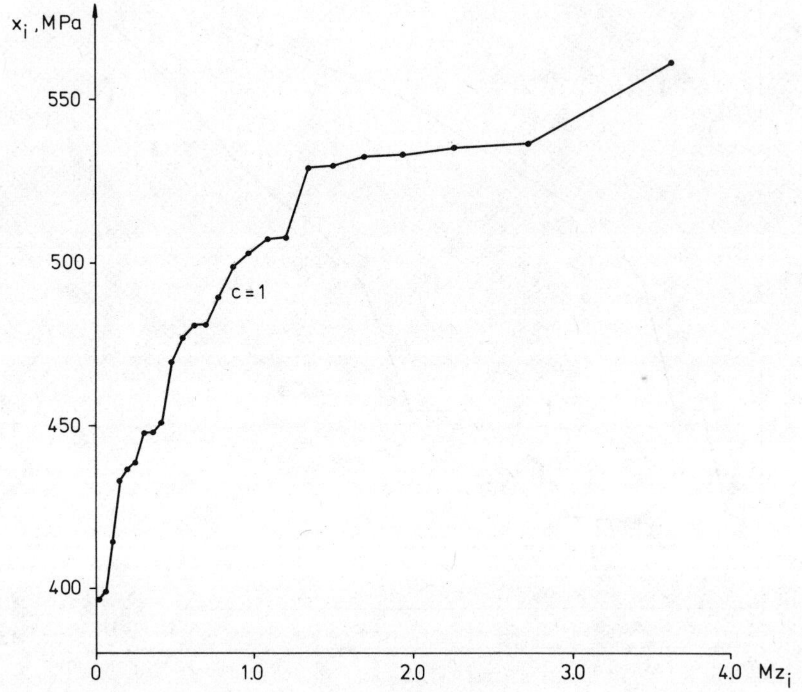

FIG. 7—*The* (x_i, Mz_i) *plot of compression test data for Laminate 1.*

nent showed that inclusion of the four highest points increased the correlation coefficient. We therefore tend to think that only the seven highest points belong to a higher component. The data for Laminate 2 plotted in Fig. 8 confirm the authors' observation that the nine highest data points belong to Mode 2 fracture.

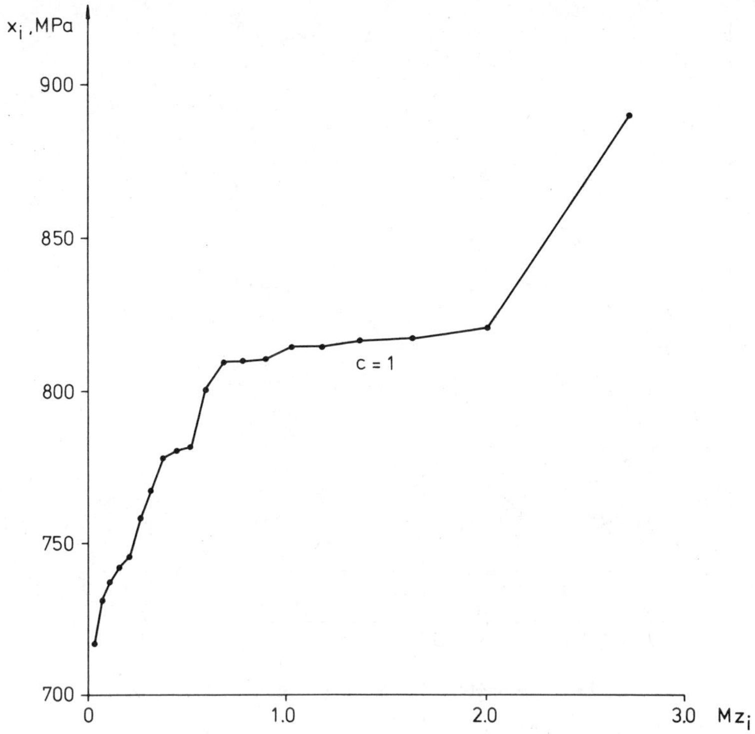

FIG. 8—*The* (x_i, Mz_i) *plot of compression test data for Laminate 2.*

The estimation procedure described in a previous section was applied to each component separately, and the estimates so obtained are listed below.

Laminate 1:
  Component 1: $a^* = 365.8$ MPa, $b^* = 138.0$ MPa, $c^* = 2.5$
  Component 2: $a^* = 422.1$ MPa, $b^* = 104.4$ MPa, $c^* = 10.0$
Laminate 2:
  Component 1: $a^* = 713.0$ MPa, $b^* = 109.4$ MPa, $c^* = 1.25$
  Component 2: $a^* = 739.8$ MPa, $b^* = 71.1$ MPa, $c^* = 7.0$

The second example taken is the fatigue life data for graphite/epoxy $[\pm 45]_{2s}$ laminate tested in shear at maximum stress of 8.143 ksi and $R = 0.1$ reported by Yang and Jones [*12*]. These data, consisting of 20 points, are plotted as x_i in kilocycles against Mz_i for trial values of $c = 1$ and $c = 10$ in Fig. 9. It is clearly seen that, except for the highest point, the data are homogeneous and have a shape parameter higher than 1, since the (x_i, Mz_i) plot for $c = 1$ curves to the right. The plot for $c = 10$ shows no more curvature to the right and confirms that the highest point is an outlier. We

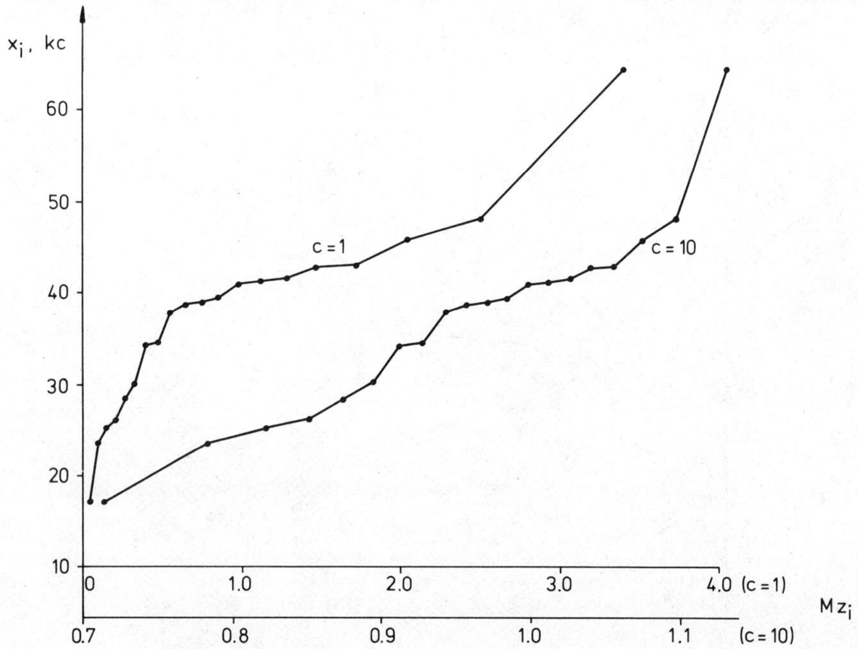

FIG. 9—*The* (x_i, Mz_i) *plots of fatigue test data.*

therefore censor this observation and estimate the parameters from the remaining data.

The estimation procedure gave a negative value of a, whereupon the constraint $a \geq 0$ was placed and the following estimates were obtained.

$$a^* = 0, \qquad b^* = 38.14 \text{ kc}, \qquad c^* = 4.21$$

These estimates may be compared with Yang and Jones' predictions of the two parameters (assuming $a = 0$), which were calculated to be

$$b^* = 41.07 \text{ kc}, \qquad c^* = 4.31$$

Conclusions

It has been demonstrated that the moment estimation and the maximum likelihood estimation methods lead to large errors in estimating the scale and the shape parameters if the location parameter is taken to be zero. The errors increase linearly with the true value of the location parameter.

The proposed method, called here the standardized variable estimation, has been demonstrated to give accurate and efficient estimates of all three Weibull parameters for low shape parameters, and more accurate and more efficient estimates than the other two methods of the scale and the shape parameters for high shape parameter. Furthermore, this method allows separation of components in a multi-component sample and detection of outliers.

Acknowledgments

The author would like to express his grateful appreciation of the inspiration and guidance he received from collaboration with Professor W. Weibull, who died recently having contributed actively to the field of statistical strength theories until his very last days.

References

[*1*] Weibull, W. and Weibull, G. W., "New Aspects and Methods of Statistical Analysis of Test Data with Special Reference to the Normal, the Lognormal, and the Weibull Distributions," Part I and Part II, FOA Report D 20045-DB, Defence Research Institute, Stockholm, June 1977.

[*2*] Gumbel, E. J., *Statistics of Extremes*, Columbia University Press, N.Y., 1960.

[*3*] Antle, C. E. and Klimko, L. A., "Choice of Model for Reliability Studies and Related Topics II," ARL-73-0121, AD 772775, 1973.

[*4*] Harter, H. L. and Moore, A. H., *Technometrics*, Vol. 7, No. 4, 1965, pp. 639–643.

[*5*] Cohen, A. C., *Technometrics*, Vol. 7, No. 4, 1965, pp. 579–588.

[*6*] Thoman, D. R., Bain, L. J., and Antle, C. E., "Inferences on the Parameters of the Weibull Distribution," *Technometrics*, Vol. 11, No. 3, 1969.

[*7*] McCool, J. I. in *Transactions*, Vol. 13, American Society of Lubrication Engineers, 1969, pp. 189–202.

[*8*] Lieblein, J., *Annals of Mathematical Statistics*, Vol. 26, 1955, pp. 330–333.

[*9*] Weibull, W., "Estimation of Distribution Parameters by a Combination of the Best Linear Order Statistics and Maximum Likelihood," AFML-TR-67-105, Air Force Materials Laboratory, 1967.

[*10*] Weibull, W. and Weibull, G. W., "High-fidelity Approximation to Median Percentage Points of Order Statistics," AFML-TR-69-317, Air Force Materials Laboratory, 1969.

[*11*] Ryder, J. T. and Black, E. D. in *Composite Materials: Testing and Design (Fourth Conference), ASTM STP 617*, American Society for Testing and Materials, 1977, pp. 170–189.

[*12*] Yang, J. N. and Jones, D. L., *Journal of Composite Materials*, Vol. 13, 1978, pp. 371–389.